Amazing Arachnids

Amazing Arachnids

Jillian Cowles

PRINCETON UNIVERSITY PRESS

PRINCETON AND OXFORD

Requests for permission to reproduce material from this work should be sent to Permissions, Princeton University Press

Published by Princeton University Press,
41 William Street, Princeton, New Jersey 08540

In the United Kingdom: Princeton University Press,
6 Oxford Street, Woodstock, Oxfordshire OX20 1TR

press.princeton.edu

Jacket image: *Phidippus apacheanus*. Photo by Bruce D. Taubert

ISBN 978-0-691-17658-1

Library of Congress Control Number: 2017956630

British Library Cataloging-in-Publication Data is available

This book has been composed in Minion Pro

Printed on acid-free paper. ∞

Printed in China

10 9 8 7 6 5 4 3 2 1

Contents

Preface

People from all over the world come to the southwestern United States in order to see this hotspot of biological diversity. From creosote flats to mountain forests, from beaches to deserts, the arid southwest is home to a multitude of extraordinary species, many of which are unique to this challenging environment. This book attempts to present the story of one group of southwestern inhabitants, the arachnids.

The need for such a book became apparent to me when I began to learn about them. There were hundreds of publications about arachnids, but most were scattered in dozens of obscure journals. Arachnids other than spiders were almost completely neglected. And finally, even field guides on arthropods focus on regions other than the southwestern United States. Clearly, the story of these amazing creatures deserved to be told. The problem was, how to tell it?

A picture really is worth a thousand words. Very little of the natural history of these creatures had been previously documented in photographs. In this book, photographs capture arachnids hunting, fighting, mating and caring for young. In many cases, these images document behavior that had never been photographed before. Fortunately, the southwestern United States (including Arizona, New Mexico and southern California) is home to all but one arachnid order (the Ricinulei), providing the opportunity to tell the story of arachnids in general through those species found in this region. The story sometimes includes a fourth dimension, time, touching upon how changes over deep time have affected the biogeography of species. In other cases, the story compares and contrasts adaptations seen in these species with related species found elsewhere in the world. This provides a broader evolutionary context for their stories.

In order to address the diversity of species in the region, more than 300 kinds of arachnids are portrayed in photographs. Common as well as rare species are included, chosen as representatives of larger groups of species. Because spiders are an extremely large order, more chapters are dedicated to them. This approach provides an opportunity to explore the rich natural history and diversity of this group.

Terminology used in arachnology can be a challenge. For example, more than one term may be applied to a particular life stage, depending on which order of arachnids is being discussed. Adding to the difficulty, different camps of arachnologists may disagree on the use of a particular term. In this book, I used a very simple criteria for deciding whether particular terminology should be used. If the term is used by arachnologists in currently accepted literature, then I followed their standard. For example, the terms "setae" and "hairs" are both used in the current literature; therefore I use both. Technical terms are explained in context as much as possible.

One of the most commonly asked questions about any given arachnid is, "Is it dangerous?" This book provides practical answers to that concern, providing photographs and information regarding potentially dangerous arachnids from this region. At the same time, photographs of the many harmless species may assist those who simply want to know, "What is it?"

Ultimately, this book has a dual focus: introducing the arachnids of the southwestern United States, and through them, telling the story of arachnids throughout the world. These small, ubiquitous neighbors have amazing stories to tell. I hope that you will enjoy them.

Prologue

A black pit opened at my feet. Staring down into the gloom, I prepared to make the descent down the vertical shaft that made up the entrance into the cave. After donning my helmet and headlamp and taking a few deep breaths, I started to descend. Others had used this path before me; pieces of cacti and rodent droppings were signs that pack rats had been present. A large black widow, plump and glossy, hung belly-up in her web displaying the unmistakable scarlet hourglass of her kind. As I continued deeper into the cave, the daylight faded into a dim twilight at the bottom of the shaft. Only a few feet further and the cave became pitch black. It was here that I searched for my quarry.

My headlamp illuminated an abundance of brown spiders, *Loxosceles sabina*, closely related to the famous brown recluse. These spiders were on virtually every rock surface, necessitating care in placing my hands when touching any rock. Their venom contains the same necrotizing component, sphingomyelinase D, found in the venom of the brown recluse spider. Although these spiders are not aggressive, one could bite if it felt threatened. These were not my quarry.

Kneeling on the uneven floor, I carefully searched under loose rocks. A surprising variety of fantastic creatures were living under the shelter of the rocks: tiny orange harvestmen, pale pseudoscorpions, booklice, swiftly scuttling scorpions, nymphs of bloodsucking kissing bugs, pillbugs, and tiny cobweb spiders. These also were not my quarry, at least not at this time.

I continued to search, carefully replacing each rock as I had found it. My knees were starting to hurt from kneeling on the rock. My husband patiently stood by, shining his light on the rocks as I laboriously lifted and searched under each one. A tiny animated speck caught my attention. Fingers shaking, I fumbled as I captured the fragile creature, using a delicate paintbrush to gently coax it into a vial. This was my quarry.

As we climbed back up the shaft to daylight, we left behind the brown spiders, scorpions, kissing bugs, and finally even the black widow. I stared into the vial at my prize. It was a tiny spider, only a millimeter in body length, with long, spindly legs. Her cuticle was suffused with a delicate rainbow iridescence, giving her a slightly shimmering appearance. We were the first to have ever seen this species; her kind of *Darkoneta* had never been collected or described before. I could not have asked for a better gift on Valentine's Day.

I had once been afraid of spiders. The journey through the camera lens changed all that. Like Alice in Wonderland, I became caught up in a world of the strange, the fantastic, and the wonderful. The world of arachnids rivals any science fiction creation.

This book is a small token of my appreciation and gratitude for these amazing creatures. I hope that you will enjoy your journey into their world.

FACING PAGE: Orb webs are just one of many types of traps and snares constructed of the phenomenal material we call silk. Western spotted orb weaver, *Neoscona oaxacensis*.

CHAPTER 1   Introducing the Arachnids

Tested by 400 million years of a changing planet, arachnids have evolved many strategies for survival. Combining the ancient with the new has produced a diversity of species of every color and shape. Despite this, one key characteristic is shared by all arachnids: two chelicerae, used for piercing, grasping, or chewing up food. In this case, each scorpion chelicera has a fixed and movable finger, giving this scorpion a somewhat toothy "smile."

If "alien" means strange, arachnids are arguably among the most alien of Earth's inhabitants. The archetypal arachnid possesses eight legs, eight beady, unblinking eyes, fangs or pincers, and venom. In addition, the hardened exoskeleton and the multi-jointed legs convey an almost mechanical quality to this living creature; arachnids seem to have stepped right out of the movie *War of the Worlds* and scuttled into our living rooms. Finally, most are small, fast, and nocturnal, and are therefore difficult to observe. Consequently, the human imagination fills in the gaps in our knowledge, creating fearsome creatures. Ironically, the arachnids that exist in the real world rival anything that our imaginations could conjure. As a group they are many things: tough, resourceful, beautiful, and incredibly diverse—but hardly terrifying. Getting to know these small neighbors who share our planet is immensely rewarding and never dull.

First of all, arachnids are arthropods. Arthropods are characterized by segmented bodies, jointed appendages, and an exoskeleton. This exoskeleton, or cuticle, is composed of layers of waxes, proteins, and chitin. Chitin, which is composed of a derivative of glucose (*N*-acetylglucosamine), is combined with other substances that enhance its function as protective armor. In many terrestrial arthropods such as insects and arachnids, the chitin is embedded in a proteinaceous matrix of sclerotin. Sclerotin imparts the brown color to many arthropods; as proteins become cross-linked in the sclerotin, it becomes tougher, harder, and darker in a process called tanning. Immediately after molting, arthropods are pale and their cuticle is soft. Over a period of hours or days, as the sclerotin proteins become cross-linked, the cuticle darkens and hardens. In other arthropods such as crustaceans, calcium carbonate is combined with the chitin in a process of biomineralization. This gives the crustacean the perfect armor, combining hardness and resiliency. The major drawback of an exoskeleton is that the arthropod must molt in order to grow in size. Molting is a risky business; the arthropod is susceptible to predation during this time, and sometimes the process of molting itself can lead to injury or death. Despite this, arthropods have been extraordinarily successful from an evolutionary perspective.

Arthropods make up more than 80 percent of all known species of animals, with more than a million described species and millions more yet to be described. Included in this group are hexapods (six-legged arthropods such as insects and springtails), crustaceans (such as crabs, lobsters, shrimp, and pillbugs), myriapods (centipedes and millipedes), and the chelicerata (including horseshoe crabs, arachnids, and sea spiders).

Arachnids have two main body sections (also known as tagmata) and six pairs of basic appendages. The front of the body is called the prosoma or the cephalothorax; it includes both the head and the thorax fused together. The back of the body is the opisthosoma or the abdomen; it contains most of the reproductive and food storage capacity. In some arachnids (such as spiders) these body divisions are clearly evident, but in others (such as mites and harvestmen) the two body divisions merge together without a clear line of demarcation.

Most arachnids have six pairs of appendages at maturity. First and foremost are the appendages that give the subphylum Chelicerata its name: the chelicerae. The most common form of chelicera is a pair of claws consisting of a fixed upper finger and a movable lower finger. Because the arachnid has a pair of chelicerae, it therefore has two of these "hands," each with its own complement of fingers. This undoubtedly is extremely valuable during the manipulation and mastication of prey. Having two hands is certainly far more effective than having only one hand, especially when attempting to cut up food. Many arachnids have some version of these clawlike chelicerae. Exceptions include some mites that have evolved specialized chelicerae for piercing and sucking and spiders that have a single fang as part of each chelicera instead of the two clawlike fingers. In addition, spider chelicerae are even more specialized, being used like hypodermic needles to inject venom into prey. Some species of pseudoscorpions have a special structure on each pincerlike chelicera called a galea. The galea is a spinneret, delivering silk used for building nests. Male mites of some species may also possess specialized structures on the movable finger of their chelicera used for transferring sperm. This may be a fingerlike projection (the spermadactyl), or an opening that receives the spermatophore (the spermatotreme).

The next pair of appendages are the pedipalps. Pedipalps are basically modified legs, and arachnids have evolved an array of variations depending on how

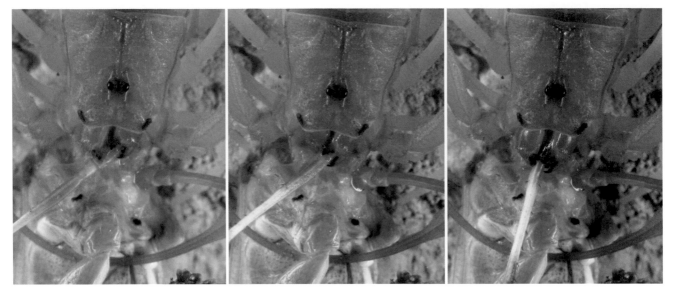

Equipped with a fixed and a movable finger, each scorpion chelicera operates like a hand but is the functional equivalent of jaws. Since the scorpion has a pair of these chelicerae, food can be cut up and masticated during preoral digestion. During this process, digestive fluids are mixed with the masticated katydid, turning it into a liquefied mush. The scorpion ingests only the liquid food: hence the term "preoral digestion." Many arachnids have some version of this form of chelicerae.

1. Instead of having a fixed and movable finger, each spider chelicera has a single fang. These fangs are specialized for injecting venom. A small hole near the tip of each fang provides an opening for the venom to pass into the prey. The subterminal position of the hole reduces the chance of it becoming plugged. The fangs of mygalomorphs such as this tarantula are mostly downward pointing, somewhat like the teeth of a saber-toothed cat.

2. In araneomorph spiders like this jumping spider, the fangs work like pincers.

Long, delicate setae called trichobothria move if there is the slightest disturbance in the air. Each trichobothrium is set in a cuplike socket in which there are dendritic nerve endings. These hairs are sensitive to sound as well as air currents, allowing the arachnid to detect moving prey at a distance. In fact, the trichobothria are so sensitive that some arachnids are capable of detecting and capturing a buzzing fly in midair. These trichobothria are from the walking leg of an amblypygid.

the pedipalps are used. In wind spiders (Solifugae), the pedipalps are elongated and resemble long legs; however, they are used primarily in a sensory capacity. In addition, the solifuge palp has a special suctorial organ at the tip of each palp, which assists it in grasping prey as well as in climbing. Other pedipalps resemble claws or pincers. Scorpions, pseudoscorpions, amblypygids, and vinegaroons have robust pedipalps that are utilized in a variety of ways. Scorpions not only use their palps to grasp prey, but the male scorpion also uses his palps to hold the claws of the female while performing his courtship dance. Some species of pseudoscorpions have venom glands in their palps; therefore, they can simultaneously grasp and envenomate their prey. Vinegaroons use their lobsterlike claws for prey capture as well as for digging. The vinegaroon must have an efficient way to carry out the soil that it has loosened as it digs its burrow; the palps function like the bucket of a bulldozer. Harvestmen (Opiliones) have a wide range of different palps. *Protolophus* shows extreme sexual dimorphism in the size and structure of its palps. The males have large palps that can be used to wrestle a female prior to mating, whereas the females have small palps, each with two delicate fingers. Harvestmen such as *Sclerobunus* and tiny *Sitalcina*

have raptorial palps with which to grasp prey. Each palp is armed with a row of spines; hence their family name Phalangodidae reflects the palp's resemblance to the line of soldiers holding long spears (the phalanx) used during Alexander the Great's campaigns. Mites have a variety of palps that may be modified for prey capture in predaceous mites, hold-fast structures in parasitic mites, or food filters in microbivorous species. The hollow specialized setae (eupathidium) on the tips of the palps of spider mites (Tetranychidae) have evolved to deliver silk instead of functioning primarily as sensory structures. The majority of arachnid palps are also armed with an array of specialized sensory hairs for the detection of water, food, disturbances in air currents, and temperature measurements. Finally, spiders present one of the most extreme examples of sexual dimorphism and specialized structure in their palps. The female has simple, leglike palps, while the male has complicated palps used for transferring sperm to the female.

The next four pairs of appendages are the legs. In wind spiders, vinegaroons, amblypygids, schizomids, palpigrades, and many mites, the first pair of legs is functionally sensory rather than being used for locomotion. In harvestmen, the second pair of legs

Given the diversity of arachnid lifestyles, it is not surprising that arachnids have evolved many kinds of pedipalps.

FACING PAGE:

1. Armed with impressive spines, the raptorial palps of the amblypygid *Paraphrynus carolynae* are its sole tool for capturing prey. These palps open on a horizontal plane.

2. The pedipalps of male spiders are highly specialized structures used for transferring sperm to the female. They must be loaded with sperm prior to courtship, a process called sperm induction.

3. Vinegaroons use their palps for digging and carrying soil. Living underground, they must dig their own burrows.

4. Pseudoscorpions have venom in the fingertips of their clawlike pedipalps. This cricket died within seconds of being stabbed with the tip of the pedipalp.

5. Some harvestmen like this *Sclerobunus* have raptorial, spiny palps used for prey capture. These palps open on a vertical plane.

serves in this capacity. These sensory legs may be referred to as antenniform legs, especially the extremely elongated and delicate first pair of legs found in the amblypygids and the vinegaroons. Specialized setae on sensory legs detect chemical traces or air movement. The sensory legs of arachnids are held out in front as they walk, tapping the ground at frequent intervals in order to "taste" or feel the substrate in search of food, water, or mates. Hard ticks (Ixodida) have a specialized sensory structure on the tarsus ("foot") of the first pair of legs referred to as Haller's organ. This organ detects chemicals, heat, and humidity, assisting the tick in finding a host. As the tick waits on vegetation, it stretches out its front legs, "questing" for a host from which it will take a blood meal. Many arachnid tarsi are endowed with specialized hairs or structures that enable them to cling to vertical surfaces. In many spiders, brushlike scopula hairs serve this function, whereas pseudoscorpions have a little pad called the arolium between the two claws on each tarsus that allows them to walk on vertical surfaces or the underside of an object. This is an extremely useful adaptation, since pseudoscorpions frequently live and hunt on the underside of rocks.

Perception of the environment is dependent on and limited by the kinds of sensory structures present. An arachnid's cuticle is analogous in function to our ears, nose, taste buds, and temperature receptors. For many arachnids, the cuticle is as important as eyes are to humans. Long, delicate trichobothrial hairs are "touch-at-a-distance" receptors, sensitive to any air disturbance as well as low-frequency vibrations (sound), alerting an arachnid to the presence of predator or prey. A spider can capture a buzzing fly even if its eyes are completely covered. Slit sensilla on the legs may also be used for detecting and locating potential prey. These narrow slits in the cuticle are covered by a thin, easily deformed membrane. Substrate vibrations deform the slit and trigger a nerve impulse, allowing the arachnid to locate moving prey. A single spider may have more than 3,000 slit sensilla, most of which are located on its legs. Although different slit sensilla serve in different capacities, it is thought that some slit sensilla may also function as the "ears" of some arachnids. It is only logical that there is some receptor for spiders to "hear" the stridulation produced by a potential mate during courtship. Specialized hairs of many kinds cover the typical arachnid, including tactile hairs that detect touch and contact chemoreceptive ("taste") hairs that have an open pore at the tip for detecting molecules. Other structures in the cuticle may also be used in olfaction, such as tarsal (foot) organs consisting of small pits that may function primarily as hygroreceptors, detecting changes in humidity. Some structures may be specific to a certain group of arachnids. Only scorpions have the comb-shaped pectines, used for mechanoreception as well as for detecting chemical traces, and only the solifuges have malleoli (racquet organs) used for chemoreception.

For many arachnids, the eyes are of secondary importance compared to cuticular sensory structures. Despite this fact, many arachnids do have eyes, and these eyes can play a significant role in their ability to function. Arachnids have two types of eyes: the main (or median) ocelli and the secondary ocelli. The first type of eyes, the median ocelli, are seen as a pair of eyes on top of the cephalothorax in many orders of arachnids, including the scorpions, vinegaroons, amblypygids, harvestmen, and solifuges. Spiders also have main ocelli, the anterior median eyes, which in jumping spiders form detailed images and may be able to detect colors as well as ultraviolet light. The secondary ocelli correspond to the lateral ocelli in many arachnids. These were compound eyes in ancestral scorpions, as is still seen in their cousin the horseshoe crab. These compound eyes evolved into the simple ocelli that are found in modern scorpions. In some groups such as scorpions, there may be up to five lateral ocelli on each side of the cephalothorax. In many arachnids such as wolf spiders, the secondary ocellus has a tapetum which enhances sight in dim light and is responsible for the "eye-shine" of reflected light from these eyes.

Arachnid eyes are covered by a thin, transparent layer of cuticle that protects their eyes against damage and desiccation. Consequently, arachnids do not have eyelids and therefore cannot blink. Instead, jumping spiders clean their eyes with a quick brush using their fuzzy pedipalps to wipe off any dust. Because the front of the eye is covered with this layer of rigid cuticular exoskeleton, the front of the eye must remain stationary. However, in jumping spiders, the back of the anterior median eye can move. These eyes are roughly conical in shape, and the narrow base of the cone rests in a harness of muscles that can move it. Consequently, the area of the retina (at the base of the eye) can be shifted, aiming the vision of the jumping spider. Of course, the

Extremely sensitive to light, scorpion eyes have their own built-in sunglasses. A larger pair of median eyes is in the center, flanked by several pairs of lateral eyes (in this case, 6 lateral eyes). Many other orders of arachnids share this general eye arrangement.

range of movement of the retina is somewhat limited, so the jumping spider orients its cephalothorax in order to better see any disturbance that the smaller eyes had detected.

Instead of forming high-resolution images as jumping spider eyes do, scorpion eyes are specialized for detecting very low levels of light. They even have pigment granules within their eyes that help shield their highly sensitive retinas during the day, comparable in function to sunglasses.

Many arachnids have other light-sensitive structures, although some of these are not fully understood at this time. Some of these are found in

unexpected areas of the body, such as the metasoma (tail) of the scorpion, or the tarsi (feet) of snake mites.

The central nervous system (analogous to our brain and spinal cord) processes all the incoming sensory input and controls the response of the animal. In general, arachnids have two major components to their central nervous system: a supraesophageal ganglion or nerve mass (the brain) and a subesophageal ganglion or mass. The brain controls and receives input from the eyes, while the subesophageal ganglion or nerve mass controls the legs and pedipalps while receiving signals from cuticular sensory structures. In the case of scorpions, additional ganglia control other

Spiders have diversified into a bewildering array of more than 45,000 species. The arrangement of their eyes is a useful tool in identifying them to family.

1. Crab spiders (Thomisidae) have 2 rows of fairly small eyes, often on tubercles, 4 eyes to a row.

2. Wolf spiders (Lycosidae) see well even at night with their large posterior eyes. The 4 anterior eyes are located in a straight row just above the chelicerae.

3. Lynx spiders (Oxyopidae) have 6 conspicuous eyes arranged in a roughly hexagonal arrangement. Two tiny eyes are facing forward, just below the conspicuous eyes.

4. Jumping spiders (Salticidae) are characterized by their large, forward-facing anterior median eyes.

body regions. The cheliceral ganglion wraps around the digestive tube between the supraesophageal nerve mass and the subesophageal nerve mass. It controls the chelicerae. Seven other ganglia are located down the length of the scorpion's body: three in the mesosoma and four in the metasoma (tail). These are all connected by nerve cords and control the opisthosoma, including the action of the aculeus (stinger). Adult spiders have only two major ganglia: the supraesophageal and subesophageal ganglia. The supraesophageal ganglion consists of the "brain" in addition to the cheliceral ganglia. Together, these control the eyes as well as the chelicerae, pharynx, and venom glands. The subesophageal ganglion receives cuticular sensory input and controls motor neurons to the legs and extremities. However, the embryonic development of the spider reveals that this compact arrangement exists only after individual ganglia migrate into the prosoma from the abdomen during development and fuse together to form the impressive subesophageal ganglion. Most other arachnids have some variation of this arrangement. Harvestmen have a large neural mass that consists of two major sections. The protocerebrum together with the deutocerebrum controls the eyes and the chelicerae, while the subesophageal ganglion controls the palps, legs, and opisthosoma. In solifuges, the dorsal cerebral ganglion controls the eyes and chelicera, while the subesophageal ganglion controls the palps, walking legs, and the structures associated with the opisthosoma. Vinegaroons, tailless whipscorpions, schizomids, palpigrades, and mites share this arrangement with slight variations. Pseudoscorpions have a single cerebral ganglion surrounding the esophagus. It is remarkable that these relatively simple arachnid central nervous systems can handle complex tasks involving learning and memory.

Together, the respiratory system and the circulatory system deliver oxygen to the various organs and also remove carbon dioxide, a waste product of respiration. There are three possible respiratory systems found in arachnids: book lungs, tracheae, and cuticular respiration.

The most conspicuous of these are the book lungs, appearing as paired whitish areas just under the cuticle on the ventral surface of the abdomen. The number of book lungs varies with the type of arachnid. Scorpions have four pairs of book lungs while vinegaroons, amblypygids, and mygalomorphs have two pairs. Schizomids and many modern spiders have only one pair of book lungs. Each book lung consists of alternating layers of lamellae and air. The thin parallel layers are stacked such that they resemble the pages of a bound book, giving book lungs their name. Contained within the lamellae is the hemolymph (blood), which picks up oxygen and releases carbon dioxide while passing through the book lung. Because the cuticle of the lamellae is extremely thin, gas exchange can readily occur by diffusion as the hemolymph flows through these hollow, flattened structures. The many stacked layers maximize the available surface area needed for gas exchange. The book lung works somewhat like a bellows, powered by the pumping of the heart. As blood

pressure increases during systole, the lamellae fill with hemolymph and the air spaces become compressed. These open up again during diastole, when blood pressure decreases. A small slit opening to the outside allows fresh air to enter the spaces between the lamellae. Although this slit can be opened or closed by muscular control, respiration in arachnids is considered passive compared with ours.

Book lungs are not very efficient; consequently, the arachnids who depend entirely on book lungs for oxygen cannot sustain a high activity level for long. Tarantulas, vinegaroons, and amblypygids are capable of extremely rapid sprints of short duration, but then they must stop while they "catch their breath," figuratively speaking. During these recovery periods, lactic acid is oxidized. A more efficient delivery system consists of tracheae. These hollow tubes open to the outside air and pass into the body, delivering oxygen directly to the interior. Arachnids and insects have independently evolved this system of respiration. Solifuges, ricinuleids (hooded tickspiders), harvestmen, pseudoscorpions, and some mites have tracheae. Many "modern" spiders have both tracheae as well as one pair of book lungs, and some species of spiders have only tracheae. Arachnids with tracheae can maintain a higher activity level for a longer time than arachnids that have only book lungs. However, even the tracheae of arachnids is still inefficient compared with our respiratory system, which employs a diaphragm and lungs.

The third method of respiration is through the cuticle. This is found primarily in tiny arachnids such as palpigrades and some mites. Cuticular respiration requires a large surface area compared with the mass of the animal, thereby limiting it to tiny animals. A thin cuticle is advantageous for this type of respiration. Palpigrades have an extremely thin, weakly sclerotized cuticle that is probably well adapted for cuticular respiration in moist environments.

The hemolymph of arachnids, like the blood of vertebrates, transports oxygen and carbon dioxide; however, instead of being red in color like our blood, arachnid blood is somewhat blue in color. This color is due to hemocyanin (containing copper) in their blood instead of hemoglobin (containing iron). Hemocyanin has been around for a very long time and is found in both mollusks and arthropods. The oxygen-carrying capacity of hemocyanin is significantly lower than that of hemoglobin. It remains a mystery why such a relatively poor transport molecule should be conserved for so many millions of years. The answer may lie in the fact that arachnids do not use specialized cells to package the hemocyanin; these enormous molecules are freely circulating in the hemolymph. In contrast, hemoglobin is contained within specialized red blood cells whose sole purpose is to transport hemoglobin. If large numbers of the red blood cells are lysed (broken open), the free hemoglobin is actually toxic to many organs of the body. In fact, free hemoglobin can kill the organism. Ultimately, there may be a greater net efficiency in having to manufacture only the transport molecule without complex cellular packaging; however, there must be a concurrent low metabolic rate for this strategy to succeed. Low oxygen use, a large food storage capacity, and long periods of inactivity may collectively contribute to a survival strategy of low energy use. This strategy may make the difference between life and death during lean times.

The circulatory system works with the respiratory system to supply oxygen and energy to all parts of the body. Humans have a closed circulatory system whereby blood is enclosed within arteries and veins. Arachnids have a circulatory system which is partly open and partly closed. The heart is tubular in shape and lies lengthwise inside the opisthosoma. It pumps hemolymph to the arteries, which in turn branch into smaller vessels. From there, the oxygenated blood is distributed throughout the body, right to the tips of the extremities. The branched arterial blood vessels are open at the end, allowing oxygenated hemolymph access to the tissues. The arachnid arterial system is primarily a closed system, open only at the tips of the smallest arterial vessels; however, the arachnid's return blood flow system is very different from ours. Arachnids do not have a venous system of blood vessels. Instead, the hemolymph flows freely within the body, traveling along a gradient of progressively decreasing pressure. Eventually the hemolymph collects in spaces called lung lacunae and from there is routed through the book lungs (if there are book lungs) before returning to the heart.

All the energy for fueling this system must come from food. Most but not all arachnids are predators. Noteworthy exceptions are the harvestmen (Opiliones) and some mites. Many species of harvestmen are omnivorous, feeding on an incredibly wide variety of plants, animals, and fungi. Mites include a vast array of different species with different lifestyles. Among mites

are predators, parasites, fungivores, herbivores, and detritivores. Harvestmen and some species of mites ingest particulate food, but most arachnids ingest only liquefied predigested food.

Wastes are concentrated and excreted by several systems. Intestinal cells within the digestive diverticula synthesize guanine, a nitrogenous waste product, and route it into the intestine to be excreted with the feces. Guanine is relatively insoluble in water and so precipitates out as a white paste, thereby conserving water. Another important system for nitrogenous waste removal consists of the Malpighian tubules. These tubules are found between the gut diverticula and empty out into the intestine. Their role is to extract nitrogenous wastes from the hemolymph and convert these wastes into the largely insoluble guanine, adenine, hypoxanthine, and uric acid, to be excreted later. Malpighian tubules are found in many orders of arachnids, including spiders, scorpions, vinegaroons, and others. A final set of excretory organs consists of the coxal glands. These glands open at the base (or coxa) of the legs and are probably important in ion and water balance. Fluid from these glands is released only during feeding and may assist in processing food for digestion.

Arachnids have a stunning variety of reproductive strategies. The transfer of sperm may involve the pedipalps (spiders), an intromittant organ (harvestmen and some mites), or spermatophores (scorpions, amblypygids, vinegaroons, pseudoscorpions, some solifuges, and some mites). Some transfer sperm directly from the male's genital opening to the female (some solifuges). In some arachnids, no mating is necessary; these species are parthenogenetic and consist only of females. Finally, in some species, males can be produced from unfertilized eggs and females are produced from fertilized eggs. The many different methods of reproduction reflect the fact that arachnids have been diversifying and adapting to changes on this planet for a very long time. The ability to colonize ephemeral habitat with only one individual (as is possible in parthenogenetic species) may be more important than the ability to recombine genes (as is seen in sexual reproduction).

Arachnids have been on Earth for at least 400 million years. In an attempt to tease out the story of their evolution, several different types of data are considered. Morphology, embryology, and DNA

1. The spermatophore of the vinegaroon was deposited by the male and picked up by the female.

2. Harvestmen transfer sperm directly via an intromittent organ. The male in this pair is on the right.

3 and 4. The pedipalp of the male spider is specialized for transferring sperm to the female. The male spider must load his pedipalps with sperm prior to courting the female. The male is the smaller of this mating pair.

Eurypterus remipes. Eurypterids, also known as sea scorpions, were distant relatives to our modern arachnids. The chelicerates split into two major groups: one clade contained eurypterids and horseshoe crabs, while the other clade included all the arachnid orders.

sequencing all contribute evidence that can be used to construct cladograms depicting the phylogeny, or evolutionary history, of a group of related organisms. The addition of fossils from extinct taxa presents additional challenges. Fossils cannot provide DNA or developing embryos; they are limited to the partial morphology of an organism. What they do provide is a glimpse of creatures that may have been extinct for hundreds of millions of years, creatures whose lives from the distant past are intertwined with the lives of their living relatives today, linked forever by their phylogenetic relationships. The inclusion of fossils may therefore provide additional pieces of the puzzle, filling gaps in the evolutionary picture.

The process of constructing cladograms may at times resemble the story of the blind men and the elephant; each cladogram may be significantly different from the others depending on what data are included and how that data are analyzed. Consequently, there is

no consensus on the exact phylogeny of the arachnids. Despite this, some interpretations of arachnid evolution do have support across a broad base.

The first challenge consists of determining where chelicerates belong in relation to other arthropods. One classification scheme placed arachnids, horseshoe crabs, and sea spiders on one branch (the Chelicerata) and all the other arthropods, including crustaceans, insects, and myriapods, on the other branch (the Mandibulata). Another analysis incorporating molecular evidence (RNA and DNA) as well as data on embryonic development has generated a different model. The Arthropoda once again split into two groups, but in this case one clade consists of the Parahexapoda (including chelicerates and myriapods) while the other clade includes the Pancrustacea (including crustaceans and Hexapoda, such as insects).

The earliest record of a possible chelicerate dates from the Cambrian Period, 542 to 488 million years ago. Better techniques for processing fossils, including improved imaging technology, have revealed detailed structures locked away in rock for millions of years. Orsten limestone nodules from Sweden contain the remains of many fossilized invertebrates. The chitinous cuticles of these tiny invertebrates were phosphatized and silicified, preserving them in a matrix of limestone. The limestone can be removed with acid, leaving the microscopic fossils intact. Electron microscopy is then used for generating images of these minute and ancient fossils. A tiny protonymph larva of a sea spider (Pycnogonida) dating from the Upper Cambrian Period was discovered in this way. This larva possessed chelicerae, also known as cheliphores in sea spiders. Only 270 micrometers in length (just over a quarter of a millimeter) and approximately half a billion years old, this larva provides us with evidence that chelicerates appeared very early in the Paleozoic Era.

The exact position of sea spiders within the arthropod cladogram remains controversial. However, many agree that the remaining chelicerates break down into two major clades. Most horseshoe crabs (Xiphosura) along with the extinct sea scorpions (Eurypterida) and Chasmataspidida (which shared characteristics with both sea scorpions as well as horseshoe crabs) form one clade. The sister group to this aquatic clade of chelicerates contains all twelve living arachnid orders as well as four extinct orders of arachnids. It is noteworthy that with the exception of some derived species of aquatic mites and one species of aquatic spider, virtually all the extant species of arachnids are terrestrial. This is in contrast to those most ancient chelicerates (horseshoe crabs, sea scorpions, etc.), which lived exclusively in a marine environment. The aquatic lifestyle of the earliest creatures reflects the story of life on Earth; the marine environment was the cradle of these early life forms in part because the land was still barren and inhospitable. The colonization of land by plants started in the Ordovician Period 488 million years ago with nonvascular plants such as mosses and liverworts. As plants continued to evolve during the remainder of the Paleozoic Era, the face of the land changed. Plants were the catalyst for the diversification of terrestrial arthropods, directly providing food for herbivores, as well as indirectly supporting detritivores and predators, including arachnids. Sometime before the Silurian Period the ancestral arachnids probably started living

on land. Their descendants diversified, filling a variety of newly available niches. Many of today's arachnid orders made their first appearance during the Silurian, Devonian, and Carboniferous periods. The following is a list of the arachnid orders along with the time period during which they are known to have appeared, as seen in the fossil record; however, this fossil record may demonstrate "sampling bias." The extremely fragile creatures such as palpigrades may only rarely become preserved as fossils, compared with more durable arachnids such as the scorpions.

A clade referred to as the Tetrapulmonata is well accepted among many arachnologists. As its name would imply, the Tetrapulmonata group includes several orders that possess four book lungs. These include

EXTANT ARACHNID ORDERS
Scorpiones (scorpions)—Silurian
Pseudoscorpiones (pseudoscorpions)—Devonian
Opiliones (harvestmen)—Devonian
Acariformes (mites)—Devonian
Amblypygi (tailless whipscorpions)—Carboniferous, or possibly Devonian
Araneae (spiders)—Carboniferous
Thelyphonida (Uropygi, vinegaroons)—Carboniferous
Schizomida (short-tailed whipscorpions)—Carboniferous
Solifugae (wind spiders or camel spiders)—Carboniferous
Ricinulei (hooded tickspiders)—Carboniferous
Parasitiformes (ticks and mites)—Cretaceous
Palpigradi (microwhipscorpions)—Tertiary

EXTINCT ARACHNID ORDERS
Trigonotarbida—Silurian-Permian
Uraraneida—Devonian-Permian
Phalangiotarbida—Devonian-Permian
Haptopoda—Carboniferous

Amblypygi (tailless whipscorpions), Thelyphonida (vinegaroons), Araneae (spiders), and the extinct orders Haptopoda and Uraraneida. The group also includes Schizomida (short-tailed whipscorpions), although they have only one pair of book lungs. The spiders considered closer to the ancestral form, such as Liphistius and the mygalomorphs, have two pairs of book lungs. Other spiders that have only one pair of book lungs (or none) are thought to be derived and

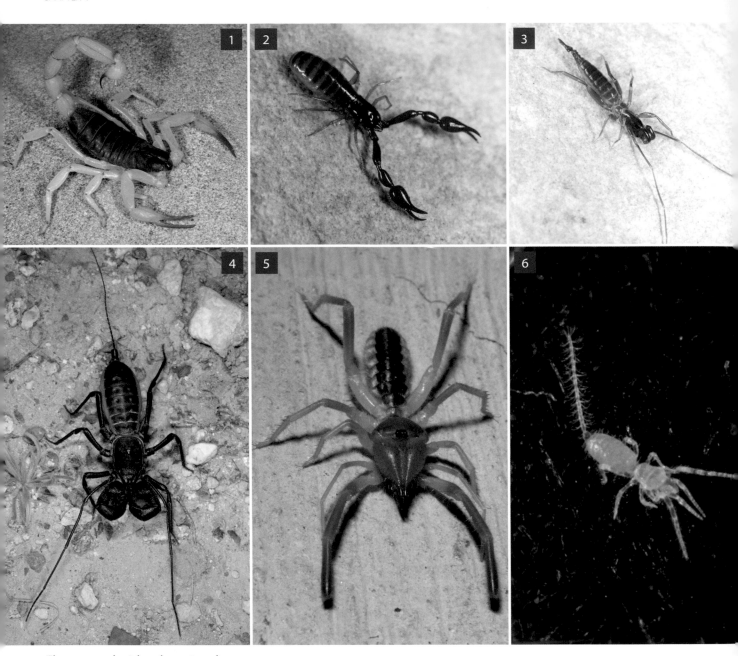

Eleven arachnid orders at a glance:

1. Scorpiones
2. Pseudoscorpiones
3. Schizomida: short-tailed whipscorpions
4. Thelyphonida (Uropygi): vinegaroons, whipscorpions
5. Solifugae: wind spiders
6. Palpigradi: microwhipscorpions

7. Amblypygi: tailless whipscorpions
8. Araneae: spiders
9. Opiliones: harvestmen
10. Acariformes: mites
11. Parasitiformes: ticks
12. Parasitiformes: opilioacarids and other mites

The Acariformes and the Parasitiformes both contain mites. However, it appears that these two major groups of mites may have evolved independently and might not be closely related.

The order Ricinulei has not been included since it does not occur in the southwestern United States. Although hooded tickspiders historically occurred in Texas, they have not been found in the United States in recent times.

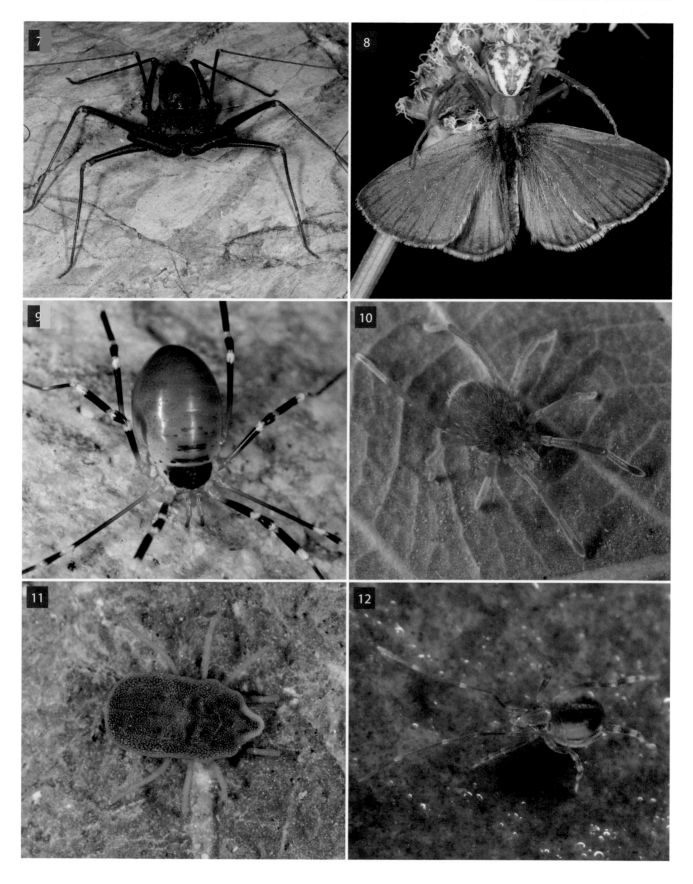

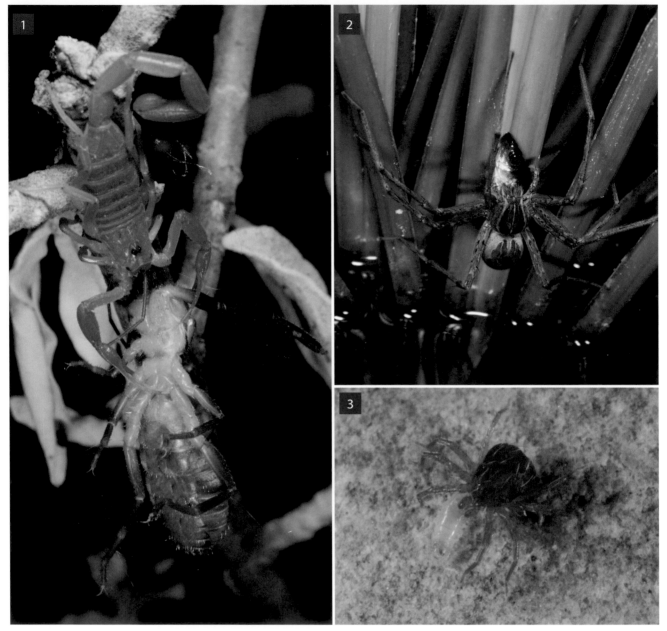

Millions of years of diversification and adaptation have produced a stunning repertoire of hunting strategies among the arachnids.

1. The business end of the scorpion is at the tip of its "tail," armed with a needle-sharp aculeus and a supply of venom. This scorpion has been able to overpower a solifuge as large as itself.

2. Fishing spiders like this *Tinus peregrinus* capture fish just at the surface of the water. After immobilizing its prey with venom, the spider must carry it up out of the water in order to feed; otherwise the enzymes needed for predigestion would be diluted out.

3. *Anystis* mites also use venom for prey capture. Known as whirligig mites, these little predators run rapidly in circles hunting other small invertebrates like this fly larva.

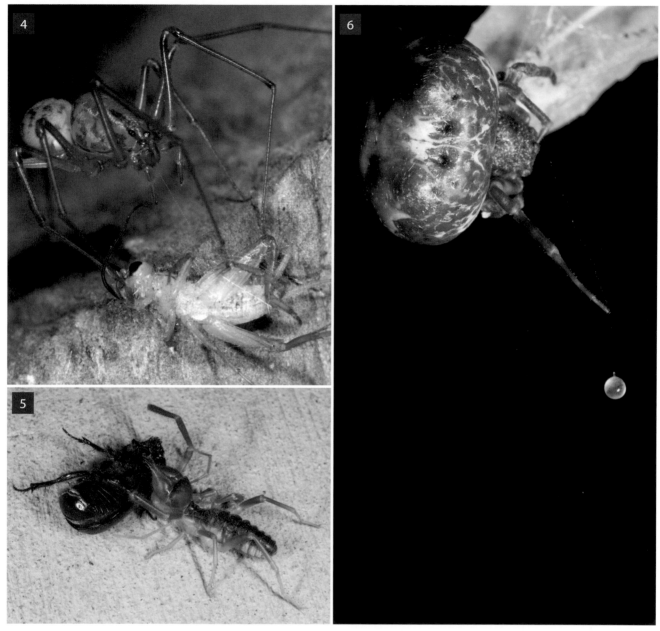

4. Spitting spiders capture prey by squirting two streams of sticky glue and silk from their tiny fangs, fastening the prey to the substrate.

5. Armed with massively powerful chelicerae, the solifuge energetically attacks and tears apart its prey. These solifuges have no venom; therefore, they must use speed and strength to overpower prey.

6. Using aggressive chemical mimicry, bolas spiders release fake female moth pheromones in order to lure male moths within striking distance. Once the male moth flies close enough, the bolas spider swings the sticky blob, hits the flying moth and captures her dinner. Photo by Matt Coors

more modern from an evolutionary perspective. The order Uraraneida is of special interest. These extinct arachnids date from the Devonian Period through the Permian Period, 416 to 251 million years ago. They share several characteristics with the Araneae (spiders), including silk glands and a naked cheliceral fang. However, this order differed from spiders in that it had a flagellum (similar to a vinegaroon's) and lacked spinnerets. Uraraneida is considered a sister group to Araneae and therefore by inference provides clues regarding the ancestors of spiders.

A controversial area of arachnid phylogeny involves the Acari (ticks and other mites). Traditionally, mites were considered to belong to a single group based in part on having a hexapodal (six-legged) larva. Within this group were two superorders: the Acariformes (also known as Actinotrichida) and the Parasitiformes (also known as the Anactinotrichida). The Acariformes is a highly diverse group containing a vast number of species found in almost every conceivable niche throughout the globe. Included within the Acariformes are the "typical" mites, such as red spider mites and velvet mites. The earliest unquestioned records of acariform mites date to the Devonian Period, 416 to 359 million years ago, preserved in the Rhynie chert of Scotland and the Gilboa Formation in New York State. Molecular data suggest that they occurred even earlier, during the Silurian Period, 444 to 416 million years ago. In contrast to the acariform mites, the earliest parasitiform mites do not appear until much later, during the Mesozoic Era, 251 to 65 million years ago (contemporary with the dinosaurs). The Parasitiformes contain the Ixodida (ticks), Mesostigmata (including many phoretic and parasitic mites), Holothryrida, and Opilioacarida (the "harvestmen" mites). Altogether the parasitiform mites make up only about one-third of the total number of described Acari.

Independently generated cladograms based on molecular data compared with cladograms that include fossil data both strongly support the concept that Acari is not monophyletic; in fact, it most likely evolved from two separate lineages of arachnids. The Acariformes are an ancient group probably most closely related to the Solifugae (camel spiders). The Parasitiformes are a more recently evolved group. They may be more closely related to Ricinulei (hooded tickspiders) or possibly to pseudoscorpions than they are to the acariform mites; however, as mentioned earlier, agreement has not yet been reached on the exact structure of the arachnid family tree, so further developments may yet modify the model.

All the extant orders of arachnids are included in this volume except for the hooded tickspiders (Ricinulei), which do not currently occur in the southwestern United States. Phalangiotarbida, Trigonotarbida, Uraraneida, and Haptopoda have become extinct, and are therefore also excluded.

From the Equator to the Polar regions, arachnids can be found on every continent of our planet in every imaginable environment. They can be found in deserts and in lakes, in jungle canopies and in underground caves. They ride the winds thousands of feet above the earth and survive buried deep in the soil. They have survived profound changes on the face of the planet during the past 400 million years. Land masses have converged forming supercontinents, only to break up and drift apart again but in a different configuration. Mountains rise up and climates change. Jungles, savannas, and deserts appear and disappear; nothing is spared from change over time. In addition to abiotic challenges, millions of other species of organisms have evolved during this period. Some were prey, some were predators, and some were competitors.

The arachnids described in the following chapters are the living legacy of these millions of years of diversification and adaptation. Their defense mechanisms, reproductive strategies, and hunting methods have been refined and perfected by natural selection over the eons, at times producing extreme solutions to challenging problems.

Ironically, natural selection has sometimes produced diametrically opposite solutions to a particular problem. For example, many arachnids use mimicry as protection against predation. This mimicry may take the form of camouflage, making the animal virtually invisible in its environment. An opposite kind of mimicry may actually attract attention with bright warning colors, imitating the appearance of stinging insects such as wasps. In another example of evolution producing opposite extremes, some spiders may be solitary and fiercely territorial, while other species even within the same family may be completely social, cooperatively catching prey and collectively raising their young. The metabolism and reproductive rates of arachnids may demonstrate opposite extremes as well. Ticks have such a slow metabolism that some have been known to survive for 10 years without food, while some mites can fit an entire life cycle into a brief 34 hours.

Arachnids present a puzzling paradox. By necessity, their brains are quite tiny, and therefore their cognitive capabilities are presumed to be correspondingly small. However, some arachnids have unequivocally demonstrated a degree of plasticity in their behavior and a remarkable capacity for storing information and problem solving that seems impossible given these minute brains.

It is in the categories of hunting and defense that arachnids have evolved almost every imaginable strategy, and even a few strategies that may be beyond the imagination. In the realm of hunting, trapdoor spiders pop up out of the ground, raft spiders capture fish at the surface of the water, pirate spiders lure their quarry by imitating prey, bdellid mites "tag" their prey with silk, oecobiid spiders race around giant ants in order to hobble them, bolas spiders lure their prey

with fake pheromones, pseudoscorpions have venom in their claws, scorpions have venom in their "tails," and spitting spiders spit out two narrow streams of glue mixed with silk. Of course, web spiders construct an ingenious variety of traps and snares using the phenomenal material we call silk. In the line of defense, vinegaroons and schizomids spray acetic acid, green lynx spiders spray venom, harvestmen release a veritable cocktail of repugnant chemicals, and crab spiders produce pigment to match the color of the flowers they sit in.

Arachnids present the perfect combination of form and function, illustrating the integration of anatomy, physiology, chemical capabilities, and behaviors. These creatures are simultaneously ancient and modern, living fossils in some respects and nimble opportunists in other ways. In the following chapters, their stories will rival the best science fiction fantasies. Welcome to the world of the arachnids.

This tiny *Bellota* from Arizona is only about a tenth of an inch (3 mm) long, the same size as the shiny black ants it mimics. Despite their small size, jumping spiders demonstrate a surprising degree of intelligence.

CHAPTER 2 Scorpions:
Scorpiones

Grasping a cricket with its claws, a scorpion (*Paruroctonus gracilior*) is able to place its stinger (aculeus) with exquisite precision. In this case, the cricket died almost instantly, thus ensuring a meal for the scorpion. The stinger may be deployed in a defensive capacity as well. In a contest between two predators, the speed and accuracy of the scorpion's sting can decide who will live and who will become food for the victor.

Four hundred million years ago, a formidable predator left its aquatic environment and crawled onto land. Perhaps this new territory freed it from the competition of those other Silurian predators, the sea scorpions known as eurypterids. The eurypterids had been paddling around the oceans since the Ordovician Period some 460 million years ago. During this period, they had diversified into many sizes and species. Some were impressive in size, reaching more than 2 meters in length. All had chelicerae, the clawlike "jaws" that allowed these predators to tear apart and masticate their prey. Some had a spine on the end of a tail-like telson.

The aquatic ancestor of terrestrial scorpions might have been something like a eurypterid, but instead of paddles for swimming, it may have had only legs for creeping along the bottom or on vegetation. Like a modern dragonfly larva, the ancient aquatic scorpion used its compound lateral eyes for prey recognition, stealthily approaching to within ambush distance. However, instead of capturing its prey by creating a suction force while opening huge, expandable jaws (as do dragonfly larvae), the ancestral scorpion used its pedipalp claws to grasp the victim while the tail-like metasoma bent forward, armed with a venomous stinger to finish it off. Like the eurypterids, it then tore apart its quarry using chelicerae.

The transition from an aquatic to a terrestrial existence required some modifications in the body of the ancestral scorpion. For the terrestrial scorpion, respiration occurred through enclosed book lungs instead of through gills, thereby reducing water loss through evaporation. Book lungs have small openings called spiracles, which can be opened or closed as needed, further reducing water loss. The scorpion's legs had to become stronger in response to the demand of holding up its full weight without the support of surrounding water.

The cuticle was protective against both desiccation and damage from the sun's ultraviolet radiation. Long-chain hydrocarbons, lipids and waxes associated with the thin, outermost layer of cuticle, called the epicuticle, also protected against water loss. This thin layer was fluorescent under ultraviolet light, possibly shielding the scorpion against the sun's damaging rays as it left the protection of the water. (A distant cousin of the scorpion, the horseshoe crab, shares this ability to fluoresce under ultraviolet light.) A thicker

Proscorpius osborni lived during the Silurian Period about 420 million years ago. It may have hunted prey on partially submerged plants.

Under ultraviolet light, scorpions fluoresce, giving off a greenish glow. Betacarboline, which forms during the hardening or tanning of the cuticle, is one substance that contributes to this effect. Coumarin (7-hydroxy-4-methylcoumarin) is a phenolic compound found in the hyaline exocuticle of the scorpion, and in some form is probably involved in fluorescence. Fluorescence might protect against the sun's radiation.

underlying layer of protein-chitin complex making up the procuticle provided additional protection as well as structural strength.

Digestion became partially external. Food was first chewed into a pulp in a preoral chamber bordered by the basal segments of the pedipalps, the first two pairs of legs, and the chelicerae. Digestive enzymes flowed from maxillary glands onto the masticated mush. As the scorpion slurped down the liquefied meal, a maxillary brush filtered out the solid particles that were expelled. An advantage of external digestion may be the reduction of ingested parasites, excluded during the filtration process. External digestion could have developed only in a terrestrial creature, since an aquatic environment would dilute out the digestive enzymes.

The morphing of the aquatic scorpions into something akin to our present-day terrestrial scorpions allowed them to exploit the increasingly complex terrestrial environment that developed during the Carboniferous Period, 360 to 290 million years ago. Lush plant growth in the form of tree ferns, giant horsetails, and lycopod trees dominated the cool, moist environment, and the earliest conifers appeared. Huge dragonflies with a wingspan more than a foot across flew through this green world, fueled in part by the high oxygen content of the atmosphere provided by the abundant plant life. Decaying plants became food for detritivores such as cockroaches, who in turn became food for predators such as terrestrial scorpions. As the primitive plants of the Carboniferous died, an occasional scorpion was trapped in the plant material. Over eons of time these layers became coal. The secrets of the ancient scorpions can now be unlocked by treating coal with fuming nitric acid. Fragile epicuticle only 10 micrometers thick can be freed from its 300-million-year-old grave, allowing a glimpse of the most delicate structures, even the fine detail of the ancestral scorpion's compound eyes.

During the Carboniferous Period and continuing into the Permian Period, the ancient continents Gondwana in the south and Laurasia in the north converged to form the supercontinent Pangaea. This allowed terrestrial scorpions to disperse across the mostly tropical landmass, providing the ancestral foundation of scorpions found on every continent except for Antarctica. The formation of the single huge landmass of Pangaea spelled the end of the lush fern forests of the Carboniferous. The interior of the continent was far away from the moderating influence of the ocean. The climate became extreme, with hot dry periods alternating with violent seasonal monsoons. Perhaps during this change in climate, scorpions became more nocturnal. Long thin hairs developed on the pedipalps. These trichobothrial hairs were exquisitely sensitive to the slightest air movement and allowed the terrestrial scorpion to locate prey in the dark. As nonvisual sensory structures evolved, scorpions depended less on their eyes for prey recognition. The compound eyes were reduced to the smaller, simpler eyes of modern scorpions.

After millions of years of evolution and diversification had produced a rich array of life forms on the planet, the slate was very nearly wiped clean. At the end of the Permian, 251 million years ago, a mass extinction event occurred. The cause of this extinction event is still the subject of speculation. One plausible theory holds that severe volcanic activity over a span of about 10 million years may have been the culprit. But one fact is certain. No other extinction event was as extreme as the Permian-Triassic "great dying." Up to 96 percent of all marine creatures and 70 percent of terrestrial vertebrates became extinct along with 83 percent of all insect genera. The marine relatives of scorpions, the eurypterids, which had been cruising the oceans for more than 200 million years, were among the casualties of this mass extinction; however, their smaller terrestrial cousins, the scorpions, survived.

Life reinvented itself. After the Permian-Triassic extinction event, an astonishing array of new species eventually repopulated all the niches. By the Cretaceous Period, 100 million years ago, an incredible diversity of dinosaurs flourished throughout the planet. The small mammals that were the ancestors to the primates persevered under the shadow of these reptiles, laying low and staying under the radar. The engine of coevolution drove the angiosperm plants to produce ever more flamboyant flowers advertising the nectar and pollen rewards for the insects (such as bees) that pollinated them. That same engine powered the arms race between plants and herbivorous insects. Almost as soon as plants could evolve a chemical defense against being eaten, insects evolved a way to sequester or utilize those chemicals, often to their own benefit. Meanwhile, scorpions remained largely unchanged in their basic morphology; however, the breakup of Pangaea into new continents during the Mesozoic did separate populations, leading to divergence and subsequent new species.

One June day 65 million years ago, everything changed. The impact of a huge asteroid off the Yucatan Peninsula formed the Chicxulub crater and initiated another mass extinction across the globe. The Cretaceous-Tertiary extinction obliterated the mighty dinosaurs, as well as countless other species. Among the survivors were flowering plants, bees, the ancestors of modern birds, the small shrewlike ancestors of modern primates, and the tough little scorpions, now survivors of two mass extinctions.

As the planet cooled about 2 million years ago, large areas of tropical forest became savanna and deserts appeared. Not surprisingly, scorpions, the ultimate survivors, have adapted to some of the most challenging environments on the planet. Modern scorpions are found in the Himalayas, the Alps, the mountains of western North America, and even under snow-covered stones 18,000 feet (5,500 m) high in the Andes of South America. Completely blind scorpions have been found deep in caves, while others are found along the shoreline of the ocean, exposed to salt water as the tides come and go. Some scorpions still thrive in tropical forests, but many others have become desert adapted. In fact, scorpions may make up one of the most significant components of the total animal biomass in several deserts relative to other arthropods and vertebrates. (However, their impact on energy flow through a system may be less than expected because of their low metabolic rate.)

A surprising diversity of species inhabits different niches in the arid and semiarid southwestern North America. Some are psammophiles, adapted to sandy deserts, having specialized bristlelike setae that allow them to walk on the sand surface with ease and agility. Others are lithophiles, living on vertical rock surfaces and using narrow crevices as retreats. Some are fossorial, living in scrapes and burrows or in layers of debris on the ground.

A number of morphological and physiological adaptations are advantageous for life in extreme environments such as deserts and caves. First of all, scorpions have a very low metabolic rate. Book lungs do not provide enough oxygen for prolonged physical activity, but by the same token, energy is conserved by animals with low oxygen consumption. Scorpions at rest have a metabolic rate only about 25 percent of a comparably sized insect at rest, which has tracheae to provide oxygen to its body. In fact, only ticks are known to have a lower metabolic rate.

The long, stiff setae on the legs and tarsi of *Smeringurus mesaensis* allow it to travel across sand with ease. This desert-dwelling scorpion is a psammophile.

In addition, when food is available, scorpions can ingest a very large meal relative to their own body weight: as much as 17 percent of their weight in some cases. The extra calories are then stored as glycogen in the hepatopancreas, a large, liverlike organ that may make up to 20 percent of the body mass. This strategy of storing calories is analogous to that used by another desert inhabitant, the Gila monster. The Gila monster stores fat in its tail and can survive on only a couple of large meals a year. This "feast or famine" adaptation is especially important for surviving in environments such as deserts that may require periods of inactivity during extreme temperatures. It is also a useful adaptation for living in caves, where the frequency of finding food may be greatly reduced.

Another adaptation for desert survival includes water conservation. Water is conserved by the excretion of almost insoluble nitrogenous waste such as guanine, xanthine, and uric acid instead of urine. The principal waste is guanine, which maximizes the excretion of nitrogen, since guanine contains one more nitrogen atom than does uric acid and is less soluble in water. The cuticle, as previously noted, protects against both desiccation as well as ultraviolet radiation. Pore and wax channels convey lipids and waxes from the epidermal glands to the cuticle. Even if the scorpion does become dehydrated, it is tolerant of high osmotic ion concentrations in its hemolymph. Sensory hairs on

Lacking pigment except for around its median eyes, a baby *Pseudouroctonus* scorpion has just left its mother's back.

the "feet," or tarsi, detect humidity, thereby assisting the scorpion in finding the most favorable microhabitat in a given environment.

In addition to the morphological and physiological adaptations for water conservation, scorpions have adopted behaviors that assist in thermal and water regulation in extreme desert environments. Nocturnal activity patterns reduce desiccation and thermal stress, and some scorpions construct burrows or scrapes that provide a cooler, more humid refuge during the worst heat.

In becoming nocturnal, scorpions' eyes have become extremely sensitive to light, so much so that they have built-in "sunglasses" protecting their eyes from excessive light. These "sunglasses" consist of pigment granules in the retinula cells of the eyes that migrate upward during the day, thereby shielding the light-sensitive rhabdoms in the retina. At night, these pigment granules migrate down to the base of the retinula cells and radially away from the rhabdoms, allowing full exposure to light. Consequently, at night the median eyes of scorpions increase in light sensitivity by up to 4 log units, and the lateral eyes by up to 1 log unit. Even so, the lateral eyes are at least 10 times more sensitive to light than are the median eyes. This allows scorpions to navigate to and from a burrow using only starlight. In fact, bright moonlight may be too intense, with the result that scorpions may avoid hunting during the full moon. In addition to migrating

pigment granules, fixed pigment in the cells that line the post retinal membrane and cells surrounding the lens form a "fixed iris" blocking all light from reaching the retina except for light passing through the lens. The pigment that surrounds the eye is evident even in newborn scorpions, which are frequently pale in coloration except for the dark pigment associated with their eyes.

While the lateral eyes may be far more sensitive to low light than the median eyes, the latter provide a better quality image. The median eyes have a well-focused dioptric apparatus consisting of lens, vitreous fluid, and retina, whereas the lateral eyes consist of rather flattened lenses and no vitreous fluid; however, even the combination of the median and lateral eyes is rivaled in sensitivity by other structures. Perhaps the most important for prey detection is an array of both mechanosensory and chemosensory structures.

The initial detection of potential prey from a distance is via the basitarsal slit sensilla on the legs that help the scorpion orient to the source of vibration on the substrate. Once the scorpion has localized the area of disturbance and is close enough to the prey, the specialized hairs on the pedipalps called trichobothria come into play. These long, thin, lightweight hairs are extremely responsive to air currents and low-frequency sound. Scorpions can pinpoint prey several centimeters away in this manner, enabling them to orient to and capture even flying insects. Once physical contact is initiated, chemosensory hairs at the tips of the pedipalps and on the tarsal leg segments identify the prey. These chemosensory hairs each have a curved tip and an open pore that allows sensory nerves access to chemical cues. Similar chemosensory hairs are found on many arachnids; however, in addition to chemosensory hairs, scorpions possess a unique structure for chemical detection in the form of pectines. All scorpions have a pair of these comblike structures attached to the ventral surface of the body. Each "comb" has from 5 to 40 "teeth," depending on species and gender of the scorpion. In turn, these "teeth" each have hundreds to thousands of microscopic peg sensilla. Each peg sensilla in turn has 10 to 18 dendritic sensory cells. Consequently, scorpions may have more than a million sensory neurons associated with the pectines. The spatial array of peg sensilla across the pectines provides a means for scorpions to follow chemical signals, and probably provides feedback to the scorpion on whether it is getting closer to or farther away from its quarry.

00 15.0kV 12.0mm x270 SE 2/5/2010 · 200

00 15.0kV 12.0mm x650 SE 2/5/2010 · 50.0 · 00 15.0kV 12.0mm x1.70k SE 2/5/2010 · 30.0

Scorpions are unique in possessing a pair of comblike pectines on their ventral surface. These pectines can detect faint traces of chemicals, permitting the scorpion to find water, prey, or mates. Electron micrographs show the peg sensilla lining the lower edge of each pectinal tooth. Each scorpion has from 5 to 40 teeth to each pectinal comb. Each tooth has hundreds to thousands of peg sensilla, and each peg sensilla has 10 to 18 dendritic sensory cells. Therefore, scorpions may have more than a million sensory neurons associated with the pectines. The spatial arrangement of the teeth on the combs gives the scorpion feedback as to whether it is getting closer or farther from the chemical trail.

Once the scorpion has located and physically contacted its prey, it must capture and kill it. Scorpions have two primary weapons at their disposal, and frequently these are used in tandem for best effect. These weapons consist of the venom, delivered by way of the aculeus, or stinger at the end of the telson, and the pincerlike chelae of the pedipalps used for grasping the prey. Venom is produced in paired glands located in the vesicle at the end of the scorpion's "tail." Contraction of circular muscles surrounding the venom glands forces the venom through two ducts that open near the needle-sharp tip of the aculeus.

The principal effective components of *Centruroides* venom are neurotoxins. These small proteins bind to ion channels (sodium, potassium, or chloride channels), and in mammals an alpha toxin prolongs the conductance of sodium ions. In *Centruroides sculpturatus* envenomation of humans, severe pain at the site of the sting and some systemic neurological symptoms— include jerking and twitching of the muscles and

The vesicle and the aculeus (stinger) of *Anuroctonus phaiodactylus*.

the eyes rolling randomly and uncontrollably—may result. *Centruruoides sculpturatus* may rarely even be lethal, especially for young children. Fortunately, an effective antivenin is produced in Mexico that has been used successfully in pediatric cases in the southwestern United States. In uncomplicated adult cases, the severe pain of the immediate site of the sting gradually fades after several hours, leaving an odd tingling in the extremities, especially the fingertips, as if they had "fallen asleep." If the venom directly hits the nerve of a finger, after the initial pain the finger may have lasting numbness for a prolonged period of time, even for weeks. Individuals who get stung repeatedly over months or years may develop resistance to the venom, presumably in the form of antibodies, leading to a much-reduced reaction.

After South America broke off from Gondwana and joined North America in the Western Hemisphere during the Mesozoic era, the buthids of the Old World and the New World started to diverge in the evolution of their venoms. Although they share some characteristics consistent with a common ancestry, the amino acid sequence, immunological properties, and pharmacological action are now different in Old World scorpions compared with New World scorpions.

The scorpion's other primary weapon, the clawlike chelae, are reinforced with heavy metals incorporated into the cuticle. The metals zinc, manganese, and sometimes iron are contained in structures that experience the most wear—specifically the cheliceral teeth, the teeth on the edges of the pedipalps, the tarsal claws, and the very tip of the aculeus (the stinger). The concentrations of metal may be as high as 25 percent of the dry mass of those structures for zinc, 4 percent for manganese, and 9 percent for iron. Zinc is predominant at the edges and tips of structures, giving increased durability through hardness, while manganese is found mostly in the shafts of structures. It is not understood how scorpions are able to concentrate heavy metals in areas of their cuticle. However, this is not entirely unique to scorpions; it is also seen in the jaw of a worm called *Nereis virens*, even paralleling scorpions in that zinc is in the outer part of the jaw structure and manganese is in the base. The buthids differ from the non-buthid scorpions in incorporating iron in addition to manganese and zinc, whereas non-buthids use only manganese and zinc.

Courtship between such formidable predators is inherently a fairly risky business. Since a mature female scorpion is usually somewhat larger than the male, she may repel or even eat a potential suitor if she is not receptive to his advances. An elaborate courtship enables her to select a fit father for her offspring—an important consideration given the huge investment of her resources in producing young.

It all begins when the male encounters chemical traces of the female on the substrate. Males of many species have more teeth on each pectinal comb than do females, and these allow him to detect even faint chemical trails. In his search for females, a male spends less time hunting for food and eating, and expends more energy in his wanderings. Thus the males become quite lean, even to the point of starvation. In addition, their extensive searches expose the males to a greater risk of mortality through predation. A surprising array of animals may capture and eat scorpions. Among vertebrate predators are lizards, frogs, toads, snakes, grasshopper mice, elf owls, burrowing owls, and bats. Elf owls and pallid bats are experts in preying on scorpions, snipping off the telson (with the sting) before devouring the scorpion. Tiny elf owls can even hover over the ground as they hunt for scorpions and centipedes. Among invertebrate predators of scorpions are black widow spiders, pholcid spiders, brown spiders (*Loxosceles* species), wolf spiders, solifuges, centipedes, ants, and especially other scorpions. After avoiding all these dangers, a wandering male has a final challenge awaiting him: his mate, the female scorpion.

Upon locating a female, the male may rapidly rock his body back and forth while holding his tarsi stationary. This "juddering" identifies his gender and

1. Courtship between scorpions may include cheliceral massage, or a "kiss."

2. Before depositing the spermatophore, the male "dances" with the female as they pull each other backward and forward. The male (on the right) holds the pedipalps of the female (on the left) with his own claws.

3. The empty spermatophore remains behind after the female has received its contents.

signals his intent to mate. If the female is somewhat reluctant, tail wrestling and tail clubbing may ensue as she attempts to push him away or even sting him. He counters with his own telson. In some species, a sexual sting by the male may help to calm the female. If the female is very willing, these preliminaries may be bypassed and the couple may proceed immediately to a "kiss" and the promenade a deux, in which the male faces the female, grasps her palps with his own chelae, and they "waltz" over the surface until the male finds a suitable area on which to deposit his spermatophore. During this courtship, the dance may be punctuated with "kisses" or cheliceral massage in which both scorpions touch and move their chelicerae together. If the male is not strong enough to lead the larger female in the promenade a deux, she may drag him, and he may not be able to successfully mate with her. Different species of scorpions may show some variation in their courtship, but all include a promenade a deux as a prelude to spermatophore deposition.

Once he finds the right substrate, he stops and deposits the spermatophore, which consists of two fused hemispermatophores deposited on an upright pedicel. When the female is positioned over the spermatophore correctly, the male nudges her backward. This forces the spermatophore to pivot at its base, and the compression assists the female in taking the sperm mass into her genital opening.

Length of gestation may be as little as roughly 3 months to 7½ months for buthids, or as long as almost 8 months to 18 months for other scorpions. Depending on species, the mother scorpion may have from 1 to more than 100 young in a single litter, although 25 to 35 young in a litter is fairly typical. Although all scorpions are viviparous, deriving at least some nutrients directly from the mother, there are two very different types of embryonic development: apoikogenic and katoikogenic.

In apoikogenic development, the embryo grows within the ovariuterine tubule, and the egg may be supplied with a moderate or even small quantity of yolk. Direct absorption of nutrients from the mother is through the embryonic integument and through the membrane surrounding the embryo (not through the mouth and digestive tract, which develop later). There is no set orientation of the embryo, so the baby may be born headfirst or breach, losing the membrane at birth. As a consequence, parturition time is quite fast, as little as one minute per young for *Centruroides* or *Vaejovis*, and the entire litter may be born in a matter of hours.

In katoikogenic development, each embryo develops in a diverticulum, or a lateral outgrowth branching from the ovariuterus. The embryo is not surrounded by a membrane and may have only a tiny amount of yolk. The head of the embryonic scorpion is oriented toward the distal end of the diverticulum where there is an appendix closely associated with the mother's hepatopancreas. The katoikogenic embryo develops a working mouth and chelicerae early. The appendix transports nutrients from the mother's hepatopancreas directly to the embryo's mouth. Because of the orientation of the embryo, all katoikogenic babies must crawl backward out of the ovariuterus at birth and are born breach. Thus, the parturition time for each baby is much longer than for apoikogenic scorpions. For *Diplocentrus*, it may be from about 7 to 62 minutes per baby, and the total time for the entire litter to be born may drag on for more than two days.

At the time of birth, the mother scorpion raises her body above the substrate in a position called "stilting." As the babies drop out of her genital opening, she cradles them with her front legs, forming a "birth basket" until each baby can crawl up onto her back by way of her head. Babies unable to negotiate this challenge are eaten by the mother—an important way for her to cull out the babies not likely to survive and to replenish some of her energy. The babies are not developed enough at this point to hunt on their own, since the telson does not have an operative aculeus and the chelae are fused together and so cannot be used as grasping claws. However, the babies do have functioning peg sensilla with which they recognize their mother's back as the desirable destination immediately after birth. As the babies cluster on the mother scorpion's back, they arrange themselves in different patterns depending on the group of scorpions to which that species belongs. In the Vaejovidae, the babies tend to line up in neat, unidirectional rows facing in the same direction as their mother faces. In the buthids such as *Centruroides*, the babies pile on in a random arrangement, and in *Superstitionia donensis*, the babies cluster in a sort of ball on top of the mother's head and mostly hold onto each other instead of holding onto their mother, unlike most other scorpions.

Depending on the species, the babies stay on the mother for just over a week to about a month as they undergo further development before their first molt. They do not feed during this time but instead tap into the abundant reserves they were born with. As the baby's metasoma (tail) elongates, the plump body becomes more slender. Most mother scorpions remain relatively inactive during this time, but *Centruroides* mothers may actively forage, crawling on vertical surfaces as the babies tenaciously cling to her back. After their first molt the babies are able to hunt on their own. They are tiny but fully functional miniatures of their parents, each possessing an operational telson and aculeus, as well as grasping chelae. It is only after the first molt that the cuticle becomes fluorescent under ultraviolet light.

Most scorpions leave their mother and disperse to live independently after their first molt. However, in a diplocentrid scorpion from Baja California, second-instar juveniles have been observed to stay with the mother and may cooperatively hunt for prey and help each other drag the capture into the burrow. In the genus *Pandinus* from the Old World of Africa and Asia, young scorpions may communally kill prey that would be too large for any single baby to kill by itself. They may share a burrow with their mother through several instars.

Centruroides sculpturatus do not display this degree of sociality, but do show considerable tolerance toward others of their own species. They may be found in groups of up to 25 or 30 individuals under rocks or other refugia during the winter, and they may be tolerant of sharing close quarters even during the warmer months.

But most scorpion species are solitary in their lifestyle and may kill and eat conspecifics, especially if other food is in short supply. Consequently, after leaving their mother, young scorpions have a high rate of mortality both from adult conspecifics as well as from other species of scorpions.

Maturation may take as little as 6 months in buthids to almost 7 years in some other groups. Most scorpions require 6 to 9 molts to mature, although males may mature in fewer molts than do females from the same species. Some species, such as *Hadrurus spp.*, may live for up to 25 years. Many species of scorpions may

Like all arthropods, scorpions must molt as they grow in size. A split in the front of the cuticle allows the scorpion to escape from its old exoskeleton, leaving behind the empty exuvia.

display considerable plasticity, maturing at different instars and at different sizes

This may present a problem with their identification, but because of a standardized set of criteria for taxonomic identification, size variation alone is not of primary importance. Taxonomists use a variety of morphological characters as well as DNA in order to classify scorpions.

Whether one is a formally trained arachnologist or an amateur naturalist, scorpions elicit respect and appreciation as one learns more about them. The morphological structures help not only with identification, but also as we understand the function of these structures we can try to imagine how this predator perceives its world. From picking up chemical trails using the peg sensilla of the pectines and chemosensory hairs, to detecting vibrations using slit sensilla and the exquisitely sensitive trichobothrial hairs, to navigating even in dim starlight with the extremely sensitive eyes, scorpions are equipped with an amazing array of sensory structures. It is no wonder that these tough little predators have survived two mass extinctions, outliving other far more formidable contemporaries such as the mighty dinosaurs. We could learn much from such enduring teachers.

The birth of a scorpion:

Scorpions are the only arachnids that give birth instead of laying eggs. After a gestation that can last for more than a year, the mother scorpion raises up her body, standing tall in preparation for the birth of her young. As the babies drop from the genital opening on the underside of the mother, she catches each one in the birth basket formed by the first pair of her walking legs. Cradling her newborn, she waits for the baby to start climbing.

The mother scorpion assists her baby in climbing up onto her back. The baby scorpion identifies its mother's back by the chemicals present on the mother's cuticle. Another baby can be seen behind this one; it is just being born.

A newborn scorpion drops from the genital opening of its mother while she is still assisting another baby. She is able to manage two in close succession. More babies are yet to come; they can be seen as white areas under the cuticle along the side of her body.

The newcomer is cradled as its predecessor gains the dorsal surface of the mother. If any baby is unable to climb onto her back, she will eat it. This provides nutrition for the depleted mother and weeds out babies that are unable to survive.

By morning, the mother has an armful of babies. Sometimes they are born so rapidly, they accumulate in the birth basket. Over the next hour or so they will make their way onto her back, where they will stay until they molt about two weeks later. Until they molt, the two fingers of the claws are stuck together and the stinger is nonfunctional.

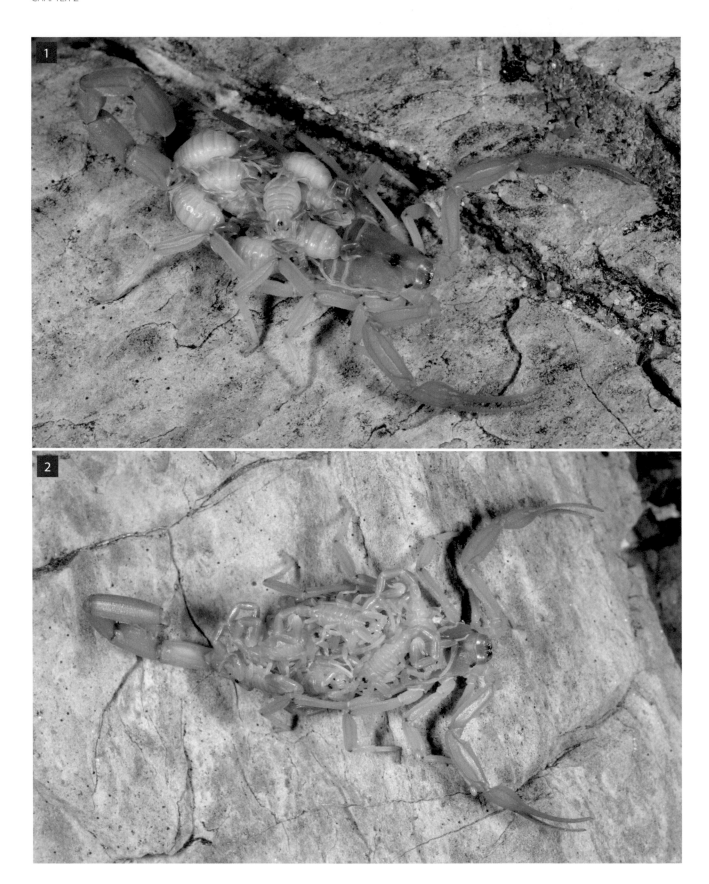

1. *Centruroides sculpturatus* mother with newborn young.

2. One week later, the baby scorpions are almost ready to molt. Note the random arrangement of the babies on the back of their mother, typical for buthids.

3. The babies of *Stahnkeus allredi* all line up facing one direction. This is characteristic of many vaejovids.

4. The babies of *Superstitionia donensis* cling to each other, forming a ball of babies perched loosely on their mother's head. Any disturbance may cause the ball to fall and the babies will scatter, but if they are given the chance, they will gather together and re-form the ball. This species takes about a month before the young are ready to molt for the first time.

Grasshopper Mice and Centruroides Venom

Scorpion venom serves a dual purpose: it is used both as a means of obtaining prey and as a weapon for self defense. Most mammals are extremely sensitive to scorpion venom, especially to the venom of the bark scorpion. A close encounter with a *Centruroides* scorpion may be an unforgettable experience, resulting in intense pain that may last for several hours. But the potentially lethal bark scorpion has more than met its match in a fierce predator, the tiny grasshopper mouse.

Weighing as little as a mere 0.9 ounces (25 g), this diminutive mouse demonstrates several remarkable characteristics. First of all, it is a predaceous, carnivorous mouse, hunting insects, arachnids, lizards, and even other mice at night. Unusual for mice, both the male and the female together raise the young. Because predators generally require fairly large areas in which to hunt, grasshopper mice are extremely territorial. The grasshopper mouse advertises its presence to neighboring mice by standing up and howling into the night. This voracious little predator attacks and kills a surprisingly large number of prey animals for its small size. The ability to add scorpions to its diet would certainly be of value, since scorpions make up a significant portion of the animal biomass in some areas of the southwestern United States. Furthermore, hunting scorpions may provide a double benefit to the grasshopper mouse. Not only does the mouse obtain an immediate reward in the form of a tasty scorpion as food, but by killing that scorpion, the grasshopper mouse is removing another predator that may compete for insect prey in its territory.

Ears back and eyes closed, a grasshopper mouse attacks and kills a bark scorpion. It then proceeds to devour all but the "tail" and claws of the scorpion. Two species of these predaceous mice live in the southwestern United States, and both species are able to hunt bark scorpions. *Onychomys arenicola* hunts *Centruroides vittatus* (at left), and *Onychomys torridus* hunts *Centruroides sculpturatus*. Killing and eating scorpions not only directly provide food for the grasshopper mouse but also eliminate a potential competitor. A mutation in the mouse's pain-transmitting neuron allows the grasshopper mouse to withstand the sting of the scorpion.

But in order to hunt and kill a scorpion, the grasshopper mouse must be able to withstand an extremely painful and potentially lethal sting. The solution to this problem is at once elegant and surprising.

Ashlee Rowe and colleagues discovered that the grasshopper mouse has evolved a mechanism that is deceptively simple but highly effective for blocking the pain induced by bark scorpion venom. Ultimately, this mechanism is dependent on a single amino acid variant expressed in a pain-transmitting neuron. This tiny mutation has an extraordinary result. The venom that normally would cause excruciating pain now binds to the amino acid variant in the nerve, blocking the pain signal and actually inducing analgesia. Consequently, in the grasshopper mouse the venom essentially checkmates itself. The story hinges upon that tiny change in the structure of pain-transmitting neurons.

Acute pain is transmitted primarily by nerves called nociceptors. These dorsal root ganglion neurons communicate with the central nervous system, transmitting information from the periphery of the body to the spinal cord and the brain. The actual signal in the nerve is generated by voltage-gated sodium channels. An action potential (nerve signal) occurs when a sodium channel gate opens and allows the movement of ions through the pore of the neuron's membrane. There are two principal types of sodium ion channels: tetrodotoxin-sensitive Nav 1.7 and tetrodotoxin-resistant Nav 1.8. The Nav 1.7 channel initiates the action potential in the pain neuron, but the Nav 1.8 is necessary for the action potential's propagation.

Grasshopper mice have a mutation that results in a single amino acid substitution in their Nav 1.8 sodium ion channel pore. This changes the sodium ion channel so that it now binds Centruroides venom peptides. Consequently, instead of propagating the nerve impulse, the transmission is blocked. Although the impulse is initiated by the Nav 1.7 sodium ion channel, it is not sustained and propagated by the Nav 1.8 sodium ion channel, so the impulse dies. It is ironic that the very venom peptides that activate the Nav 1.7 channel are the same peptides that block the Nav 1.8 channel. Instead of causing pain in the grasshopper mouse, the venom actually provides an analgesic effect after an extremely brief initial reaction. Despite being stung repeatedly, the grasshopper mouse continues to hunt, impervious to the scorpion venom. After killing its prey, it devours the scorpion with apparent relish, leaving only the tail (metasoma), telson, and claws uneaten.

Because the analgesic effect is triggered only by scorpion venom, the mouse still has a functional nervous system for pain detection in regard to other injuries. This has important survival value, since the detection of pain serves a protective function. Pain is frequently triggered by a harmful stimulus that may damage the animal; therefore, pain detection is necessary to avoid the dangers inherent in any environment. Presumably, the selective advantage gained by adding scorpions to the diet of the grasshopper mouse outweighs the disadvantage of a temporary reduction in pain detection.

1. *Anuroctonus phaiodactylus* occurs in areas of well-packed sandy soils where it can dig a burrow. This scorpion waits in its burrow until an insect enters, then uses its heavy claws to capture and subdue the prey. This species is found in Nevada, Utah, and the California deserts east of the coast ranges.

2. *Anuroctonus pococki* is found in the coast ranges of southern California and south into Baja California.

3. *Hadrurus anzaborrego* occurs in southern California and into Baja California. This impressive scorpion is named for the Anza-Borrego Desert where the type specimen was collected.

4. The giant desert hairy scorpion, *Hadrurus arizonensis*, has many setae, especially on the last segment of its "tail," the vesicle. This very large scorpion is found in low desert sandy flats where it can dig a burrow 10 feet (3 m) or more in depth. This species is found in Arizona, California, Utah, Nevada, and Mexico.

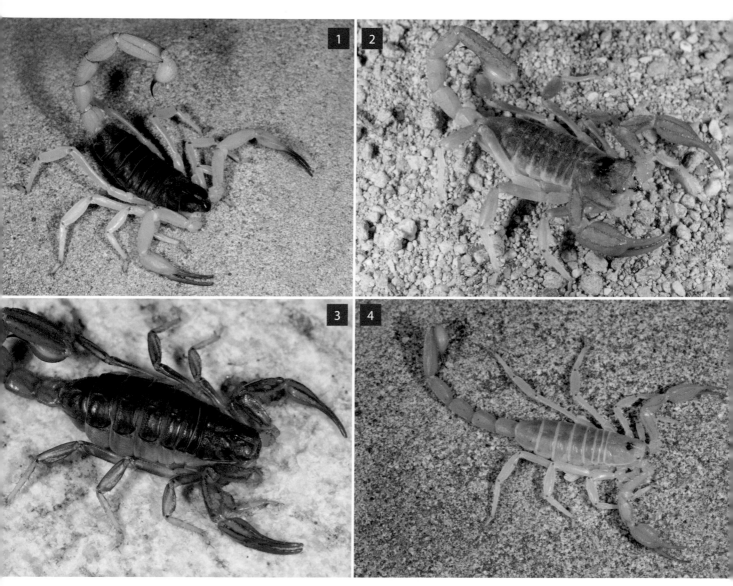

1. *Hadrurus spadix* is found in the Great Basin, northern Mojave Desert, and Colorado Plateau of eastern California, Nevada, Oregon, Utah, northern Arizona, and western Colorado. It favors canyon lands, where it can be found in vertical sandstone cliffs. It is easily identified by its black body, hairy vesicle, and large size.

2. *Paruroctonus gracilior* is a medium-sized Chihuahuan Desert "marker species" found in rocky or soft sandy soils in New Mexico, west Texas, southeastern Arizona, and throughout the Chihuahuan Desert into northern Mexico. In Mexico, it is also found in dunes, where it has reduced pigment. It lives in burrows, frequently at the base of a creosote bush or other plant.

3. *Paruroctonus silvestrii* is a common, medium-sized scorpion found along the Pacific coast from Baja California to as far north as the San Francisco area.

4. *Paruroctonus utahensis* is adapted for walking on sandy substrates, with numerous stiff setae on the tarsi and basitarsi. It is found in the Rio Grande drainages of western Texas, northern Chihuahua, and New Mexico, the Colorado River drainages of southern Utah and northeastern Arizona, and also the dunes of southeastern Arizona.

1. *Pseudouroctonus apacheanus* mature scorpion.

2. *Pseudouroctonus apacheanus* immature scorpion. This small species is found at higher elevations starting at 1,000 ft. (305 m) in elevation in Del Rio, Texas, to over 8,500 ft. (2,590 m) in southeastern Arizona, usually near streambanks, north-facing slopes, cliffs, or hidden canyons. It becomes easily desiccated and so is restricted in the arid southwest to microhabitats that have more available moisture.

3. *Pseudouroctonus* species near *apacheanus*. Some populations of these delicate scorpions are found in the protection of caves, even at lower elevations. Because cave populations are genetically isolated, they may evolve into distinct species. The babies of this species are born pure white, except for black pigment surrounding the median eyes.

4. *Stahnkeus allredi* is a tiny, very fast scorpion of the low desert of southern Arizona. It is almost never found on the surface by using a black light; instead, it can most easily be found in pack rat nests.

1. *Serradigitus wupatkiensis* occurs from northern Arizona to Idaho, and is associated especially with the canyon lands of Zion and Moab in southern Utah. *Serradigitus* typically have thin, serrated claws.

2. *Centruroides sculpturatus* lives in rock crevices and under the loose bark of trees. Unfortunately, it is also adept at getting into houses. This is one of the few North American scorpions with a potentially dangerous sting. This species of *Centruroides* is found in Arizona, California, Nevada, Utah, New Mexico, and Sonora, Mexico.

3. *Smeringurus vachoni* is found in rocky habitats where it takes refuge in small, protected crevices. Frequently, only its claws are visible as it hides in rock crevices. It is found in California and Nevada.

4. *Smeringurus mesaensis* adult male. This species is found in open desert and dune habitat in southern California and Arizona as well as into Mexico. The tarsus and basitarsus of this species have stiff setae that enable it to easily walk on the surface of sand without sinking in.

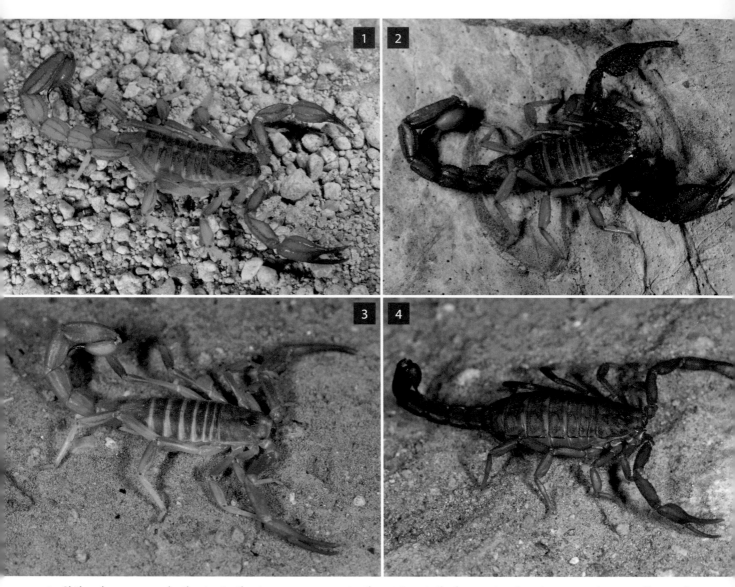

1. *Chihuahuanus coahuilae* is similar in appearance to the stripe-tailed scorpion but is somewhat smaller and found primarily in the Chihuahuan Desert. It lives in burrows over a wide range of elevation, from low desert up to 7,000 ft. (2,130 m) in elevation.

2. *Uroctonites huachuca* is a "Sky Island" species, found only at higher elevations in the Santa Rita and Huachuca Mountains of southern Arizona. It is similar in appearance to *Pseudouroctonus apacheanus* but is somewhat larger.

3. *Vaejovis jonesi* is found in juniper woodlands in northern Arizona and southern Utah on the Colorado Plateau. It typically resides in sandstone crevices.

4. *Vaejovis electrum* is one of many small brown scorpions inhabiting the higher elevations of the Madrean Archipelago, also known as the Sky Islands. As the last Ice Age ended 10,000 years ago, the climate became warmer and drier, isolating populations of these scorpions living in the cooler mountains. Consequently, these isolated populations have evolved into distinct species. *Vaejovis electrum* is limited to the Pinaleño Mountains of extreme eastern Arizona.

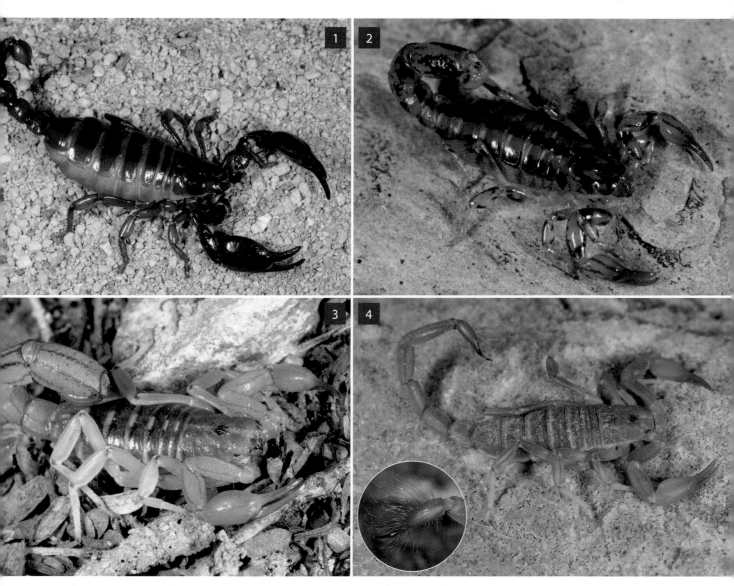

1. *Diplocentrus spitzeri* lives in scrapes that it constructs under large, heavy rocks at intermediate elevations in the oak zone. It also constructs deep burrows in mesquite groves along washes. This species of *Diplocentrus* lives in southern Arizona.

2. *Superstitionia donensis* is the only epigean (surface dwelling) species in its family. All other species in the family Superstitioniidae are found in caves. *S. donensis* has a large range including much of the southwestern United States and parts of Mexico and can be found in the lower desert living in scrapes under rocks as well as in leaf litter under trees.

3. *Paravaejovis spinigerus*, the famous stripe-tailed scorpion. These robust scorpions are common throughout the Sonoran Desert, living in burrows in open ground and scrapes under rocks. They may be encountered over a wide range of elevations, from creosote flats up to the oak zone of the mountains and occasionally to lower pine forests.

4. *Kochius sonorae* presumptive from southern Arizona. Another species, *Kochius hirsuticauda*, occurs in southern California and into southwestern Arizona, Nevada, southeastern Utah, and south into Mexico. The name *hirsuticauda* refers to the long, silky setae on the last segment of the "tail," the vesicle (INSET).

CHAPTER 3 Pseudoscorpions:
Pseudoscorpiones

Clinging to their mother, newly emerged baby pseudoscorpions require another day before they are ready to live on their own. Once they leave their mother, these diminutive predators must be able to capture prey that may be as large as themselves. In the case of this *Dinocheirus* species, the robust claws are utilized in grasping and killing the prey. The hand of each pedipalp contains a venom gland; the pointed claws deliver the lethal dose of venom.

For their small size, pseudoscorpions are associated with some of the most extraordinary and interesting natural history of any living creature. Among their attributes is the ability to produce silk from their "jaws," venom from their claws, and "milk" from their ovaries. In addition to this, not only are they are adept at traveling on their own stubby little legs, but they also hitchhike on larger flying arthropods. Naturally resourceful, pseudoscorpions make a living in a variety of habitats, including (but not limited to) caves, tree bark, beehives, mammal nests, bird nests, ant colonies, beaches, and libraries. Being small sometimes does have its rewards. As long as a pseudoscorpion can find a nice little crevice in which to live and an occasional prey animal, it can survive.

Like their more familiar arachnid cousins, the spiders, most pseudoscorpions produce both silk and venom. But unlike the spiders, pseudoscorpions produce silk from glands in the prosoma. The spinneret, called the galea, is located at the tip of the movable finger of each chelicera. Hence they deliver silk from their "jaws." In most species of pseudoscorpions, silk is produced to build molting chambers, brood nests, and hibernation chambers.

Chelicerae are also utilized in the important task of grooming. After touching anything with its pedipalps, a pseudoscorpion will almost always fastidiously clean them using the comblike serrula on the chelicerae, especially the exterior serrula on the movable finger of the chelicerae. Finally, the chelicerae are used in the typical arachnid fashion to tear into and masticate prey in the process of preoral digestion.

A hunting pseudoscorpion uses both lyriform organs and trichobothria to locate prey. Lyriform organs, found over the entire body surface, detect surface vibrations. Trichobothria are the long, delicate hairs anchored to a cuplike socket. As the hairs move with the slightest air disturbance, they allow the pseudoscorpion to precisely locate prey up to 0.6 inches (15 mm) away. The pedipalp of a mature pseudoscorpion has a total of 8 trichobothria on the hand and fixed finger, and 4 on the movable finger. As a hunting pseudoscorpion walks forward, it holds its pedipalps ahead of it and slightly above the substrate in a manner similar to that of an ambulatory scorpion, allowing the sensory structures on the pedipalp to "see" ahead of it. If movement is detected behind the pseudoscorpion, it rapidly turns to face the disturbance, and it can retreat while walking backward faster than it can walk forward. Some pseudoscorpions can even jump backward.

The pedipalps assist the hunting pseudoscorpion not only in the detection of nearby prey, but also in the identification and capture of its quarry. Chemosensory setae on the palpal fingers may identify an object once the chela makes physical contact. Once the pseudoscorpion has grasped a prey animal, it must be able to subdue it. The multipurpose pedipalps deliver venom via the tips of the chelae, allowing the pseudoscorpion to subdue animals larger than itself. The venom is produced in the hand portion of the pedipalp of most species (although there are a few species that produce no venom).

The preoral digestion used by pseudoscorpions is similar to that employed by many other arachnids. One of two methods is utilized, depending on the species of pseudoscorpion. Some species masticate the entire prey animal while digestive fluids are worked in, and the subsequent liquid is then ingested. In this case, the remains of the prey become an unrecognizable, compact pellet by the end of the process. Other species tear a small hole in the cuticle of the prey with the chelicerae and inject digestive fluids into it. The pseudoscorpion can then ingest the predigested liquefied meal, leaving the cuticle of the prey behind. The latter method is superior to the former in several ways. First, because the pseudoscorpion does not have to masticate the entire animal, it is more efficient. The pseudoscorpion processes only the useful parts of the prey, leaving the indigestible cuticle mostly intact. Second, it allows the pseudoscorpion the opportunity to feed on much larger prey. Consequently, it can utilize a greater size range of potential prey animals. Surplus calories from a large meal can be stored in the diverticular midgut and the peritoneal epithelium (lining) as glycogen and lipids, much the way a human stores extra calories in the liver as glycogen. After a large meal, a pseudoscorpion can survive weeks or even months without food.

The pedipalps, like the chelicerae, are multipurpose in function. In addition to being used to capture and kill prey, pedipalps are also used for transportation purposes, enabling the pseudoscorpion to grasp hairs or even the leg of a larger arthropod and hitchhike rides. Hitchhiking, or phoresy, on larger arthropods is a characteristic of pseudoscorpions (as well as some mites). Wood-boring cerambycid beetles frequently have these little hitchhikers clinging to them, usually under the elytra (wing covers). As the beetles fly to new

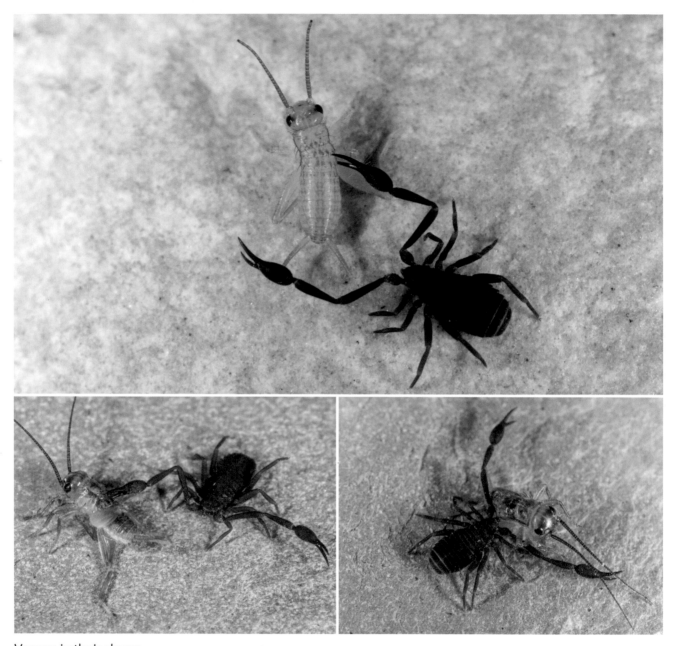

Venom in their claws:

With a touch from its claw, or chela, a *Parachelifer* pseudoscorpion can capture a cricket as large as itself. Within seconds, the cricket is dead. Each finger on the claws of *Parachelifer* can inject venom. If even the tip of one claw penetrates the cuticle of the cricket, the pseudoscorpion will soon have a meal. The venom gland is located in the "hand" of the pedipalp.

Grasping the killed cricket with its chelicerae, the pseudoscorpion triumphantly carries its prey to a secluded spot in order to feed. After it tears a small hole in the cricket, digestive fluids are regurgitated into the prey. The predigested, liquefied contents are then ingested by the pseudoscorpion, leaving behind the empty exoskeleton of its victim. It may be weeks or even months before the pseudoscorpion obtains another meal.

Some pseudoscorpions such as neobisiids masticate the prey into a mush, leaving behind just a small pellet. Neobisiids have relatively larger chelicerae, a necessary requirement for mastication.

trees, the tiny phoretic pseudoscorpions can disperse to new locations where they can live under tree bark and hunt small arthropods. In the Sonoran Desert they have also been observed to hitch a ride on cactus flies whose larvae live in the rotting corpses of fallen saguaro cacti. These environments are rich in tiny arthropods, making a good hunting ground for the opportunistic little pseudoscorpions.

Finally, the pedipalps may be used during social interactions including fighting and courtship.

The reproduction of pseudoscorpions involves the transfer of a spermatophore from the male to the female, similar to many other arachnids such as scorpions, amblypygids, and uropygids. However, the courtship involved in the transfer may range from deposition of the spermatophore in the absence of a female, to an elaborate courtship with extensive physical contact between the male and the female. In general, the more complicated the courtship, the more complex in structure is the spermatophore.

The simplest of the reproductive strategies is found in the cheiridiid family. The males of this group deposit sufficient spermatophores in a small area to form a small forest of them, whether or not a female is even present at the time. Each spermatophore may contain only 6 to 15 large spermatozoa in a sperm package. This sperm package sits on top of a stalk, upon which a droplet of pheromone attractant is added. Once a female wanders by, she may encounter the sperm packet and take it up after stepping over it. Upon contact with moisture, the sperm packet swells and ejects the spermatozoa into the female's genital atrium. If no female chances to accept the spermatophore, the male destroys it and replaces it with a fresh one. This method works only in a humid environment and in gregarious species, increasing the probability that a female will encounter the spermatophore soon after deposition, before it can desiccate. It is costly to the male since he invests energy and resources in many unproductive spermatophores. But since only a few spermatozoa are dedicated to any one spermatophore, he can afford to waste a number for the chance of occasional success.

A more efficient strategy involves depositing a spermatophore only in the presence of a female, and a further refinement involves the deposition of a spermatophore only in the presence of a receptive female. The latter scenario may be accomplished by incorporating a mating dance into the courtship.

Mating dances for pseudoscorpions are similar to those of scorpions and vinegaroons, starting with the male and female facing each other. In all species the dance is highly specific and ritualized, but a common component includes the male grasping the chelae of the female with his own pedipalps, after which they pull each other forward and backward. In the Cheliferidae the male may display his ram's horn organs. These are a pair of long, slender, hollow tubes attached to the genital atrium. Together, they form an approximate V shape when everted by an increase in hemolymph pressure. They emit an attractant pheromone for the female. Only after the female shows interest in him does he deposit a spermatophore. He then proceeds to physically contact the female. He even assists her in taking up the spermatophore. A spermatophore of a cheliferid is complex in structure and may contain several hundred spermatozoa. A greater efficiency in sperm transfer is achieved by way of the courtship dance, allowing the male to invest more energy and resources into one large successful effort, rather than many small wasted efforts. The courtship dance of pseudoscorpions lasts from only a few minutes to more than an hour in length.

In the Cheliferidae and Chernetidae families, the females can store sperm for future use in a structure called a spermathecae. Consequently, multiple broods may be produced over several months after a single mating. The female (of most species) constructs a brood chamber in which she stays while her embryos are developing. A natural crevice or sheltered spot is selected for the nest site, and the nest may be constructed of tiny pieces of wood and pebbles lined with silk, or it may consist entirely of silk, depending on the species of pseudoscorpion. The brood sac is composed of a membranous pouch carried under the female's opisthosoma that opens to her genital atrium. If she is disturbed, the pseudoscorpion can drop the pouch and abandon it.

The eggs are small and have little yolk, but the mother pseudoscorpion produces a nutrient-rich fluid, a sort of "milk" with which she nourishes the developing embryos. The ovarian cells that did not produce follicles transform into glandular cells after the eggs are shed. Initially flat, these epithelial cells elongate into columnar cells that secrete fluid into the lumen of the ovary. This fluid is rich in nutrients including phospholipids, proteins, and polysaccharides. The female releases the nutritious fluid into the brood pouch at a specific time during the embryos' development depending on the species of pseudoscorpion. In some species, the production and release of the nutritive fluid is synchronized with the formation of an embryonic structure called the pumping organ. The pumping organ allows the embryo to take in the nutritive fluid, and the subsequent growth of the embryo is rapid. The embryo actually undergoes a molt into a second embryonic instar before hatching. While it is a second embryonic instar, it continues to grow in size as well as developing the organs it will need to survive as a free-living instar. During this time, the pumping organ transforms into the mouth.

The release of the nutritive "milk" and the development of the embryo vary from species to species. Perhaps the most dramatic example of the perfect synchronization between the mother's release of the nutritive fluid and the uptake of this fluid by the embryo is seen in *Chelifer cancroides*. In this species, the second embryonic instar actually breaks through the brood pouch membrane, leaving its dorsal surface extending outside the membrane while the remainder of its body is still inside the brood sac. This is fortuitous, because as soon as the pumping organ has formed, the mother releases a large amount of nutritive fluid that is absorbed by the embryos in a mere 3 to 6 seconds. The embryos at this point resemble tiny water balloons. Since they had already extended their dorsal surface through the brood pouch membrane, their bodies have room to swell enormously, and they grow very rapidly after this point. Just before hatching, the embryos grow a sawlike "egg tooth" to assist them in emerging as protonymphs.

In species in the genus *Cheiridium*, the mother pseudoscorpion drops the brood pouch once the young have molted into second embryonic instars, after which they continue to develop and eventually hatch without further maternal assistance.

The entire development of the embryos from the production of the brood sac until their emergence as protonymphs is fairly rapid, taking only 3 to 5 weeks for most species. In many species, the newly emerged young ride clinging to their mother's opisthosoma for a day or two before striking out on their own. Pseudoscorpions undergo three nymphal stages before reaching maturity. These three stages are protonymph, deutonymph, and tritonymph. Prior to molting, they become swollen and torpid. Many species build a molting chamber out of silk where they can retreat during this vulnerable period. With each molt, the number of trichobothria on the pedipalps increases until they attain the full complement of 12 trichobothria per pedipalp at maturity.

Pseudoscorpions do not molt after reaching maturity. They may live 3 or 4 years as adults. As they reach old age, they stay in their little crevice home but will readily capture any passing prey as long as it comes within reach of their palps. Pseudoscorpions are predators to the very end of their lives.

As a consequence of their tiny size, pseudoscorpions are easy to overlook; however, they may be a significant component of an ecosystem. They may reach population densities of as many as 900 individuals per square meter in some soils, and may attain comparable densities under the bark of some trees. They are a common component of cave ecosystems, in part due to their ability to survive long periods without food. This attribute "preadapts" pseudoscorpions to life in caves, where meals may be few and far between. Despite the overall paucity of organic material in most caves, bats do deposit droppings, which in turn grow mold or fungus. Tiny insects called psocids, or booklice (not related to parasitic lice), feed on the mold growing on the bat guano. In turn, the psocids are preyed upon by pseudoscorpions. In addition to psocids, pseudoscorpions capture beetle larvae, springtails, or other arthropods as the opportunity presents. Pseudoscorpions have adapted so well to life in caves that some species have become true troglobites, living their entire lives only in caves. These species may demonstrate morphological adaptations associated with cave existence, which may include loss of pigment or elongation of their pedipalps.

Two species of pseudoscorpions, *Chelifer cancroides* and *Cheridium museorum*, are cosmopolitan in their distribution, having been dispersed by humans around the globe.

Sociality in pseudoscorpions follows the pattern seen in a number of other arachnid orders. Although the vast majority of pseudoscorpions are solitary predators, several species demonstrate some degree of sociality. Sociality ranges from the merely gregarious *Neobisium* to the fully social *Paratemnoides*. Both species of *Paratemnoides* (*P. nidificator* and *P. elongatus*) form multigenerational colonies under the bark of trees. These colonies may consist of only a few pseudoscorpions up to almost 200 individuals. The advantages of sociality become evident even in small colonies. Cooperative hunting enables these small predators to capture and kill prey significantly

1. The early embryos are carried in a thin, membranous pouch under the female's abdomen. The mother must tip her abdomen upward in order to accommodate the bulky brood pouch. If the mother is disturbed, she can drop the pouch and abandon it.

2. The membranous brood pouch is so thin that the developing embryos can easily be seen within it.

3. Four days later, the rapidly developing embryos show the snoutlike pumping organ. At this stage, the mother releases a nutritive fluid into the brood sac. This fluid is produced by special cells in her ovaries. The growth of the embryos is extremely rapid after ingestion of the nutritive "milk" from their mother. In just a few more days, the pumping organ will have become the chelicerae of the baby pseudoscorpions.

4. After emerging from the brood sac, these baby *Dinocheirus* will ride on their mother for only a day or two before dispersing. Once they leave their mother's abdomen, they will lead solitary lives.

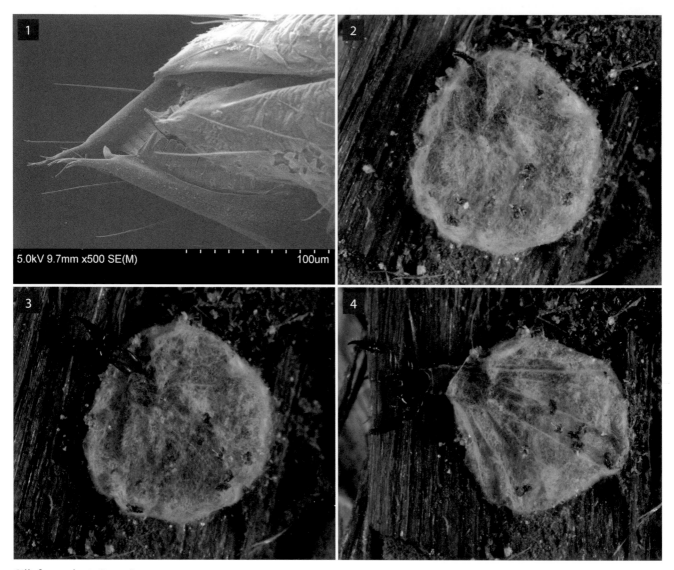

Silk from their "jaws":

1. Pseudoscorpions construct silken brood nests, molting chambers, and hibernation chambers. The silk spinnerets, called galeae, are located at the tip of the chelicerae. Each galea is seen here as a small, branched structure extending from the tip of the chelicera of a chernetid pseudoscorpion. The comblike serrula used for grooming can also be seen in this image.

2, 3, and 4. Tucked under the bark of a juniper tree, a tiny garypinid pseudoscorpion rests in its silken chamber. Once disturbed, it emerges from its refuge. These tiny pseudoscorpions are only about 0.08 to 0.12 inches (2–3 mm) in body length, giving them the ability to live in small spaces, such as under the bark of trees.

larger than themselves, such as the ant *Cephalotes atratus*. A single ant of this species not only weighs 30 times as much as a single pseudoscorpion, but it is also heavily armored and can capably defend itself. The pseudoscorpions wait just under the edge of their bark refuge. As an ant passes close by, a number of pseudoscorpions reach out from under the bark and grasp the ant's legs with their chelae (claws), pulling the ant against the refuge's entrance. While their quarry is pinned in this way, other pseudoscorpions can emerge from the protection of their refuge and proceed to attack and feed on the ant. The entire colony benefits from this strategy, since cooperative hunting enables them to capture a far greater size range of prey.

In a more surprising aspect of their sociality, *P. nidificator* parallels the honey bee in that individuals perform certain duties based upon their age and gender. In the case of the social pseudoscorpions, older females protect the youngest instars of the colony, while the younger females "babysit" older juveniles. Juveniles work together to construct the silk nesting areas. Meanwhile, the males provide food for the entire colony, capturing and killing prey. Consequently, mature females living within the colony are able to invest significantly more effort toward reproduction than a solitary female is able to do. Colonial females therefore have a rate of reproduction 3 times as high as solitary females. Additionally, nymphs benefit from extended parental care until they reach adulthood, enhancing their survival. Under situations of food scarcity, the mother pseudoscorpion is consumed by the juveniles, thus reducing cannibalism among the nymphs and preserving colony cohesion.

These social pseudoscorpions are found in tropical South America and the southeastern United States, where the warm temperatures and high humidity provide the abundant prey base necessary for colonies of predators to exist. The division of labor and the cooperative hunting and sharing of prey are unexpected in a non-web–based arachnid. Its presence suggests the possibility of sociality that may exist in other orders of small, obscure arachnids.

Hitchhiking on a longhorn beetle, these tiny *Cheiridium* pseudoscorpions can be transported over long distances. The larvae of many longhorn (cerambycid) beetles bore into wood; therefore, the adult wood-boring beetle must oviposit on trees. As newly mature beetles emerge from the tree, pseudoscorpions can catch a ride and disperse to new habitat. Once they reach their destination, they disembark from the beetle and can make their living hunting tiny invertebrates under the bark. Hitchhiking is also called phoresy.

Many pseudoscorpions ride tucked under the hard wing covers (elytra) of beetles and are therefore hidden from view, unlike this species.

1. *Albiorix parvidentatus* (family Ideoroncidae) is an extremely common and widespread pseudoscorpion throughout Arizona and southern California. It is found under rocks even in extremely hot, dry, low-elevation desert.

2. *Albiorix anophthalmus* has been found only in two adjacent caves in the Rincon Mountain foothills of southern Arizona. This troglobitic species has elongated, thinner pedipalps compared with its surface relative, *Albiorix parvidentatus*.

The elongation of appendages is one adaptation to life in a cave.

3. This pseudoscorpion may be in the genus *Hesperochernes* (family Chernetidae). Compared with surface species of *Hesperochernes*, this cave species is pale. Loss of pigment is another adaptation associated with troglobitic species.

4. This specimen of *Archeolarca* near *rotunda* (family Larcidae) was collected in yet another cave in the mountains of southern Arizona. Pseudoscorpions are frequently found in caves, where they feed on booklice and other tiny arthropods.

1. The southwestern United States has a variety of species within the family Chernetidae. *Dinocheirus* can be found in lower-elevation desert. However, unlike *Albiorix parvidentatus*, *Dinocheirus* is not commonly found under rocks, but instead is often associated with areas of dead vegetation and other organic debris.

2. This species of *Parachernes* (family Chernetidae) was found in vegetation in extreme southern Arizona. It may have been hunting under the bark of a juniper tree.

3. In the arid southwestern United States, *Hesperochernes* (family Chernetidae) are found in the foothills and mountains where conditions are more cool and moist. They can readily be found under rocks and rotten logs, where tiny invertebrate prey is abundant and where there is some moisture.

4. This female *Hesperochernes* can move surprisingly fast, considering that she is carrying a full brood pouch.

1 and 2. *Lustrochernes grossus* (family Chernetidae) is an extremely common species found under rocks and in leaf litter. They occur over a range of elevations, from low desert to as high as 4,000 ft. (1,220 m) in the oak zone.

The females are seen with brood pouches in late spring and after summer monsoon rains.

3. This chernetid pseudoscorpion was found in a cave in one of the sky island mountains of southern Arizona. Since this species is not restricted exclusively to life in caves, it would be classified as a troglophile rather than a troglobite.

4. *Cheiridium* species (family Cheiridiidae) are frequent fliers, riding on beetles and other arthropods to disperse to new habitat. These tiny pseudoscorpions hunt other invertebrates under the bark of trees.

Members of the family Neobisiidae are generally found in cool, moist habitat. Consequently, populations of these pseudoscorpions may be restricted to mountains, coastal areas, and riparian areas in the arid southwestern United States. These pseudoscorpions may be relicts from the last Ice Age, when the climate of the southwestern United States was cooler and wetter than it presently is.

1. *Globocreagris* species found in a canyon riparian corridor in the Santa Catalina Mountains of southern Arizona.

2. *Microcreagris* species found in the Santa Rita Mountain foothills of southern Arizona along a riparian corridor. This individual was 0.2 inches (5 mm) in body length, truly enormous for a pseudoscorpion.

3. *Parachelifer* species (family Cheliferidae). Armed with venom in both fingers of each claw, this little predator is able to kill animals as large as itself, including the pseudoscorpion *Dinocheirus*. It can be found foraging on dead, rotting vegetation as it actively searches for prey.

This species seems to have little tolerance for its own kind. When two meet, they rapidly and briefly touch each other with their pedipalps and then usually separate, going in opposite directions.

1. *Garyops* or *Idiogaryops* (family Sternophoridae). This tiny pseudoscorpion is found under the bark of ironwood trees in the low, hot creosote desert of Arizona. The specialized microhabitat provides this delicate little pseudoscorpion with both protection against the sun as well as a place to hunt for food.

2. *Serianus* species (family Garypinidae). A fallen, rotten saguaro cactus provided moisture and a wealth of invertebrate prey for this little pseudoscorpion.

3 and 4. *Solinus* or *Serianus* species (family Garypinidae). The Superstition Mountains of Arizona may seem inhospitable, but these tiny pseudoscorpions have found the perfect niche under the bark of juniper trees. You can see the pedipalp of a second pseudoscorpion reaching out from under the crevice next to the silken nest. This particular species seems tolerant of conspecifics, frequently being found in numbers under small strips of bark.

Life on a beach:

Pseudoscorpions live in a surprising variety of habitats, including caves, tree bark, beehives, mammal nests, bird nests, ant colonies, libraries, and even beaches. *Garypus californicus* (family Garypidae) is found along the coast of California and Baja California, living in the supralittoral zone where debris is deposited by the high tide. Dead, decaying seaweed, as seen in this photo, accumulates as wrack along the high tide mark, providing habitat for a wealth of small detritivore crustaceans including isopods and amphipods. The detritivores in turn make up a prey base for a variety of predators including spiders and pseudoscorpions. Pseudoscorpions may disperse to new habitat as they are swept out to sea on driftwood and later wash up on a new beach. Photo by Alice Abela.

1. Sheltering in a silken brood nest with her developing embryos, this female *Garypus californicus* is at home on the beach. The outer surface of the silken nest is camouflaged with sand.

2. This little olpiid pseudoscorpion (family Olpiidae) is so tiny that the grains of sand from the beach almost look large in comparison to it.

3 and 4. These chernetids probably belong to the genus *Dinocheirus*, which demonstrates considerable sexual dimorphism in some species. Males have significantly larger pedipalps than do the females. The pale, recently molted male later darkened in color.

Photos by Alice Abela.

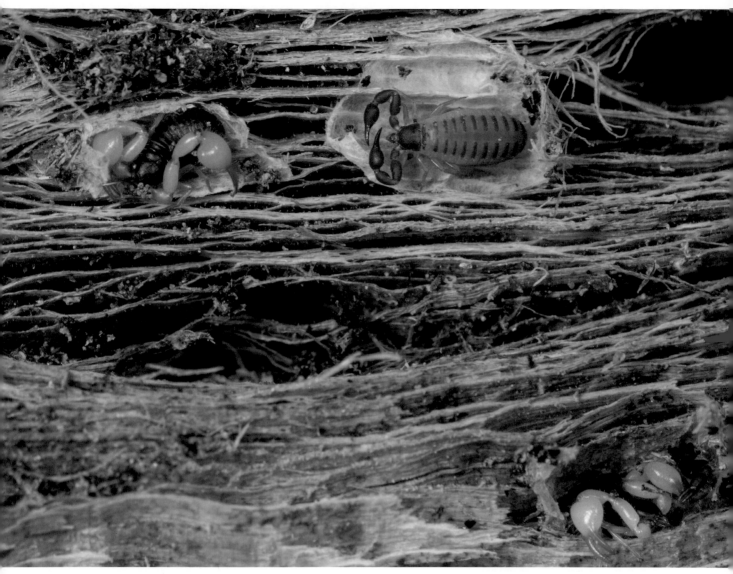

Dead vegetation forms the perfect shelter for these chernetid pseudoscorpions. Native to sandy backdune habitats along the coast of California and Baja California, giant coreopsis (*Coreopsis gigantea*) is a woody perennial in the sunflower family that grows 3 feet (1.0 m) high and has a thick trunk several inches (centimeters) in diameter. The death of this plant provided numerous protected nooks and crannies for these chernetid pseudoscorpions to build silken refugia. *Dinocheirus* is frequently found in association with dead vegetation, which provides not only shelter but also a variety of tiny invertebrate prey. Photo by Alice Abela.

CHAPTER 4 Vinegaroons:
Thelyphonida (Uropygi)

A vinegaroon can aim its defensive spray with considerable accuracy. The principal component of the spray is almost pure acetic acid. Photo by Bruce D. Taubert.

Armed with heavy, lobsterlike claws at the front end of their bodies and shooting almost pure acetic acid out of their rear ends, vinegaroons seem to have stepped straight out of a science fiction novel. Despite their fantastic abilities, vinegaroons are perhaps the most poorly understood of the large arachnids. This may be the result of their nocturnal habits, dark nonfluorescent coloration, and the fact that they live most of their lives underground. However, the story of these enigmatic creatures is well worth the cost, albeit paid for in sleepless nights. Their story rivals and even surpasses the creations of fiction.

The common name "vinegaroon" is well chosen. The defensive spray of the vinegaroon *Mastigoproctus giganteus* of the southwestern United States consists primarily of acetic acid (up to 84 percent), water (10 percent), and caprylic acid (5 percent). Acetic acid is, of course, the component that gives vinegar its characteristic odor. Hydrophilic "water-loving" acetic acid in pure form simply beads up on the lipid-containing cuticle of most arthropods. But with the addition of the lipophilic "lipid-loving" caprylic acid, the spray spreads easily and penetrates into the cuticle. The caprylic acid derives its surfactant properties from a chain of 8 carbon atoms in the molecule, as compared with only 2 carbons contained in acetic acid. The acetic acid spray is produced in a pair of pygidial glands in the abdomen of the vinegaroon. Contraction of muscles in the outer layer surrounding the gland discharges the mixture as a spray from a knoblike structure called the pygidium at the base of the "tail" (called the flagellum). By bending the abdomen and rotating the knob, the vinegaroon can direct the spray with considerable accuracy, even if the target is almost directly in front of it.

Vinegaroons may spray repeatedly (as many as 19 times) before depleting their reserve of defensive chemicals. It takes about a day for them to recharge their reservoir. The spray has proven to serve as a deterrent to the most formidable arthropod foes such as ants. It also repels vertebrate predators such as the fierce little predaceous grasshopper mice. In contact with human skin, it may cause a burning sensation, and of course the eyes of a potential vertebrate predator such as a bird or a grasshopper mouse would be highly vulnerable to the effects of the acid.

The acetic acid is used purely as a defense weapon—not for capturing prey. A hunting vinegaroon employs tools similar to those used by scorpions for

1. Strong, clawlike pedipalps and a tough cuticle provide vinegaroons with their own armor.

2. As added protection, vinegaroons can defend themselves by spraying almost pure acetic acid from the pygidium, located at the base of the flagellum. By rotating the pygidium, the vinegaroon can aim the spray in almost any direction, even almost immediately in front of it. The flagellum assists the vinegaroon in accurately aiming the spray. Photo by Bruce D. Taubert.

detecting prey. A combination of sensilla (to pick up substrate vibrations) and trichobothria (to detect airborne vibrations) on the uropygid's legs allow it to narrow down the general location of its quarry. The tiny hairs on the flagellum might also assist in this task. At the same time, the antenniform legs are extended forward, tapping the surface as the uropygid seeks out prey. Chemosensory hairs on the antenniform legs provide chemical clues as to the identity of any objects it encounters. As soon as the vinegaroon has positively identified a potential prey animal, it charges forward, grabbing with its heavy, clawlike palps. If it misses with the first try, it excitedly feels around with the antenniform legs, searching until it has once again located its quarry.

Like an arachnid linebacker, the vinegaroon is an unhesitating and powerful predator in action, tackling animals that may be as large or even larger than itself. All kinds of other invertebrates are taken as well as the occasional small vertebrate. Once the prey is captured, the vinegaroon masticates and liquefies the meal in the typical arachnid method of preoral digestion. It may also scavenge carrion when the opportunity presents itself.

The courtship and mating of uropygids follows the same general pattern of many other non-spider arachnids. Like scorpions, vinegaroon courtship may include some wrestling, a waltz or promenade a deux, and spermatophore transfer; however, it does differ in some respects, notably in the length of time for the entire process. Vinegaroons may require 12 to 16 hours to complete their courtship and mating.

The breeding season starts with the summer monsoons and continues into fall. Courtship may be initiated with an episode of wrestling, in which the male grasps the female with his pedipalps and may actually lift her off the ground. This "chase and grapple" stage may be extremely brief or essentially nonexistent if the female is very receptive, which she signals to the male by vibrating the tips of her antenniform legs in front of him. He then grasps the tarsomeres (the tips) of her antenniform legs in his chelicerae and may nibble at them as he holds them. The presence of secretory cells on the tarsomeres of the female suggests the possibility that chemical signals or pheromones may be conveyed to the male as he nibbles them. Likewise, chemosensory hairs on the female's antenniform legs may be detecting chemical signals from the male.

As the male holds the female's tarsomeres, they "dance" facing each other as they slowly walk backward and forward in search of a suitable substrate for the deposition of the spermatophore. During the entire slow waltz, the male continues to hold the female's antenniform leg tarsomeres in his chelicerae. Sometimes he appears to be dragging her forward, and sometimes she appears to be dragging him. This stage may last for just over an hour to more than 6 hours in length. As in scorpions, this stage may assist the female in determining the "fitness" of her prospective mate—an important consideration given the huge investment she makes in producing young.

Once a suitable spot has been chosen for spermatophore deposition, the couple stops dancing and the male turns 180 degrees all the while still grasping the female's antenniform legs. The female stands directly behind the male during the next phase, known as the generation phase. She faces the same direction as the male as she gently holds his opisthosoma with her palps. The two remain perfectly still except for the occasional shudder of the male's antenniform legs during the 3½ to 6 hours it takes for the male to produce the spermatophore. The spermatophore consists of a pair of mirror-image sperm packets held together "back to back" perched on a short stand, the whole structure resembling a squat letter Y. As soon as the male deposits the spermatophore on the substrate, the female advances and lowers her gonopore over it. As her opisthosoma is lowered, the "stand" acts like a pivot (similar to the structure of scorpion spermatophores), and small, fingerlike structures within the gonopore pick up and hold the spermatophore in position for sperm transfer.

1. During the early stages of courtship, vinegaroons perform a "dance." Over a period of several hours, the courting pair continuously pull each other back and forth, presumably testing each other's strength. This may assist each party in determining the "fitness" of their potential mate. Grasping the female's antenniform legs in his chelicerae, a male vinegaroon maintains constant contact with the female during the dancing stage of courtship. The female's legs contain both secretory cells as well as chemosensory hairs.

2. After the dance, the female stands behind the male while he generates a spermatophore. He continues to hold her antenniform legs with his chelicerae during this phase, which can last 3½ to 6 hours.

At this point, the male turns around to face the female once again, but this time he overlaps her body from above so that he can reach his palps in under her opisthosoma, "hugging" her with his palps. During this final stage of courtship and mating, the male presses the spermatophore farther into the female's gonopore and then in a series of strokes with his palps assists in the transfer of sperm from the spermatophore into the female. After the transfer of sperm is complete, the empty spermatophore is abandoned. The pressing stage may last from 2 to 7 hours. Thus, the entire courtship and mating lasts an average of 13 hours but may exceed 15 hours in length.

The male may be ready to mate again the following night, but most males cannot successfully mate on three consecutive nights.

In the fall, the female digs a winter chamber 30 to 50 centimeters deep. She is an accomplished excavator as she diligently digs her burrow. Looking like an animated bulldozer, she carries out the excavated soil with her massive palps. After the chamber is complete, she seals herself in with soil. She stays in this chamber for the entire winter and spring, emerging only with the arrival of the summer rains. It is during this period of confinement that she produces an egg sac in the spring. A mass of large white eggs is deposited in a membranous clear sac attached to her gonopore. There may be as many as 68 eggs in the sac, a truly prodigious investment in resources for the mother vinegaroon. She must tilt her opisthosoma upward in order to accommodate the bulky egg sac. After 5 weeks of incubation, the eggs hatch and pinkish white larvae emerge and climb onto the mother's opisthosoma. The larvae are helpless in virtually all respects except for their ability to climb up onto their mother and hold on. To this end, their tarsi have rounded, suckerlike tips that help them hold onto the relatively smooth cuticle of their mother.

During the next 5 weeks, they undergo further development, tapping into the energy and protein stored in their fat little larval bodies. They do not move at all during this developmental stage, staying in exactly the same position for the entire 5 weeks. At the end of this period, they molt to become free-living protonymphs, with shiny black bodies and scarlet pedipalps.

With the start of the summer rains, the mother and her babies are ready to emerge from the burrow. The mother vinegaroon emerges ravenously hungry. She

has seriously depleted her own energy reserves not only by the long underground fast lasting from November until July, but also by investing tremendous resources in the production of her offspring. The collective weight of the newly emerged protonymphs may actually exceed the weight of their mother. It is no wonder that she aggressively charges and captures any available prey. It is also noteworthy that the mother vinegaroon avoids mistaking her offspring for food. The ability to differentiate between a cricket and a similarly sized protonymph vinegaroon is compelling evidence of the exquisite sensitivity of the uropygid's sensory apparatus.

In fact, the famished mother may even share her food with her babies. As the mother masticates the captured prey, nearby protonymphs may swarm near her, practically reaching right between her chelicerae in order to steal some food. At times, a number of young may mob the mother, at which point she may adjust her position slightly. The babies hastily back away from their mother, only to return a few minutes later. The young seem able to share their mother's food with impunity at this stage.

A truly extraordinary level of maternal care was observed in one captive mother vinegaroon. Even 4 months after the mother and young had emerged from the maternity burrow, the mother vinegaroon was observed sharing food with her offspring. After capturing a large cricket, she turned away from her own burrow entrance and instead approached the tiny burrow of one of her babies. There, she reached down with her antenniform feelers into the baby's burrow, and up the baby came, immediately grasping hold of the captured cricket. At this point, the mother and baby stood still for several minutes as they gently tapped and caressed each other with their antenniform legs. Then the mother vinegaroon turned 180 degrees and walked to her own burrow entrance, all the while holding the cricket with her baby still clinging to it. She descended into the burrow, and by morning, the cricket was gone and both the baby and mother vinegaroon were still in the mother's burrow. By that evening, the baby had returned to its own burrow. Several times after that, the mother was observed approaching the same baby's burrow and reaching down it with her antenniform legs after having captured a cricket. The mother's active sharing of food with her protonymph babies gives her offspring a tremendous advantage, since the adult vinegaroon is capable of capturing relatively large prey.

1. The spermatophore is produced after several hours of generation time. Once it is deposited on the substrate, the female is led over it by the male. She proceeds to pick it up with her gonopore.

2. After the female picks up the spermatophore, the male turns to face her and proceeds to embrace her. His pedipalps reach under the female's abdomen; during this pressing stage he assists in transferring the contents of the spermatophore to the female.

3. The "thumb" or apophysis of the male pedipalp helps to push the contents out of the sperm packet.

4. The transfer of the spermatophore contents requires several hours. After completion, the empty spermatophore is abandoned.

apophysis

Protonymph instars behave like miniature adults. They dig their own little burrows and capture and kill prey. If the young vinegaroon obtains enough food, it will undergo one molt a year until it reaches maturity. The molt normally occurs in the spring while the overwintering vinegaroon is still entombed in its underground cell. Damaged or missing legs, including antenniform legs, and the flagellum may be regenerated before the molt. There are 4 free-living instar stages before maturation, and since the young uropygid may not molt following a poor year, it may take from 5 to 7 years to reach maturity. Vinegaroons do not molt again once they are mature, and so their normal lifespan in the wild is probably in the range of 6 to 9 years. Eventually, this magnificent predator slows down due to old age, as joints stiffen and lost appendages cannot be regenerated. Perhaps even in the wild, it may actually die of old age, still a formidable predator to the end.

Protonymphs stay underground in their burrows during the winter and frequently molt before emerging the following spring. After molting, the pedipalps become larger and darker in color, as seen above. The mother vinegaroon is intolerant of her 1-year-old offspring; therefore they must live solitary lives from this point on.

A female vinegaroon that mated in the fall stores the sperm until spring, at which time she lays her eggs. The eggs are deposited into a clear, membranous sac attached to the underside of the female's abdomen at the gonopore. The mother vinegaroon must tilt her abdomen upward in order to accommodate the bulky egg sac. The number of eggs may vary considerably, from just a few to more than 50.

1. After 5 weeks of incubation, the deutembryos molt, and the pink-and-white larvae crawl onto their mother's abdomen. There they will stay while they undergo further development.

2. The vinegaroon larvae have nonfunctional chelicerae and claws and are unable to survive independently at this stage of their development. Their tarsi have broad "sucker pads" that assist them in holding onto the smooth cuticle of their mother.

3, 4. After 5 more weeks of development, the larvae are ready to molt into free-living protonymph instars. The larval feet hold fast to the mother's abdomen, allowing the young vinegaroon to extract itself completely from its old cuticle. If it were to fall to the ground before completing its molt, the young vinegaroon would die.

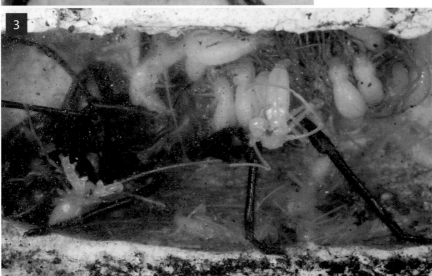

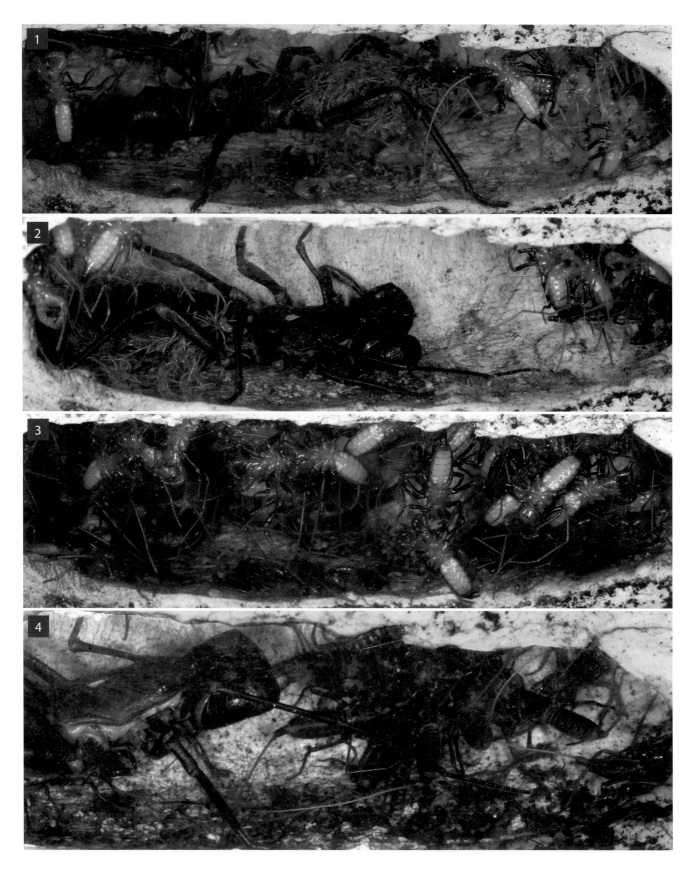

FACING PAGE:

1. After molting, the empty larval exoskeletons, called exuviae, still hold fast to the mother's abdomen.

2. The mother vinegaroon scrapes the exuviae off her abdomen. Her offspring crowd together at the ends of the burrow, staying clear of their mother while she actively removes the empty larval exoskeletons.

After their mother has finished scraping off the exuviae, her offspring freely move about within the burrow.

3. The free-living protonymph instars remain in the burrow for almost a month following their molt. During this time, their exoskeleton becomes darker as it sclerotizes. Their palps go from pale pink to a rich red color, while the dorsal surface of their bodies becomes black.

4. After a month, the mother vinegaroon has begun to venture out of the burrow to hunt for food. She brings some of the captured prey back down into the burrow. Sometimes she eats the food there, while her babies swarm around her, tearing off chunks of masticated prey from between her pedipalps. Other times, she deposits the dead insect in the burrow and steps back while her babies attempt to tear the prey into pieces. Here the babies have converged on a dead cricket that the mother deposited in the burrow.

5. Reaching between its mother's claws, a baby vinegaroon steals food from her. The mother vinegaroon shares her food, despite the fact that this is the first meal she has had in more than 7 months.

The mother might not emerge from her burrow until her babies are also ready to do so. It is remarkable that the ravenously hungry mother can differentiate insect prey from her own babies.

Once emerged from the burrow, the famished mother vinegaroon hunts insect prey. The babies swarm around their mother as soon as she has captured her quarry.

The young vinegaroons tear off chunks as their mother masticates a cricket into mush.

Unable to capture large prey by themselves, the tiny vinegaroons benefit from their mother's ability to take down relatively large prey and her subsequent willingness to share it with her offspring. This extended maternal care of the young gives offspring a tremendous advantage. The enhanced survival of the tiny protonymphs may translate into selection for maternal care of the young if these offspring are better able to reach maturity and produce their own offspring.

CHAPTER 5 Short-Tailed Whipscorpions:
Schizomida

Small size and a flexible body are advantages for living under logs, rocks, or in the soil. The shield covering the prosoma has several divisions, thus giving the order its name Schizomida, meaning "split in the middle." These divisions give flexibility to the 0.24-inch (6 mm)-long *Hubbardia pentapeltis* female.

One of the most unlikely inhabitants of the southwestern United States is a tiny arachnid, the short-tailed whipscorpion, also known as the schizomid. Schizomids resemble miniature vinegaroons in many ways. The overall body shape and proportions of the schizomid are comparable to the vinegaroon. In both groups of arachnids, the first pair of legs is elongated and covered with sensory setae. These antenniform legs are used strictly in a sensory capacity rather than for locomotion. Both vinegaroons and schizomids employ similar chemical defense systems when threatened, spraying acetic acid or a closely related chemical compound from the end of their abdomens. It is no coincidence that these two groups of arachnids appear to be close relatives; DNA and morphological characteristics indicate that Schizomida and Uropygi are indeed sister taxa. Some classification schemes even place Schizomida in the same order as Uropygi.

However, schizomids differ from vinegaroons in several important ways. First, schizomids are much smaller than vinegaroons. Most are only 0.16 inch to 0.28 inches (4–7 mm) in body length, not including the flagellum (although those in the genus *Agastoschizomus* may reach a length of almost half an inch or 12.4 mm). In comparison to the diminutive schizomids, the vinegaroon *Mastigoproctus giganteus* reaches a body length of about 2.5 inches (64 mm). Second, schizomids possess slender raptorial pedipalps oriented in a vertical plane instead of the heavy, clawlike palps of the vinegaroon, which are oriented in a horizontal plane. In addition, the schizomid has enlarged hind legs (in particular the femora) that enable it to jump. Next, the dorsal surface of the schizomid prosoma has several platelike divisions; hence the name schizomid means "split in the middle" in reference to these divisions. In contrast, the prosoma of the vinegaroon has a single shieldlike plate covering the entire dorsal surface. Finally, the flagellum of the schizomid is short and stubby compared with the long flagellum of the vinegaroon. This short little flagellum fulfills a different functional role in the schizomid, at least in regard to the male schizomid.

With the final molt, the male schizomid has a markedly different flagellum compared with that of the female. In fact, schizomids exhibit such strong sexual dimorphism that males and females from a single species were sometimes initially described as separate species. Whereas the female has a slender flagellum consisting of 3 to 6 segments, the mature male has an enlarged, unsegmented flagellum. This may be bulbous, spatulate, triangular, or elongated in shape. In some species, the male flagellum even has holes in it. Both the male and female flagella have a number of protruding setae. Species groups of schizomids are differentiated in part by the characteristic shape of the male flagellum for each species group. The functional significance of the male's flagellum becomes evident during courtship and mating. After a preliminary "dance" during which the male courts the female by vibrating and shaking his legs (without physically touching the female), the female grasps the male's flagellum with her chelicerae. At this point, there may be a contest in strength as the male attempts to pull the female forward over his deposited spermatophore and the female may try to pull the male backward. Considering the huge investment in resources required from the female for reproduction, she may be testing the male's "fitness" during this contest. Only a male strong enough to pull the female forward over the spermatophore can succeed in mating. Once the female is in position over the spermatophore, she picks it up and the sperm are transferred to her internal sperm storage structures, the spermathecae.

In southern California, mating takes place during the middle of the cool, rainy winter season. Schizomids are highly susceptible to desiccation; therefore, they are found near the surface only following times of abundant rainfall or in naturally moist areas. In the spring, the female schizomid *Hubbardia pentapeltis* excavates an underground maternity burrow before laying her eggs. The burrow may be a few centimeters in depth and may take several days to excavate. After the maternity burrow is completed, the female lays her eggs. Schizomids produce from 6 to 30 eggs in a single clutch. These are attached as a roughly spherical mass to the underside of the female's abdomen. The brooding female *Hubbardia pentapeltis* angles her abdomen up over her body such that the mass of eggs more or less faces upward. After a little more than a month's incubation, the larvae hatch and position themselves on all sides of their mother's abdomen. All the larvae face in the same direction, toward the mother's prosoma, and neatly overlap one another. There they stay immobile for a month, during which they undergo further development. The mother continues to hold her abdomen upright during this time. At the end of this period, the offspring molt and drop to the floor of

A subadult male *Hubbardia pentapeltis* (left) resembles a female in size and in morphology. The flagellum at this stage is slender, short, and simple. During the spring, this male will excavate a molting chamber underground and then remain there for several months. After his final molt, the mature male (right) emerges with the triangular flagellum characteristic of *Hubbardia pentapeltis* males. The mature male may survive for more than one breeding season.

the chamber. They stay in the maternity chamber until the mother breaks through to the surface, releasing the protonymphs. The young schizomids are lighter in color than are the adults. It may take 2 or 3 years and 5 stages (beyond the larval stage) to reach maturity.

In tropical areas that are warm and moist year-round, schizomids may not construct an underground brood chamber but instead stay on the surface of the ground. The female may not even hold her abdomen angled over her back during the incubation of the eggs and is therefore able to walk around in a relatively normal manner during this time. Young are produced throughout the year in these tropical climates.

Some species of schizomids do not require males in order to produce offspring. These parthenogenetic species consist solely of females. Other species may or may not have males to mate with; these species are facultatively parthenogenetic.

Like their closest relatives, the vinegaroons and the tailless whipscorpions, schizomids are predaceous. Probing with their antenniform legs, schizomids hunt for small invertebrate prey. They can move rapidly forward or backward, seizing prey with their raptorial palps. They capture a variety of small invertebrates, including isopods, collembolans, termites, psocids, and worms. They may take prey even as large as themselves. Like many other arachnids, each schizomid chelicera has a fixed and movable finger armed with teeth. This pair of jawlike claws enables the schizomid to masticate the food while introducing digestive fluids. Only liquefied food is actually ingested. Some evidence suggests that

schizomids excrete water-soluble ammonia instead of uric acid. The excretion of ammonia would be more efficient from an energy standpoint, but would require plenty of water available to the schizomid. Perhaps this is a factor in their distribution.

Most of the schizomids of the world are found in a band of tropical or subtropical moist lands lying relatively close to the equator. For example, no schizomids naturally occur in Europe (although a few populations have become established in greenhouses through unintentional transplantation). However, several species of schizomids are found in the temperate climate of the southwestern United States, and a few species are even found in the arid regions of the southwest. The historic distribution of schizomids in the southwestern United States consists of two main species groups. The *pentapeltis* group (characterized by a triangular-shaped flagellum in the males) occurs in the most southern part of coastal California, extending north to the San Bernardino Mountains and east, including the Peninsular Ranges. The *briggsi* group (characterized by a rounded, club-shaped flagellum) occurs in the United States starting near Los Angeles and extends as far north as Fresno, California. Isolated populations of the *briggsi* group have historically been recorded farther east, including Joshua Tree Monument, the Anza-Borrego Desert, Organ Pipe Cactus National Monument, and even Tucson, Arizona. To solve the mystery of why these schizomids have been found in such an apparently inhospitable region, we need to go back in time to the Tertiary Period of the Cenozoic Era.

The Cenozoic Era started one June day 65 million years ago following the impact of the Chicxulub meteorite off the Yucatan Peninsula. This event marked the end of the dinosaurs and the Mesozoic Era. For the first 10 million years of the Cenozoic Era (the Paleocene Epoch), the climate was warm and mild. This was fortuitous, since this moderate climate enabled plants to recolonize the southern part of North America following the devastation wrought by the meteorite impact. Ferns left a record of this recolonization in the form of a fern spore spike. After the Paleocene was the Eocene Epoch (from 56 to 34 million years ago). North America during this time was frost-free and had high rainfall. In fact, crocodiles lived in what is now Wyoming. Following the Eocene was the Oligocene Epoch (34 to 23 million years ago). Although the Oligocene was somewhat cooler than the Eocene, it was still warmer compared with our current climate. New Mexico was covered with subtropical broad-leaved forests. But during the Miocene Epoch (23 to 5 million years ago), a major change in climate began to occur. Although temperatures were a bit warmer in the early Miocene compared with the Oligocene (both Florida and the Yucatan Peninsula were under water as a consequence of higher ocean levels), the interior of North America dried out and became savanna, resembling the modern-day Serengeti Plains of East Africa. The uplifting of several mountain ranges—including the Rocky Mountains, the Sierras, and the Cascade Mountains—created a barrier that blocked moisture from the Pacific Ocean from reaching the interior of the continent. This rain-shadow effect in conjunction with a cooling trend later in the Miocene produced an arid climate in southwestern North America for the first time since the Jurassic Period. During this time, Baja California split off from the Mexican mainland and moved northward, eventually to connect to southern California. The last epoch of the Tertiary Period of the Cenozoic Era was the Pliocene (5 to 1.8 million years ago). Magma upwelling beneath the Four Corners area of the southwestern United States lifted the Colorado Plateau by more than a mile in elevation. It was during the Pliocene that southwestern North America became true desert.

The geologic and climatic history of the region provides clues that might explain how, why, and when populations of schizomids occurred in this region. Starting with the Eocene Epoch, the mild climate and abundant rainfall in southwestern North America would have been favorable for these delicate arachnids. Fossil evidence confirms that they were indeed present in the area during the Tertiary Period. In 1944, fossil schizomids were discovered in onyx marble pen bases mined from the Bonner Quarry near Ashfork, Arizona. Three different genera of schizomids were described from the Bonner Quarry onyx. The estimated age of these fossils ranges from 50 million years old to a few million years old, placing them in the Tertiary Period of the Cenozoic Era. Presumably, the increasing aridity of much of the southwestern region during the late Miocene and continuing into the Pliocene became unfavorable for schizomids. Habitat probably became fragmented, and some populations could have become completely isolated.

However, another epoch was yet to come, a dramatic epoch called the "Ice Age." The "Ice Age," also known as the Pleistocene Epoch, started 1.8 million years ago and ended 11,500 years ago. This epoch consisted of 17 glacial periods, each of which lasted an average of 90,000 years, alternating with warmer interglacial periods that lasted an average of 8,000 to 16,000 years. Glacial periods occurred as the result of a combination of factors, including the wobble of the Earth, the elongation of the Earth's elliptical orbit, the tilt of the Earth's axis, and the presence of land masses near the Polar Regions that could support thick, heavy ice sheets. The glacial periods had greater rainfall than what is seen today, including in southwestern North America. The extensive ice over the northern half of the continent diverted the jet stream and its attendant storms southward. During this period of greater rainfall, much of the downcutting of the Grand Canyon occurred. Interglacial periods had less rainfall and the temperatures were more extreme; the winters were actually colder than during the glacial periods, and the summers were hotter. The last major glacial period ended roughly 11,500 years ago. Evidence from pack rat middens indicates that the southwestern United States was cooler and moister some 12,000 years ago than it currently is. We are now in an interglacial period.

Two species groups of *Hubbardia* currently live in the southwestern United States: the *briggsi* species group and the *pentapeltis* species group.

The *briggsi* species group, found over a greater area of the southwest, probably either was present in the region for a very long time or migrated here from central Mexico. This group is considered Madro-Tertiary in origin. It includes the robust populations of *Hubbardia belkini* and *Hubbardia briggsi* found in California, as well as the relictual populations of *Hubbardia joshuensis*,

Hubbardia borregoensis, and *Hubbardia wessoni*. These last three species are restricted to the marginal habitats of tiny desert oases that are like islands in a sea of desert; schizomids have no means of dispersing out from them. Schizomids cannot ride on the wind or hitchhike on another animal. Both the eggs and the adults are susceptible to desiccation; therefore, they are effectively stranded in favorable pockets as the climate changes and a region becomes more arid. In Arizona, *Hubbardia wessoni* may even be extinct at this point in time.

The *pentapeltis* species group probably migrated to southern California via Baja California. The action of plate tectonics split Baja California off mainland Mexico during the Oligocene Epoch. Either *Hubbardia pentalpeltis* came along for the ride or migrated up along the coast afterward. This group is considered Neotropical-Tertiary in origin.

Schizomids provide a remarkable example of survival in a region that has undergone dramatic changes in geography and climate. They have adapted to surprisingly cool climates. *Hubbardia briggsi* is found at a higher latitude than almost any other schizomid, and it is even found in the foothills of the Sierra Nevada Mountains where winter snowfall occurs. Schizomids are found in southern California, where for much of the year conditions are extremely dry and they must retreat down into the soil in order to survive. Finally, schizomids live in desert oases, albeit in low numbers, indicating that this habitat is marginal at best. The presence of these delicate little arachnids in southwestern North America is both surprising and thought-provoking. Their survival, despite all challenges, is a tribute to their adaptability and resilience.

FACING PAGE:

1. During courtship, the male *Hubbardia pentapeltis* (left) presents his triangular-shaped flagellum to the female (right).

2. A receptive female will grasp the male's flagellum with her chelicerae. Following this, a spermatophore will be deposited by the male for the female to pick up. This female had produced a clutch of babies from the previous breeding season but, with abundant food, was ready to mate again. The cool, rainy winter months are the breeding season for *Hubbardia pentapeltis*.

3 and 4. In March, the female excavates an underground burrow, which may range from just under the surface to several centimeters underground. Once the burrow is completed, she closes the passageway to the surface, sealing herself underground. She then produces a spherical cluster of eggs which is attached to the underside of her abdomen. Unlike vinegaroon eggs, these are not held in a membranous sac; instead, *Hubbardia* eggs adhere to each other to form a roughly spherical mass of up to 30 eggs.

The female freely moves about in the brood chamber, even climbing on the ceiling of the burrow. She will remain sealed in the brood chamber for 2 months, until her offspring molt into free-living instars.

5. By mid-April the deutembryos show developing legs. Three days later the deutembryos molted and climbed onto their mother's abdomen as prenymphs (larvae).

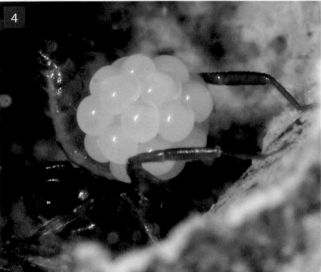

1. Immediately after molting from deutembryos, *Hubbardia* larvae position themselves in layers on their mother's abdomen. The mother eats any larva that fails to climb on board, thus replenishing her own strength to some degree. If the mother were to die before the larvae molt into free-living protonymphs, all the babies would die.

2

3

2. About 5 weeks later, the larvae molt into free-living protonymphs. This mother clung to the ceiling of her burrow during the entire time her offspring were molting. The mother *Hubbardia* proceeded to open the passageway out of the burrow the very next day.

3. Tiny but self-sufficient, protonymphs emerge from the burrow the day after they molt.

CHAPTER 6 Tailless Whipscorpions:
Amblypygi

Paraphrynus carolynae. Two long, slender antenniform legs characterize amblypygids, giving them the name "whip spider." These legs are no longer used for locomotion but are sensory in function. The antenniform legs are also essential in intraspecific communication.

Imagine a world of giant horsetail trees, ferns, and clubmosses. Everything looks oddly familiar, but dramatically out of scale. Plants that you think of as being only a few inches in height are now tall trees. Huge dragonflies with 2-foot (0.6 m) wingspans hunt in the air, while 4-inch-long (10 cm) cockroaches and 2-foot-long (0.6 m) millipedes feed on the abundant decaying plant material on the forest floor. Small reptiles forage among the vegetation, and occasionally one is drowned in the water-filled stump of a clubmoss tree. Lurking in the swampy pools are large, salamander-like amphibians. This is the world of the Carboniferous Period 300 million years ago, and in this world another predator lived. This flat, spiderlike creature stalked its prey on vertical surfaces, tentatively bending its antenniform feelers around curves as it hunted. These feelers were in fact modified legs but were no longer used for locomotion. They had become long and thin, and articulated with many joints, forming the elegant structure that gives this arachnid its common name "whip spider."

Today, in a cooler, drier planet than was present during the Carboniferous Period, whip spiders are found as remnants of that ancient time, living in warm, mostly tropical regions and some deserts.

According to some studies, whip spiders, also known as amblypygids, are a sister clade to the group containing the vinegaroons and the schizomids. All three of these taxa have evolved modified first legs that are now antenniform in nature. In the whip spider these slender legs have become extraordinarily elongated and thin, giving them a more delicate appearance than seen in the vinegaroon or the schizomid. These antenniform legs are equipped with a variety of specialized hairs, or setae, that allow the whip spider to taste and smell as well as detect vibrations and locate prey. In a way, the antenniform feelers are the eyes, ears, nose, and even voice of the amblypygid. Almost every aspect of a whip spider's life involves the antenniform legs in some fashion. They function both as receptors, picking up chemical and vibrational signals, as well as transmitters, communicating aggressive or amorous intentions. They are also absolutely essential for the amblypygid in regard to its ability to orient within its environment. Without antenniform legs, a displaced amblypygid cannot find its way "home" from even a short distance away. These antenniform legs have glomeruli in the nerve ganglia comparable to the antennal glomeruli

found in crustaceans and insects. In addition to these nerve ganglia, each antenniform leg has 2 giant interneurons whose cell bodies reside in the periphery. These receive input from the setae and bristles, which convey vibrational and other mechanoreceptive data to the whip spider.

Whip spiders possess a diverse array of cuticular sensory structures on their antenniform legs. Among the hairs (or setae) are bristles, club sensilla, porous sensilla, rod sensilla, leaflike hairs, and trichobothria. In addition to the setae, there may be other structures present, including a pit organ, a plate organ, and slit sensilla. Among the most numerous of these cuticular sensory setae are bristles. Bristles are most likely contact chemoreceptors, able to "taste" chemical traces via an open pore at the tip of each bristle. These bristles are arranged in 5 evenly spaced rows around the circumference or the tarsi and may range in number from almost 500 in the protonymph to more than 1,500 in the adult. Approximately 500 club sensilla may also be involved in chemoreception, primarily in olfaction, as are the porous sensilla consisting of hairs perforated by numerous pores. Among mechanoreceptors are the trichobothria, the leaflike hairs, and the slit sensilla. In whip spiders, the long delicate trichobothrial hairs are found on the tibia of the whip as well as on the tibia of the walking legs. These, as well as the slit sensilla, provide long-distance mechanoreception, extremely important in detecting moving prey. In fact, the walking legs provide an important backup in this crucial aspect of hunting. Even if the whip spider has lost both antenniform legs, it can still find and capture prey successfully by using the trichobothria on the walking legs; however, the whip spider absolutely must have trichobothria in order to locate moving prey. Once potential prey has been detected, the amblypygid orients itself facing toward it and unfolds its raptorial palps in preparation for the capture as it approaches the quarry. With a sudden lunge, the prey is grasped with the armed palps. Perhaps because whip spiders possess no venom and are somewhat delicate in physical structure, the creatures they capture are usually smaller than themselves. Crickets, moths, small lizards, and small frogs have been documented as prey of whip spiders in the wild. Once the prey has been captured, the whip spider uses its chelicerae to tear a hole in the body wall and regurgitates digestive juices into the opening. The chelicerae continue to masticate the prey into an amorphous mush while the digestive fluids break down the tissue. Like most other arachnids, the whip spider ingests only liquefied food, filtering out solid particles as the powerful phyrangeal pump and stomach suck in the predigested meal.

The digestive system and food storage structures of the whip spider allow it to survive for long periods of time between meals. Two types of cells in the epithelium (lining) of the whip spider's gut assist in digestion. Secretory cells produce digestive enzymes that are regurgitated into the preoral chamber. Resorptive cells develop food vacuoles that digest the food further intracellularly and then pass it along to a structure called the fat body. The fat body surrounds the diverticula of the gut and fills much of the space in the opisthosoma. This is where digested food reserves are stored for weeks or even months.

Nitrogenous waste products from protein metabolism are excreted as guanine, adenine, uric acid, and their derivatives. Since these compounds are largely insoluble in water, the whip spider can conserve water, an important factor in the ability of some amblypygids to survive in some deserts.

In the deserts of the southwestern United States, amblypygids are strictly nocturnal, another important strategy for water conservation as well as a good defense against both temperature extremes and predators. Their eyes are extremely sensitive to light, although the resolution is probably poor. Both the lateral and median eyes are simple; however, the lateral eyes, consisting of a cluster of 3 eyes on each side of the prosoma, are homologous to compound eyes. These eyes are extremely sensitive to light due to the presence of a tapetum, the reflective surface that gives many nocturnal animals "eye shine." The light-sensitive parts of the retina are directed away from incoming light, but toward the tapetum, resulting in the lateral eyes being inverted. The single pair of median eyes has no tapetum and these eyes are everted, meaning that the light-sensitive rhabdomeres point directly toward the light source. The eyes are used in a defensive capacity, assisting the amblypygid in avoiding predators by detecting dark crevices. If the amblypygid feels threatened, it takes refuge in rock crevices or rodent burrows, scuttling sideways with lightning-fast speed that is almost too fast for the human eye to follow.

Like most arachnids, amblypygids depend on hydraulic pressure to extend their legs. This is achieved

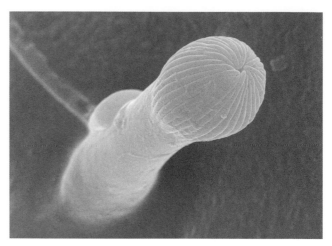

Abundant club sensilla are located on the antenniform legs. These microscopic setae may be involved in olfaction. A diverse assemblage of specialized setae are found on the "whips" of the amblypygid, assisting it in detecting and identifying potential prey, enemies, or mates.

by increasing blood pressure to each leg, regulated by the dorsoventral muscles. It is thought that a narrow constriction between the prosoma and opisthosoma, called the pedicel, protects internal structures within the opisthosoma from potentially damaging high blood pressure. Any injury that compromises the integrity of the cuticle poses an immediate and serious threat to the amblypygid. A break in the cuticle not only presents the risk of bleeding to death but may also instantly immobilize the amblypygid once the high internal blood pressure can no longer be generated to extend the legs for walking. Luckily, small wounds can be sealed by the coagulation of blood. Like that of many other arachnids, amblypygid blood is bluish because of the presence of hemocyanin for transporting oxygen from the book lungs. Hemocyanin contains copper instead of iron, giving it a distinctly blue color. (The most famous "blue-blooded" arthropod is the horseshoe crab, from which we obtain blood to make an extract called limulus amoebocyte lysate, used to detect bacterial endotoxins.)

Amblypygids are extremely fastidious creatures. A whip spider may spend from 1 to 3 hours a day grooming. Particular care is taken to groom the most important sensory structures, including the antenniform legs, the walking leg tibia, and the pedipalp tarsus. Two types of grooming apparatus are employed by the whip spider:

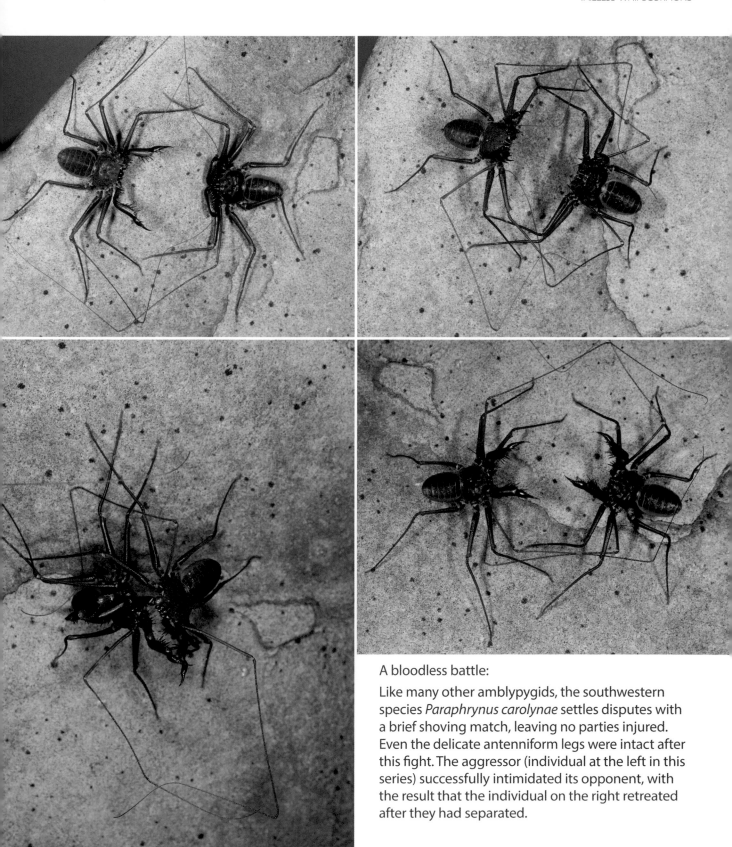

A bloodless battle:

Like many other amblypygids, the southwestern species *Paraphrynus carolynae* settles disputes with a brief shoving match, leaving no parties injured. Even the delicate antenniform legs were intact after this fight. The aggressor (individual at the left in this series) successfully intimidated its opponent, with the result that the individual on the right retreated after they had separated.

a cleaning brush on the pedipalp tarsus, and fringe hairs found both on the chelicerae and on the coxal endites of the pedipalps. The cleaning brush on the pedipalp tarsus consists of rows of setae analogous to a comb used to brush off dirt. Dense brushes of setae are also found on the chelicerae. The whip spider draws the distal part of its antenniform legs between the two chelicerae in order to clean them. These antenniform legs are not only important for locating and capturing prey, but also necessary during contests between males and for courtship and mating.

Amblypygids with shorter, more robust pedipalps such as *Paraphrynus carolynae* share a common fighting behavior between rival males. This combat is highly ritualized and in some ways reminiscent of a pair of people fencing. First, the two opponents stand obliquely facing each other. Each then extends an antenniform leg and stiffly taps his rival while the other antenniform leg is extended sideways in the opposite direction. One can imagine two fencers signaling "en garde" to one another. After a few minutes of this preliminary behavior, the contest escalates into the actual fight. Facing each other directly, each combatant rushes against the other with his pedipalps unfolded and proceeds to engage in a pushing match. This lasts for only a few seconds and culminates with the loser either crouching down or leaving the premises.

In species that have elongate, delicate pedipalps, fighting involves pushing at each other with open chelicerae. In one species, the dominant male may even go as far as to lift his opponent and toss him off the vertical surface. In all cases, the carefully ritualized fighting results in no real physical damage sustained by the combatants as long as the loser can retreat. The fighting is a contest of strength between the rivals and is very different from the behavior employed in killing prey.

The antenniform legs are also utilized during the courtship dance. The courting male taps the female with his antenniform legs in a pattern unique to each species. He alternates this with periodically stepping forward and grasping or stroking the female with his pedipalps. If the female is receptive, she crouches low to the surface and vibrates her antenniform legs. This behavior is repeated but with greater intensity as the courtship progresses. This phase may last from 1 to 8 hours, depending on the species of whip spider. The dance allows the couple to become synchronized and

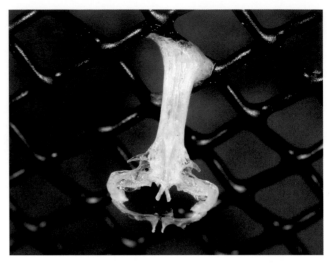

Courtship in amblypygids takes place on vertical surfaces or even on the "ceiling" of their living space. In this case, an empty spermatophore is testimony to the successful courtship of a male amblypygid during the night. A total of 6 empty spermatophores appeared over the course of several nights on the underside of the terrarium lid.

attuned to one another. At the end of this stage, the female is ready to pick up the spermatophore.

During spermatophore formation, the male turns around, facing away from the female, and she stands behind him facing in the same direction. Despite having his back to the female, the male maintains constant contact with her. He extends his antenniform legs behind him, vibrating them and tapping his mate. Spermatophore formation may take 10 to 20 minutes in phrynids. It begins with the male touching the substrate with the tip of his spermatophore organ and depositing the base of the spermatophore stalk. As he raises his body, secretions are released that continue to form the stalk. Finally, the head of the spermatophore is formed, and the male releases it from the spermatophore organ by raising his body and stepping forward. He then turns to face the female in order to lure her over the spermatophore.

All the prior courtship is a mere prelude to this moment. It is critical that the female step over the spermatophore and pick it up with her gonopods. In some species, the male must pull the female over the spermatophore. Therefore, in these species there is selection for larger, stronger males who can succeed in this endeavor.

All female amblypygids carry the developing eggs in a sac attached to the underside of the abdomen. This individual is an African *Damon diadema*.

The spermatophore consists of two small sperm packages contained deep within a highly sculptured and complex structure. The sperm packages sit on small plates, and as the female presses down, levers within the structure of the spermatophore slightly lift up these plates. In the neotropical Phrynidae, females have gonopods with hard, clawlike sclerite appendages. She uses these clawlike "grabbers" to pull out the sperm packages and then proceeds to draw the sperm up into the seminal receptacles. These seminal receptacles are lined with glandular pores that release nutrients into the receptacle, maintaining and nourishing the spermatozoa. At this point, the sperm are coiled and inactive.

Amblypygids mate more than once. Some males may deposit two spermatophores in a single night. For the females, oogenesis and vitellegenesis (i.e., the formation of eggs) occurs next. By the time the eggs are formed, the female is no longer receptive to mating, but by then, she may have mated several times.

The female lays the eggs weeks after mating. They are fertilized at the time of laying. As the eggs and egg sac fluid pass through the female's genital atrium, the encysted spermatozoa activate. As each spermatozoon uncoils and elongates, it resembles a corkscrew with a flagellum. At this stage it is able to fertilize an egg.

The female chooses to oviposit at night. First, she positions herself with her head facing upward as she clings to a vertical surface. As the eggs and fluid are extruded, they flow by gravity down the ventral side of the opisthosoma. The female then turns to face downward, while holding her opisthosoma almost horizontally in a concave shape. The eggs and egg sac fluid fill this depression on the ventral surface of her opisthosoma. She maintains this position for 1 or 2 hours until the external surface of the egg sac has hardened.

Pale and almost completely helpless, the prenymphs cling to their mother's abdomen while they undergo further development. After about 10 days, the young will molt to become free-living protonymphs.

This hardening is made possible by a filamentous substance secreted by the ovaries into the egg sac fluid that hardens to form the protective exterior of the egg sac. After a couple of hours, once the exterior of the egg sac has become stabilized, the female whip spider can take up her customary position facing upward on the vertical surface. The egg sac is held firmly in place, glued to the ventral surface of her opisthosoma by the filamentous substance. This "belly bag" darkens with time to become "tanned" or brownish.

An amblypygid may lay anywhere from 6 to 50 eggs, depending on the size of the female. The female broods the eggs for 90 to 106 days.

After fertilization, the protoembryonic phase of development commences with cleavage, and continues with germinal band development and formation of the body. After about 20 days, this first phase ends with the protoembryo molting into the deutembryo. Using sawlike teeth on its chelicerae, the deutembryo cuts through the protoembryonic cuticle and the egg shell;

however, the embryo is still contained within the brood sac. During the deutembryonic phase (lasting 70–80 days), organs and tissues are formed. The deutembryo shows no external change during this period. Roughly 3 months after the eggs were laid, the deutembryo molts into the prenymph, hatches from the egg sac, and climbs onto its mother's opisthosoma. She elevates her opisthosoma in order to accommodate the babies clinging to it. The mother amblypygid does not hunt or feed while her babies are clinging to her.

Despite having already undergone 2 molts, the prenymph is still basically a larva. The cuticle has not yet sclerotized and so is pale in color. The legs are articulated but not yet functional. The prenymph is almost completely helpless at this point but able to cling to its mother's back with the large arolium located between the claws of each tarsus. This arolium is a pink, almost circular pad on each foot that enables the prenymph to hold fast to its mother's cuticle. For the next 10 days, the prenymph uses the yolk in its gut for further development. During this period, the mother remains motionless unless disturbed.

At the end of 10 days, the young molt into protonymphs—the first free-living stage of their lives. One by one, the prenymphs move to the posterior end of the female, molt, then step off their mother. The exuviae from the babies accumulates on the mother's opisthosoma until all the young have molted and stepped away. Then the mother discards the molts. The babies are on their own from now on. In contrast to the ebony black of their mother, the protonymphs of *Paraphrynus carolynae* are pale with a green opisthosoma. Their antenniform legs are extremely long in proportion to their tiny bodies, and they have double the number of trichobothria of the adult.

These graceful little hunters are lightning fast as they capture prey. It seems incongruous that such fragile, delicate creatures can be lethal predators.

The young whip spider grows rapidly in the first year, molting several times to accommodate the increase in its size. Every time it molts, it must find a vertical or overhanging surface. The amblypygid takes a firm grip on the surface and leans back, rocking its body as the old cuticle splits, and allows gravity to assist in drawing its body free. Extracting the walking legs and long, delicate antenniform legs from the old cuticle requires extreme finesse. The old cuticle must hold firm to the substrate during the entire process. If

the amblypygid falls to the ground during the molt, it will usually die.

Immediately after molting, the whip spider is pale in color and appears somewhat milky and slightly translucent. The body darkens fairly quickly, but the legs remain pale for some time. After a few days, the fresh cuticle is shiny and black, and the amblypygid is ready to hunt again.

Molting not only accommodates growth but also permits the whip spider to regenerate lost limbs. Regeneration is possible only if the leg is broken at certain sheer points, specifically at the patella-tibia joint, which is adapted for easy breakage. The loss of a limb is an active process; thus, the amblypygid must be awake and conscious in order to successfully autotomize a limb. If the tibial portion of a leg is grasped, 3 muscles associated with the femur and the patella contract suddenly. This allows the tibia to be released from the patella with the tension exerted by the outside force grasping it. If the leg has been injured, the whip spider can autotomize the leg by contracting the 3 muscles while simultaneously pressing the injured limb against the substrate. The ability to "drop" limbs at will has an obvious survival advantage. If a predator manages to grasp the leg of a whip spider, the whip spider may escape with its life at the expense of a leg.

Since trichobothria are absolutely essential for the whip spider to locate moving prey, it is no surprise that they are present on both the walking leg tibia as well as the antenniform leg tibia. This redundancy is no accident; it provides insurance that if one of these legs is lost, the amblypygid will already have a backup in place. As long as the limb is severed at the patella-tibial joint, it can be regenerated. This includes the antenniform legs. The freshly regenerated antenniform leg may be shorter than the original and have fewer sensory setae and cuticular sensilla, but with another molt it very nearly matches an unbroken leg. If the femur is severed, the leg will never regenerate. Many amblypygids lose a leg or two in the wild, but because they continue to molt and regenerate limbs even after reaching sexual maturity, the loss of limbs is not cumulative. In addition, they can renew worn cuticle and continue to grow. After a female produces a litter of babies, she usually molts. At this point, she is once again a "virgin" (because of molting the lining of the seminal receptacles as well as their contents) and must mate again before she can produce more eggs and young.

With additional molts, females can grow in size and therefore be able to produce more eggs. Larger males also have a selective advantage over smaller males during contests between males and may subsequently obtain more opportunities for mating with females. Given the benefits of continuing to live and grow larger, it is puzzling that more arachnids do not molt after reaching sexual maturity.

It is not known for certain how long amblypygids live after reaching maturity. It is thought that individuals may live for 10 years or longer.

As a group, however, whip spiders are ancient. During the Carboniferous Period, amblypygids inhabited what is now North America and northern Europe. The neoamblypygids probably evolved in Gondwana before that continent fragmented during the Cretaceous Period. That fragmentation separated the oriental Charontidae from the western Phrynida. The subsequent separation of South America from Africa may have split the Phrynida into two groups, the neotropical Phrynidae and the Phrynichidae. Finally, the Phrynidae are split into two groups: Phryninae, found in Mesoamerica and the southern United States (including the genus *Paraphrynus*), and the Heterophryninae, found in South America.

Until recently, it was assumed that amblypygids were completely solitary; however, a study by Linda Raynor and Lisa Anne Taylor at Cornell University casts doubt on this assumption. In this study, it was discovered that two species of amblypygids (*Phrynus marginemaculatus* and *Damon diadema*) demonstrate some degree of sociality. Under captive conditions, young amblypygids cluster together forming aggregations. This behavior seems to be correlated to their age: the younger the amblypygid, the more likely it is to seek out the company of others. This behavior may provide some defensive advantage to the young. If suddenly disturbed, the early instars seek out their mother and attempt to huddle under her body, analogous to chicks under a mother hen. In the African species *Damon diadema*, this reliance on their mother for protection may be warranted. The mother amblypygid actively defends her brood, attempting to stab an attacker with the sharp terminal claw at the tip of her palp.

Even when not clustered together, amblypygids appear to seek each other out, and they frequently remain within "whip-distance" of their siblings. Interactions between individuals are mediated through the antenniform legs. Loaded with chemosensory and mechanosensory setae, these feelers are critical in all aspects of communication. The tips of the antenniform legs delicately touch any other amblypygid, or may even approach to almost connect without actually making physical contact. The siblings remain tolerant of each other until they approach maturity.

Although the behavior of *Paraphrynus carolynae* has not been studied in detail, several adults may share a common refuge and appear to be tolerant of each other. This may provide a survival advantage, given the challenge of a dry, hot environment for such a delicate creature. Fighting over a scarce resource may result in physical damage or stress to the participants, whereas sharing a refuge might provide mating opportunities or other advantages.

Sociality in arachnids has been largely defined by social spiders. Although the interactions and groupings of other arachnids such as harvestmen and amblypygids may not readily fit the spider model of sociality, the interaction and communication between tolerant individuals within a group may justify broadening the definition of sociality. Amblypygids not only provide a window into the past but also challenge our assumptions and models regarding sociality in arachnids.

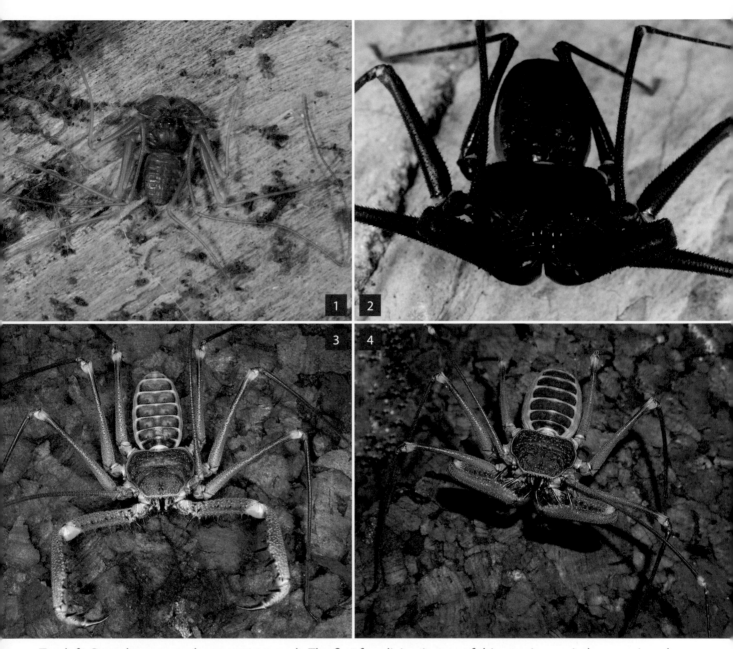

Top left: *Paraphrynus carolynae* protonymph. The first free-living instars of this species are jade green in color. As they grow older, they will develop the black cuticle characteristic of this species.

Top right: *Paraphrynus carolynae* occurs in southern Arizona and northern Mexico. Populations from Sonora, Mexico, demonstrate some color variation, such as the burgundy-colored pedipalps of this individual.

Bottom left: *Acanthophrynus coronatus* male. Photo by Michael Seiter.

Bottom right: *Acanthophrynus coronatus* female. Photo by Michael Seiter.

Acanthophrynus coronatus occur primarily in Mexico but have been reported in extreme southern Arizona and California. This species is among the largest amblypygids, reaching a body length of almost 2 inches (5.1 cm) and a leg span of about 9 inches (22.9 cm).

Microwhipscorpions:
Palpigradi

Tiny and delicate, palpigrades are so weakly sclerotized that they are partially translucent. Small size and flexibility enable this diminutive arachnid to travel through spaces in soil.

Tiny, fragile, and hidden from sight, palpigrades are among the most mysterious of all the arachnids. This order of arachnids is also known as microwhipscorpions, a name inspired by their resemblance to their larger relatives the uropygids (also known as whipscorpions or vinegaroons). The operative part of the name is "micro"; these diminutive arachnids average only a millimeter or two in length. Almost every aspect known of their natural history is "micro" as well, consisting of incomplete fragments of information supplemented by conjecture and extrapolation. Perhaps it is precisely this lack of data that lends an appealing air of mystery to these delicate little creatures.

Palpigrades share a few characteristics with their distant cousins the uropygids. Probably the most conspicuous of these is the flagellum (tail), which in palpigrades is long and whiplike, made up of 14 or 15 segments. Each segment of the flagellum has a circle of setae radiating out from it. Consequently, the flagellum resembles a bottlebrush or perhaps a horsetail plant. When harassed, the palpigrade rapidly and repeatedly lashes this flagellum up and down. Both vinegaroons and palpigrades spend much of their time underground. Whereas vinegaroons use subterranean burrows as retreats, many palpigrades live continuously in the interstitial spaces within soil. Perhaps the flagellum is evidence of convergent evolution for an underground existence. It gives the arachnid a sensory structure that detects disturbances behind it in the confines of a burrow or in the tiny passages within soil. Palpigrades also have setae on each side of their carapace that are almost certainly sensory in function, analogous to the setae found in some predaceous soil mites such as rhagidiid mites.

Palpigrades use their first pair of legs primarily in a sensory capacity comparable to the antenniform legs of vinegaroons and schizomids. These legs each possess 7 trichobothria and other sensory structures, and may function as the "eyes" of the blind palpigrade. The palpigrade carries the antenniform legs in a slightly elevated position as it walks, frequently touching the substrate. In fact, the scientific name of the order Palpigradi means "touching with each step" in Latin (*palpare*, "to touch," and *gradus*, "step," from which we also get words like "palpate" and "gradual").

Palpigrades have a pale opisthosoma and a colorless, translucent prosoma. Their legs are also colorless and clear, but their chelicerae are pinkish in color. When first seen with the unaided eye, palpigrades appear to be animated white specks against dark soil. The tiny flagellum is the characteristic that distinguishes a palpigrade from other tiny white soil predators, such as rhagidiid mites. The lack of color is due in part to the lack of sclerotization of the exoskeleton. Palpigrades are pale, fragile, and susceptible to desiccation partly because of their extremely thin, poorly sclerotized cuticle. At the same time, the lack of a rigid, armorlike exoskeleton gives the palpigrade a degree of flexibility that may be necessary for traveling through the convoluted interstices of soil. The prosoma is composed of platelets, similar to the prosoma of schizomids. Perhaps it is no coincidence that both schizomids and palpigrades have these platelets on the dorsal surface of the prosoma. This may provide yet more flexibility for these small creatures as they traverse narrow and tortuous soil passages.

The thinness of the cuticle makes it more permeable to gas exchange. Palpigrades have no book lungs or tracheae and therefore obtain all the oxygen necessary for their survival through the cuticle. The palpigrade's tiny size is another advantage in this regard; a high ratio of surface area to mass is a prerequisite for cuticular respiration. It is remarkable how well this respiratory system works, enabling this tiny arachnid to demonstrate speed, agility, and a surprisingly active lifestyle. They are even capable of jumping.

One family of palpigrades, the Prokoeneniidae, has three pairs of ventral sacs lined with soft cuticle that can be everted. It is thought that these opisthosomal sacs may be involved with respiration or possibly used to absorb water from the atmosphere. Other organisms such as some booklice (psocids) have structures that absorb water from humid air, thus enabling a tiny, easily desiccated animal to live in an environment that may lack free-standing water.

It is thought that ancestral palpigrades were aquatic. One genus of modern palpigrades, *Leptokoenenia*, consists of two species that are both found living in the intertidal zone of beaches (one species is found in Saudi Arabia, and one comes from the Congo). These species apparently can swim and may provide a glimpse of how palpigrades used to live prior to adapting to a terrestrial existence.

So little is known about palpigrades that it is not even known for certain whether all palpigrades are predators. Their large, three-segmented chelicerae are certainly suggestive of a predatory lifestyle. In fact, the fossil palpigrade *Paleokoenenia mordax* was given the name *mordax* meaning "biting" (in Latin) in reference to its formidable chelicerae. Palpigrades have been reported to hunt and capture springtails and other tiny invertebrates; however, recent research on the European palpigrade *Eukoenenia spelaea* suggests an alternative lifestyle.

As its name implies, *Eukoenenia spelaea* is found in European caves. Researchers examined the gut contents of specimens collected from Slovak caves using special sectioning and stains. Surprisingly, round structures that appear to be cyanobacteria were found to be the predominant gut contents of all the tested palpigrades. Cyanobacteria are among the oldest and most resilient of all Earth's living organisms. Strong phenoplasticity lends cyanobacteria the ability to respond to highly variable environmental conditions. In the 3.5 billion years that cyanobacteria have been on Earth, they have adapted to surprisingly challenging environments. They can withstand low light, extreme temperatures, and low oxygen levels. They can be found in the complete darkness of caves and have even been found living within sandstone as cryptoendoliths. Cyanobacteria contain the energy-rich starch glycogen, which may provide nutrition for cave-dwelling palpigrades. In addition to the cyanobacteria, glycogen granules were observed as part of the palpigrades' gut contents.

It is possible that palpigrades ingest the cyanobacteria incidentally by feeding on other small invertebrates that may have previously ingested the cyanobacteria; however, one piece of evidence supports the possibility that palpigrades feed directly on the cyanobacteria. The chelicerae of *Eukoenenia spelaea* were examined using scanning electron microscopy. Each chelicera has a fixed and movable finger, similar to many other arachnids. The inner edges of both the fixed and movable fingers are lined with 7 to 10 serrated teeth. Each serrated tooth resembles a comb, lined with many smaller flattened teeth along its margin. These cheliceral teeth may be used for scraping cyanobacteria off of surfaces. Perhaps the relatively resource-poor environment of caves has provided a selective advantage to palpigrades that can utilize these cyanobacteria, whether ingested directly or indirectly.

As food is ingested, it is filtered by rows of sclerotized projections lining the buccal cavity (the mouth).

After digestion, free cells called hemocytes transport nutrients and wastes. Guanine is a nitrogenous waste product carried within the vacuoles of the hemocytes.

The excretory system consists of coxal glands and storage cells called nephrocytes. The coxal glands are similar in structure to those found in other arachnids. They originate in the third abdominal segment and exit at the coxa of the first pair of legs. Unlike spiders, schizomids, and uropygids, palpigrades have no Malpighian tubules.

Reproduction in palpigrades is largely a mystery but probably involves the transfer of a spermatophore. The two sexes may use pheromones in species and gender recognition, implied by the difference between males and females in the number and arrangement of the glandular setae on the ventral surface of the abdomen. Although prelarvae and larvae have never been observed, it is thought that the life stages are similar to most arachnid life histories, including a prelarva, larva, protonymph, deutonymph, tritonymph, and adult; however, only 2 nymphal stages have been observed in the family Eukoeneniidae.

Eukoenenia mirabilis is composed almost exclusively of females, with males comprising less than 1 percent of collected specimens. Parthenogenesis is almost certainly a factor in the success of this cosmopolitan species. In theory, a single female could start a colony if transported to a favorable environment. Although the type locality of *Eukoenenia mirabilis* is Sicily, it has a worldwide distribution, including Algeria, South Australia, Chile, Egypt, France, Greece, Israel, Italy, Madagascar, Majorca, Malta, Morocco, Portugal, Romania, the Canary Islands, and Tunisia. Their widespread occurrence throughout the Mediterranean region may be the consequence of unintentional transport by seafaring peoples dating back to the days of classical Greece or even earlier, perhaps thousands of years ago. Modern trade has continued the tradition of transporting these tiny arachnids around the globe.

Eukoenenia florenciae has also been transported to widely separated locations. It has been documented in Argentina, Queensland, France, Colombia, Nepal, Paraguay, Texas, Louisiana, Hawaii, and Arizona.

Contrasting with these widely distributed species, other palpigrades are strictly limited to specific caves. Approximately a third of all described species are troglobitic. The constant temperature and humidity of caves is favorable to this delicate arachnid. They are also found in the soil and leaf litter of moist tropical and subtropical areas of the world.

In the mainland United States, modern palpigrades have been documented primarily in Texas, California, and Oregon. Palpigrades have also been found in Arizona caves and in riparian areas in the foothills of the Santa Rita Mountains in southern Arizona. These latter palpigrades have adapted to an area that is analogous to an oasis in the desert. Surrounded by desert scrub and grassland, palpigrades are found in the soil adjacent to a stream under sycamore and live oak trees. During the dry season, the surface flow of the stream may disappear completely, but the presence of the sycamore trees is testimony to the fact that moisture still exists down in the soil. Palpigrades simply move to deeper soil levels as the surface dries.

Fossils from the Bonner Quarry onyx in Arizona dating from the Tertiary Period (65 to 2 million years ago) demonstrate that palpigrades lived in southwestern North America in the past, when the area had significantly more rainfall than it now receives. Populations of palpigrades in Arizona caves and in riparian areas may be relictual populations, surviving in isolated pockets of favorable habitat and surrounded by a sea of arid desert. That such a fragile and easily desiccated creature could survive at all in the arid southwestern United States is eloquent testimony to the adaptive advantage of being small. The palpigrade's ability to migrate through the interstitial spaces in soil has enabled it to survive in an otherwise harsh and unforgiving environment. The palpigrade is a living paradox; its tiny size and delicate cuticle make it susceptible to desiccation and damage but also permit it to specialize in living and traveling within the protected environment of the soil.

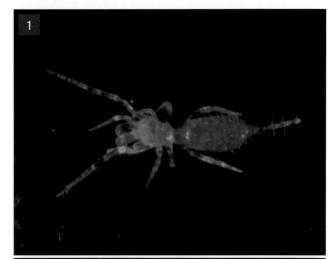

1. *Eukoenenia florenciae*. This species has been found in the Santa Rita Mountains of southern Arizona. This individual had a short, broken flagellum even before it was captured.

2. *Eukoenenia florenciae*. Palpigrades are extremely flexible because of their thin, poorly sclerotized cuticle. While grooming themselves, as this one is doing, they can readily bend and contort their bodies.

3. Palpigrades use their slender pedipalps as walking legs, and the first pair of legs is used primarily in a sensory capacity. The flagellum is also an important sensory structure.

CHAPTER 8 Harvestmen:
Opiliones

Eurybunus species. Desert harvestmen can be surprisingly colorful. This morphotype is found in southern Arizona in low desert. These harvestmen mature during the cooler winter and early spring months.

Four hundred million years ago, near a hot spring in what is now Scotland, a long-legged harvestman died. As its small body lay along the marshy margin of the spring, water rich in minerals from the spring flowed over it, silicifying the cuticle of the harvestman and preserving it. Eventually, the body of the harvestman was entombed in a hard, translucent matrix of chert, preserving the anatomy of the long-dead arachnid in exquisite, three-dimensional detail. Remarkably, this harvestman from the early Devonian Period is almost identical to the harvestmen we find today. A classic example of evolutionary stasis, harvestmen have retained their basic morphology despite having survived several mass extinctions, changing environments, the rise and fall of the dinosaurs, and the arrival of modern flowering plants, mammals, and other modern organisms. This ancient group radiated across the Earth before the Devonian, and it has come to us today fundamentally unchanged from that time, truly a living fossil.

Opiliones, or harvestmen, are perhaps the most puzzling of the arachnid orders. Like other arachnids, they possess 8 legs, 2 pedipalps, and 2 chelicerae, but in many ways everything else about them is different from most other arachnids. Their respiratory system is strictly tracheal, their feeding involves ingesting chunks and particles instead of predigested liquefied prey, and their reproduction (for most) involves an intromittant organ. Their adult lives may involve clustering into large aggregations of unrelated individuals, unlike the typically solitary lives of most other arachnids or the kinship-dependent relationships of the few subsocial and social arachnids. Not only is parental care seen in some species of Opiliones, but a few species even have paternal care of the eggs and young—unique among arachnids and rare among arthropods in general. It is ironic that a small, common, and unglamorous creature provides such a rich chapter in the story of evolutionary biology.

If one word were to be chosen to characterize harvestmen, it might be "opportunistic." Perhaps this is the key to the survival of an otherwise somewhat delicate creature. Although harvestmen are found on every continent except for Antarctica, they are largely restricted to areas that provide shade and moisture. Harvestmen, especially nymphs, die if they become dehydrated; however, because many species are flexible in their dietary requirements, they can opportunistically move from one microhabitat to another, taking different food in each location. Thus, their dietary flexibility allows them to maximize the potential of any particular habitat.

Alone among arachnids, most harvestmen are omnivorous in their diet. At times predators and at other times scavengers, harvestmen defy classification into a neat category. They have been observed to ingest soft-bodied arthropods such as collembola (springtails), spiders, insects of many kinds (especially larvae), mites, isopods, earthworms, nutty seeds, dead vertebrates, millipedes, small plants, bird droppings, gastropods, pollen, mushroom juices, fungus, slime molds (which are actually amoebae), lichens, hickory nuts, red raspberries, psocopterans (booklice and bark lice), apples, and pears. They can be a significant beneficial predator of crop pests, feeding on both eggs and caterpillars of lepidopterans (butterflies and moths). Although a few species of harvestmen are extremely restricted in their diets (such as gastropod predators), many species are flexible in their preferences, taking what is available. The primary requirement is that the food needs to be soft enough for the chelicerae to penetrate and tear off portions. Consequently, adult hard-bodied insects are rarely eaten.

Some species of harvestmen have even developed the ability to neutralize the defensive toxic chemicals produced by potential prey. For example, *Mitopus morio* detoxifies ingested alkaloids of the chrysomelid beetle larva *Oreina cacaliae*. The leaf beetle's defensive chemical is detoxified by N-oxidation in the gut of the harvestman and is eliminated in the feces. This ability to detoxify alkaloids by N-oxidation is unique among nonsequestering arthropods.

Harvestmen bite off bits of food and ingest actual particles after fluid from the mouth is mixed with them. Similar to many other arachnids, each chelicera has a fixed and movable finger armed with teeth along the inner margins that help to dismember the food. Unlike most other arachnids, however, harvestmen lack a sucking stomach. Instead, they have muscles that dilate the preoral chamber and pharynx, expanding the pharyngeal lumen, thus allowing passage of particulate, solid food into the gut for further digestion. The only other arachnids known to ingest particulate food are some mites. The ingestion of particulate food does have its drawbacks, among which is exposure to parasites. Harvestmen may ingest eggs or oocysts of parasites either while directly feeding or possibly even while grooming. If a harvestman has acquired oocysts on its legs while walking, it may accidently ingest them while grooming itself. Some internal parasites, such as gregarines, mature in the gut of the harvestman and release thousands of infective oocysts in the feces, which may in turn infect other harvestmen through environmental exposure. These gregarine endoparasites are common in opiliones and somewhat rare in other arachnids, demonstrating the benefits of preoral digestion for the arachnids that ingest only liquefied, filtered food. Perhaps it should come as no surprise that harvestmen become parasitized, given that more than half of all the described animals on the planet are parasites; more surprising perhaps is the apparent effectiveness of preoral digestion in protecting against exposure to internal parasites.

Once particles of food reach the gut, they are broken down and digested. The gut of the harvestman contains 3 cell types: resorptive, digestive, and excretory cells. A fourth type of cell, the ferment cell, is found only in the diverticula of the gut. The diverticular and ventricular cells of the gut contain small, concentrically structured mineral spherites that sequester heavy metals and minerals such as calcium and phosphorous, which are then depleted during molting. Coxal organs control the elimination of metabolic wastes and regulate concentrations of ions and water in the hemolymph (blood) of the harvestman.

Opiliones have a tracheal respiratory system. Despite the fact that tracheal respiration is inherently no more costly in water use than are book lungs, Opiliones are ultimately susceptible to desiccation because of the completely passive nature of their respiratory system. Harvestmen lack the musculature associated with spiracular control, and thus cannot close the spiracles that allow passage of air in and out of the tracheae. Some species have a grill of cuticular spines partially covering the spiracles, but this filter does not stop airflow enough to prevent water loss through respiration. Both the diversity of species and overall numbers of harvestmen are greater in moister environments than in dry habitats. Consequently, the arid North American southwest has a relative paucity of species in the order Opiliones compared with moister, cooler environments.

Although they are perhaps less formidable predators than some of the other arachnids, harvestmen do hunt and capture invertebrate prey. Prey is probably detected through mechanical stimuli using a combination of specialized hairs and slit sensilla. Although harvestmen lack trichobothria (the incredibly sensitive hairs found on scorpions, pseudoscorpions, and spiders), their legs and pedipalps are covered with an abundance of hairs called sensilla trichodea (or trichoid sensilla). Some species have been documented as being able to capture flying insects, localizing the vibrations and successfully catching the prey even with their eyes covered. Distance mechanoreception may be provided by slit sensilla, alerting the harvestmen to the presence of potential prey.

Setae are concentrated at the tips of the palps and the legs, especially the second pair of legs, which are more tactile than locomotor in function, analogous to the antenniform legs of uropygids and amblypygids. A harvestman may detect food as it forages, tapping the long second pair of legs in front of it as it walks, or it may remain stationary, ambushing prey as it is detected. Once the potential prey is located, identification is accomplished with chemoreceptor hairs on the distal segments of legs I and II, called sensilla chaetica.

The pedipalps are especially important for prey capture in several groups of opiliones. The pedipalps of nemastomatids have glandular hairs with spherical tips. An adhesive "glue" is secreted from these specialized setae, assisting the harvestman in capturing small prey such as collembola.

Armed with formidable spines, the raptorial palps of Laniatores may have been the inspiration for the family name Phalangodidae. *Phalangodes* was the Greek word for phalanx, referring to a row of soldiers standing side by side with long spears held pointing forward. This military formation was used in antiquity by the likes of Alexander the Great. The name is apt for opiliones in this suborder, since many use their spiny palps in the capture of prey and also for defense. After the prey is captured with the legs or the pedipalps, it is passed to the chelicerae for dismembering.

Since no harvestmen are venomous, and their offensive weaponry is somewhat limited, they tend to avoid large, potentially dangerous prey. Harvestmen are in turn preyed upon by amphibians, reptiles, birds, rodents, insectivores, spiders, scorpions, other harvestmen, ants, predaceous beetles such as carabids, and predaceous bugs such as reduviids, *Lygus*, and pentatomids. With so many enemies, harvestmen have

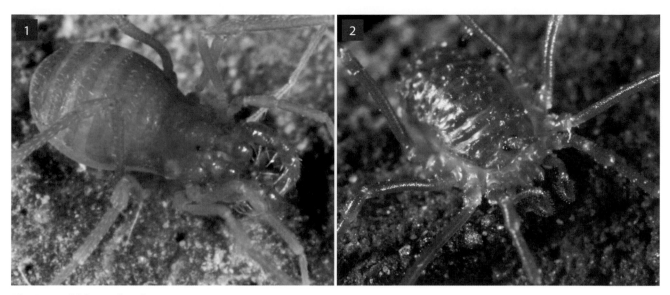

The incredible pedipalp:

1. *Sitalcina* has spiny raptorial pedipalps used for capturing prey.
2. *Ortholasma coronadense* has glandular setae on its pedipalps. Experiments have shown that the tiny drops of glue effectively hold prey such as springtails, even if the pedipalp only just touches the springtail. This is a subadult, but adult *Ortholasma* also have glandular pedipalps.

evolved an array of defensive mechanisms, some of which are unique among arachnids.

First of all, the majority of harvestmen in temperate climates are cryptically colored. The sclerosomatids tend to so closely resemble their surroundings that they are inconspicuous unless they move. Some species of harvestmen take crypsis a step further, covering themselves with a coating of dirt particles. This layer of dirt serves a dual role in camouflaging the harvestman: not only does it blend into the environment, visually protecting against hunters such as birds and lizards, but it may also mask the chemical identity of the harvestman, thus protecting against other arthropods such as ants. If crypsis fails to shield the harvestman from the attentions of a predator, several other defense mechanisms may be deployed. Rapid bobbing up and down may confuse a predator; this is also employed by long-legged cellar spiders, the pholcids. Thanatosis, or feigning death, may also be used. The harvestman may drop to the ground and lie motionless until the threat has passed. If all these have failed, the harvestman may secrete a noxious chemical to repel the attacker. Scent glands secrete different defensive chemicals in a variety of harvestmen. Eupnoi species tend to secrete alcohols, ketones, and aldehydes; Laniatores secrete alkylated benzoquinones and phenols; Travunioidea secrete terpenoids; Phalangiidae secrete naphthoquinones; and Cyphophthalmi secrete methyl ketones and naphthoquinones. The chemicals are produced in glandular epithelium and released to the surface via an ozopore; they may be employed in a variety of forms, depending on the species of harvestman. Some coat their entire body with a chemical shield that spreads over the harvestman as it is channeled along grooves in the surface cuticle. Others may produce a droplet of noxious chemical at the ozopore that may be transferred to a leg and applied directly to an attacker. Some squirt liquids, others spray a gas, and some species may direct the spray in any direction, even in front of the harvestman. These chemical shields appear to be effective protection against ants and spiders. Some spiders avoid harvestmen even before fresh defensive chemicals are released; perhaps the integument of the harvestman contains intrinsically repellant chemicals.

Harvestmen are known to form large aggregations of unrelated individuals that may accentuate the effectiveness of their chemical defenses. Laniatores form loose aggregations, and Eupnoi, especially the sclerosomatids including the daddy longlegs, form extremely dense clusters. There is no kinship requirement to join these aggregations, which are made up almost entirely of mature, unrelated individuals. For sclerosomatids living in temperate climates, these aggregations may form under conditions of low temperatures and low humidity; consequently, this behavior may have evolved primarily as protection against desiccation and cold and only secondarily as a defense strategy. Nonetheless, the collective release of scent gland defensive chemicals as well as the rapid scattering in all directions from a large group may increase the odds of survival of any single harvestman from that group.

Perhaps the most novel defense mechanism of harvestmen involves the loss of a leg. Many arachnids share the ability to lose a leg voluntarily. This is technically called autospasy, since the arachnid cannot simply shed the leg but requires some tension to be applied to the leg for it to break off. In harvestmen, lines of weakness at the base of the femoral leg segment predispose the leg to break easily at that point, and muscles of the trochanter pull the remaining tissue inward, closing the wound and preventing the harvestman from bleeding to death. Immature spiders and many other arachnids can regenerate a missing leg as long as it is lost far enough in advance of the next molt to allow time to regenerate. However, not only are harvestmen different from other arachnids in not regenerating missing legs, but when they do lose a leg, the leg provides a unique diversion; it kicks and twitches on its own accord after being separated from the body. Two pacemaker nerve ganglia within the leg are activated as soon as communication with the central nervous system is severed. These two pacemaker ganglia operate independently of each other, causing the leg to spontaneously twitch and jerk. The legs even have their own tracheae and accessory spiracles which supply the disembodied leg with oxygen. Consequently, a detached leg may continue to twitch for up to an hour (whereas if the spiracles are sealed, thus blocking the air supply, the leg may twitch for only about forty seconds). Experiments have demonstrated that the twitching leg distracts predators such as ants and spiders while the harvestman can slip away unnoticed. A similar defense mechanism is seen in a completely unrelated group of animals. In a surprising example of convergent evolution, some species of lizards can drop their tails when grasped by a predator and can escape while the tail continues to twitch spasmodically.

Thick and "furry," an aggregation of harvestmen consists of hundreds of individuals clustered tightly together. These extremely tight aggregations occur during the cold winter months. Photo by Philip Kline.

The reproductive strategies of harvestmen provide models of both natural selection as well as sexual selection. With the exception of small fossorial species in the suborder Cyphophthalmi that mate using spermatophores, male opiliones use an intromittant organ for mating. This organ may be referred to as either a penis or a spermatopositor. In the suborder Eupnoi, which includes the familiar daddy longlegs in North America, males may stake out a rock or other suitable egg-laying site, and any female seeking to lay her eggs there encounters the amorous male. During courtship, almost all communication between the male and female is thought to be chemical and vibratory signals communicated through the legs. Mating harvestmen may stroke or rub each other using their legs, chelicerae, and pedipalps. The male hooks his pedipalps near the base of leg II of the female. In some species, the male even has a femoral spur that holds the female's leg securely, preventing her escape. Consequently, it is easier for a female to accept the male's advances and mate with him than attempt to fight him off; however, the female holds a card up her sleeve. Harvestmen sperm are aflagellate and thus immobile. As a result, sperm do not seek out and fertilize the eggs on their own power. Instead, the female harvestman receives and stores the sperm in seminal receptacles at the tip of the ovipositor, and she fertilizes only the eggs at the apex of the ovipositor immediately before oviposition. Thus the female harvestman may exercise cryptic sexual selection at the time of egg laying when she chooses whether or not to use a male's sperm.

In the Dyspnoi, the males rely less on coercion and force and more on gifts in order to persuade the female to mate with them. Although the male's chelicerae are enlarged, neither the chelicerae nor the pedipalps are used to grasp and hold the female. Instead, in some species of Dyspnoi, the male produces secretions from his pedipalps which are then taken into the female's mouth as a nuptial gift prior to mating.

Protolophus singularis presents an intriguing puzzle in regard to courtship and mating. Male *Protolophus singularis* occur in two morphotypes. One morphotype has robust, heavy pedipalps, while the other has gracile, slender pedipalps. (In addition, the females have completely different pedipalps from the males; their pedipalps are small and delicate, branching into two slender "fingers.") During courtship, a robust male leaps on top of a female, grasps her with his heavy pedipalps, and wrestles her into submission before mating with her. Immediately after mating, this male may drag her front legs between his chelicerae and palps, presumably "marking" her with chemical pheromones; however, if the female is large and strong enough, she may resist the male's efforts to subdue her. Even after repeated attempts to overpower her, the female may be able to break free of the male and escape without mating. In a curious twist, such a female may immediately accept a gracile male without a struggle. After mating, the gracile male and the female may remain facing each other, making physical contact "mouth to mouth" for 90 minutes or more. It is possible that an exchange of some sort is taking place during this time; perhaps the male is sharing a postmating nuptial gift with the female. These observations are preliminary and will require confirmatory research but are certainly suggestive. It seems logical that a species containing dimorphic males would demonstrate two significantly different mating strategies. The robust male uses the "caveman" approach, overpowering the female and forcing her to mate with him. As long as he is stronger than the female, this is a successful strategy. But in situations where the female is strong enough to repel the male, an alternative courtship strategy may succeed. The gracile male presumably is using the "candy and flowers" approach, perhaps conveying the promise of a nuptial gift immediately following mating. Each strategy succeeds to some degree, thereby selecting for two morphotypes of males along with two different mating behaviors.

In harvestmen, the ovipositor of the female is the counterpoint of the male's penis. The harvestmen in the suborder Eupnoi possess a long ovipositor, allowing the female to probe potential egg deposition sites in crevices and cracks found on rocks and fallen trees. She lays only 1 egg in each little nook, hiding it well but not guarding it afterward. In some species, after the male has mated with the female, he guards her by wrapping the tarsi of his leg I in several loops around one of her femurs and follows her as she searches for suitable egg deposition sites. She is able to explore the terrain at her leisure with the male in tow, accepting the male's protection against other males as she lays her eggs. Egg hiding is a logical strategy for these harvestmen that live in temperate climates; the adults do not live very long after mating and so would not be present to guard the eggs. Most Eupnoi are small and lack conspicuous weaponry, in contrast to Laniatores, which may possess substantial spines and other armament. In addition, most Eupnoi secrete aldehydes, ketones, and alcohols as defensive chemicals, which are not as effective for egg guarding as the quinones of Laniatores. As a result, egg hiding has evolved as the most successful strategy for Eupnoi, and with it, selection for a long ovipositor that can reach deep into hidden crevices and recesses.

In contrast, the female Laniatores has a short ovipositor (which necessitates laying her eggs on more exposed surfaces), while the male's penis has developed diverse and complex "ornamentation." This may be analogous to the sexual selection seen in other arachnids such as jumping spiders, resulting in complex courtship displays. However, since the courtship in harvestmen is not communicated visually or acoustically and depends more on tactile communication, the male intromittant organ has undergone sexual selection for complexity. Since female Laniatores are restricted to laying their eggs on the surface of a rock, fallen tree, or vegetation, they must protect the exposed eggs against predators. Although some Laniatores lay only one egg at a time and hide it under debris, laying eggs in batches and guarding or brooding the eggs is very common in this suborder. The female may even guard the newly emerged nymphs for a couple of weeks.

Since tropical Laniatores may live longer than one year, they are available for egg brooding. In turn, those that developed spines as weapons and quinones and

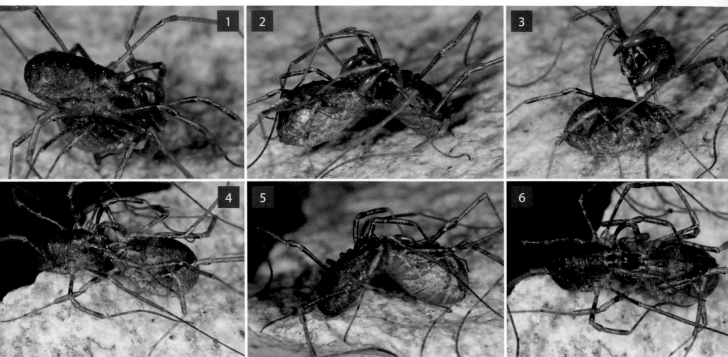

The puzzling case of *Protolophus*:

1. A large male *Protolophus singularis* leaps on top of a female and uses his heavy pedipalps to wrestle her into submission.

2. The pair mate. The female is on the left.

3. Immediately after mating, the male drags the front legs of the submissive female between his chelicerae and his pedipalps. He might be marking her with his pheromones.

4. A smaller male *Protolophus singularis* with slender pedipalps encounters a strong female that had successfully fought off a large male with heavy pedipalps. No wrestling takes place between this pair.

5. The large female immediately accepts the small male and they mate. Mating takes only a few minutes.

6 and 7. After mating, the pair stay "mouth-to-mouth" for more than 90 minutes without moving. Afterward, they separate. The male never dragged the female's legs between his chelicerae and his pedipalps.

Protolophus singularis is formally described as having two male morphotypes: one with heavy pedipalps and one with slender pedipalps. These preliminary data suggest that there could be two courtship strategies for this species as well. One strategy appears to be coercive, and one strategy might involve some sort of gift or exchange after mating. In each case, a male succeeded in mating with a female.

phenols as chemical defenses may have had a selective advantage for this task. In all species in which the mother guards the eggs, she broods one large batch rather than scattered small batches. In some species of Laniatores, if the egg-brooding female disappears, the male holding that territory may step in and take over the job of guarding the eggs for a couple of weeks. This extension of the male's defense of egg-laying sites may constitute the basis of the evolution of paternal care in harvestmen. In some species, the male is the exclusive egg-guarding parent, a unique situation among arachnids but not unknown among other arthropods. Thus, sexual selection may be the driving force for paternal nest-guarding behavior, while natural selection is the driving force behind maternal egg brooding. Females compete with each other for the privilege of mating with a male that is guarding a nest site, and a male without a nest site would stand little chance of finding a mate.

Embryonic development can take as little as 30 days or may stretch over months in temperate climates where eggs may have to overwinter. In some species the developing embryos go into diapause (a state of suspended development) if the eggs are laid in summer and don't hatch until the following spring. Eclosion, or hatching, is facilitated by the use of one or two egg teeth in the suborders Cyphophthalmi, Eupnoi, and Dyspnoi. The freshly hatched larva undergoes further development, and a short time after eclosion it molts into a first-instar nymph.

The number of nymphal stages may be from 4 to 8 but is usually 6, with the final nymphal stage referred to as a subadult. Once sexual maturity is reached, opiliones do not molt again, with the curious exception of some species in the family Grassatores (in the suborder Laniatores), in which two sexually reproductive instars may be found: both the adult and the subadult. This is not seen anywhere else in arachnids but is seen in mayflies.

The arid southwestern United States has relatively few harvestmen compared with wetter environments. The majority of these belong to the family Sclerosomatidae (in the suborder Eupnoi), typified by the familiar daddy longlegs. Several different species fall into this category, including the surprisingly tough *Trachyrhinus marmoratus*. This harvestman is a common sight even in the harsh environment of lower-elevation deserts. Sheltering in a protected niche by day, *Trachyrhinus* is active during the night, when temperatures are more moderate. Other harvestmen are found either in riparian areas or at higher, cooler elevations. *Protolophus singularis* is found almost exclusively in low desert canyons or near streams. This species grows and matures rapidly during the milder winter and spring months, thereby synchronizing its growth and activity with the most favorable temperatures of the year.

Perhaps surprisingly, the arid southwestern United States is home to several species of small harvestmen belonging to the suborder Laniatores. Only a few millimeters in body length, both *Sclerobunus* and *Sitalcina* are particularly susceptible to desiccation. Therefore, these two genera are restricted to cooler, moister mountain areas or to humid caves. These habitats are separated by dry, inhospitable desert, deadly for such delicate creatures. Because these small harvestmen have no means of ballooning or hitchhiking, populations are effectively isolated. The lack of gene flow between isolated populations has resulted in high levels of genetic divergence. This divergence probably started even before the Pleistocene Epoch. Despite the relatively poor morphological differentiation between these populations, the genetic difference as well as their geographical isolation may justify their status as distinct, though cryptic, species.

In an interesting note, *Sitalcina* fluoresce a pale yellow-green under ultraviolet light.

The name "daddy longlegs" may reflect a friendly affinity we feel for harvestmen that is absent for other arachnids. After all, harvestmen are not venomous, lack formidable weaponry, and have only 2 eyes. They ingest particulate food and most are omnivorous, eating a wide variety of fruits, plants, and animals. They form large aggregations of unrelated adult individuals, and cannibalism is rare among these tolerant creatures. They mate using an intromittant organ, and maternal or even paternal care is demonstrated by a number of species. The characteristic flexibility seen in harvestmen gives them a "toolbox" of options in any given environment that may maximize their chances for survival. Flexibility as a survival strategy has withstood the test of time and change, forged ever stronger by natural selection. Consequently, harvestmen provide a model for survival that parallels human survival in some respects. Perhaps it should not be surprising that we are tolerant of these inoffensive little arachnids, given that in some ways they are not so very different from us.

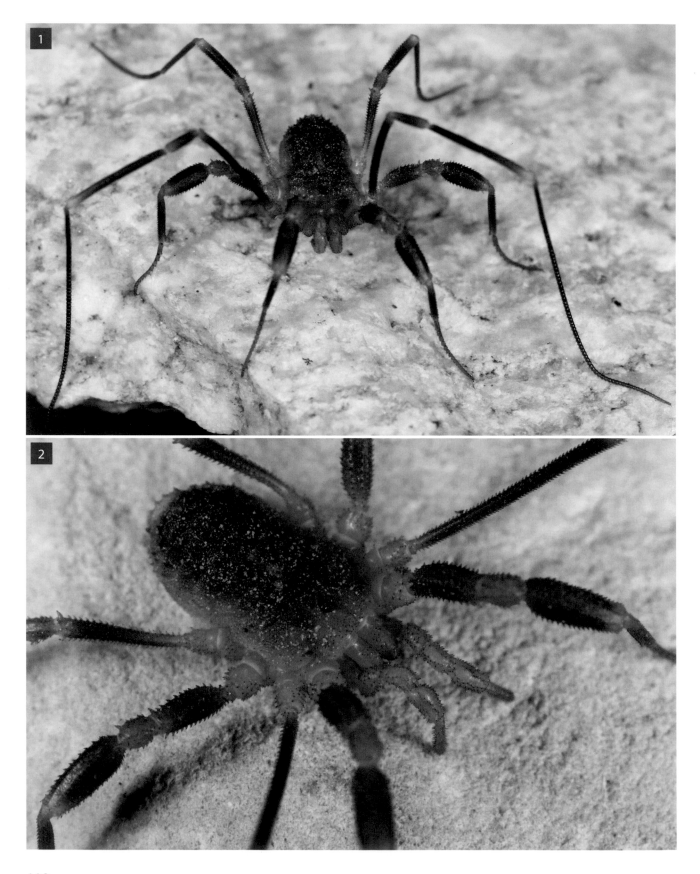

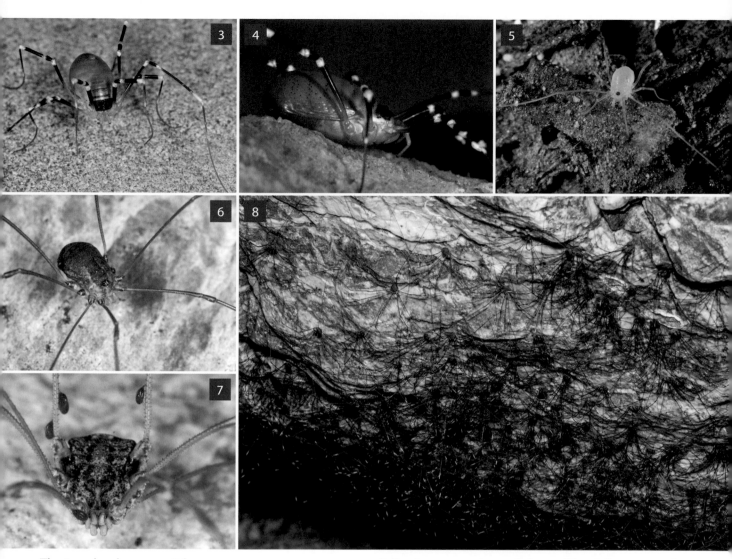

The tough sclerosomatids:

FACING PAGE:

1 and 2. *Dalquestia rothorum.* This species occurs in Arizona. Related species occur in Texas and California.

Distinctively adorned with bumps and spines, the tough cuticle helps to protect this sclerosomatid from desiccation and may also confer some protection against predators. Sclerosomatids belong in the suborder Eupnoi.

3 and 4. *Eurybunus* species, adult. This particular desert harvestman may be undescribed and occurs in Maricopa County, Arizona. The adults mature in late winter.

5. *Eurybunus* species early instar. The delicate nymphs can desiccate easily, and so are found in protected areas such as under rocks.

6. *Leuronychus pacificus* (presumptive). *Leuronychus* are found in southern California and in riparian areas in southern Arizona.

7. *Trachyrhinus marmoratus.* Easily the toughest of the desert-adapted harvestmen, this species can be common in low-elevation, hot, dry desert. *Leptus* mites are frequent parasites on these harvestmen.

8. *Trachyrhinus* may form large aggregations of unrelated individuals. These had gathered in a sheltered spot during an exceptionally severe heat wave with temperatures reaching over 115 °F (46 °C). Photo by Timothy A. Cota.

Living sculptures: *Ortholasma*, Nemastomatidae:

1. *Ortholasma coronadense* adults. Although this species occurs in dry chaparral along the coast of southern California, it survives in moister microhabitats, such as under rotten logs. Several individuals are frequently found in close association with each other. These harvestmen are in the suborder Dyspnoi, family Nemastomatidae.

2. *Ortholasma coronadense* adult. It is only after the final molt that the characteristic hood is seen. Vaguely resembling the nose-guard on a Spartan helmet, this peculiar structure might be sensory in function. The hood is present on both sexes.

3. *Ortholasma coronadense* adult. Like a living Fabergé egg, the cuticle of this freshly molted adult is covered with a sparkling fine lattice-like network.

4. The subadult *Ortholasma* has a projection at the front of the body which will become the hood with its next molt. The pedipalps are thin, long, and covered with glandular setae.

5. A subadult *Ortholasma* has captured a springtail (collembola). The pedipalp has only to touch the springtail in order to capture it. Glandular setae on the pedipalps secrete sticky droplets of glue. This glue is actually most effective at holding an object that suddenly pulls the glue, such as a springtail attempting to jump free.

Spiny harvestmen:

Laniatores have spiny, raptorial pedipalps for capturing prey.

1. *Sclerobunus robustus* mature male.

2. *Sclerobunus robustus* mature female. This species is sexually dimorphic. In the arid southwestern United States, *Sitalcina* and *Sclerobunus* populations are genetically isolated in cooler, moister habitats. Because of limited dispersal capabilities, gene flow does not occur between populations, and so different species arise.

3. *Sitalcina lobata* is found in chaparral in coastal southern California.

4. *Sitalcina peacheyi* is found in the Santa Rita Mountains as well as protected humid caves in southern Arizona.

5. The early instar nymphs of *Sitalcina* are extremely delicate and less than 0.04 inches (1.0 mm) in body length, not surprising given that the adults are only about 0.08 inches (2 mm) long.

6. *Sitalcina catalina* is found in the Catalina Mountains of southern Arizona.

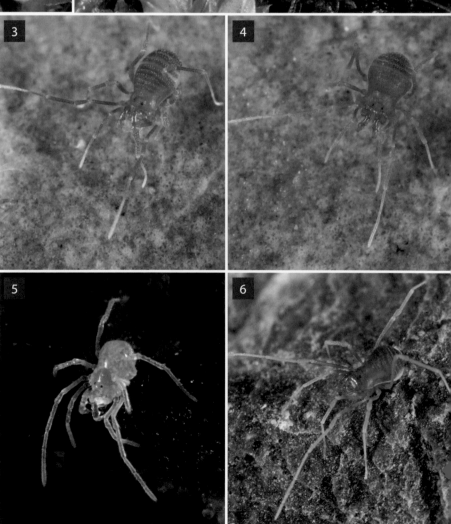

CHAPTER 9   Wind Spiders:
Solifugae

Two close-set beady eyes and massive chelicerae characterize the solifuge. These arachnids have no venom; instead, they have evolved large and powerful chelicerae for subduing and dismembering their prey.

Chelicerae gaping and pedipalps held aloft, a solifuge epitomizes an arachnid with attitude. Its ferocious appearance has inspired names such as *matavenado*, meaning "deer killer" in Spanish. Certainly, close-set beady eyes and a bristly face reinforce an impression of implied threat. But it is when this creature is in motion that the imagination is suddenly electrified by the essence of living speed; impossibly swift, it truly lives up to the name "wind spider." In sharp contrast to their distant cousins the scorpions, solifuges are arachnids that have evolved for a life in the fast lane.

Paradoxically, solifuges possess both primitive and surprisingly advanced characteristics. By some criteria, solifuges represent an archaic form of arachnid. The prosoma has 3 to 5 sclerites (the propeltidium, mesopeltidium, and metapeltidium) that may parallel the segmentation seen in the prosoma of ancient groups such as Xiphosura. In addition, the prosoma is joined to the opisthosoma by a broad connection instead of the narrow pedicel seen in spiders. The ten abdominal segments are readily distinguished by their chitinous plates (tergites). Expansion of the abdomen is accomplished by the presence of a thinner, more flexible membrane between the tergites, a necessary requirement given the voracious appetite of the solifuge.

While solifuges may be primitive as evidenced by their visible segments, they are simultaneously revolutionary in their insectlike adaptations. These adaptations include using only tracheae to supply their bodies with oxygen, and running using only 6 legs.

Oxygen is supplied directly to the interior of the body by branching tracheae, which open to the outside via stigmata. Tracheae are faster and more efficient than book lungs at transferring oxygen to the interior of the body. In an instructive example of convergent evolution, insects and arachnids (such as solifuges, spiders, harvestmen, pseudoscorpions, and ricinuleids) have independently evolved this structure for respiration. Tracheae allowed insects to adapt to new niches (such as the air by way of flight) in an explosion of adaptive radiation. In solifuges, tracheae allowed sustained activity as is seen in their style of hunting, which consists of running on the ground and changing direction every few seconds as they randomly search for prey. The energy for this sustained activity is supplied by the tracheal system of respiration, but the speed and agility are achieved by using only 6 legs for running.

As leg length increases, greater speed and efficiency may be gained by reducing the number of legs involved in locomotion. This allows each leg to maximize its stride while minimizing the chance that 2 legs may overlap and hence interfere with each other. Most solifuges not only have long legs and consequently long strides, but their legs also have a unique joint arrangement that enhances their speed and agility. Additional segments in their legs (both a divided trochanter and a prefemur and postfemur) impart greater flexibility as well as a sort of rocking motion. This permits the tarsal claws more complete contact with the substrate, hence greater traction, and therefore greater speed.

From an ecological standpoint, a solifuge is closer to a predaceous ground beetle than to a scorpion. Of course, the downside of greater oxygen use is a faster metabolism and consequently a shorter lifespan—usually only 1 or 2 years—in contrast to the slow metabolism and relatively long lifespan of scorpions, which have book lungs.

To a casual observer, a solifuge appears to use 10 legs as it races across the ground. But both the palps and the first pair of legs are used for sensory purposes rather than for locomotion while on horizontal surfaces; however, on vertical surfaces, the palps can function in locomotion. Solifuges have a unique structure called the suctorial organ (or palpal organ) at the tip of each palp. This strange, balloonlike structure is hidden within the palp most of the time. When the solifuge needs to climb or grasp an object, the chitinous covers at the end of the palp open up, and out pops a small, white structure.

This structure has microscopic ridges that permit it to cling to smooth surfaces, allowing solifuges to climb even glass. As the solifuge pauses between climbs, the suctorial organ may be almost completely hidden except for a rhythmic pulsing, showing a hint of white at the tip of the palp at regular intervals. The most likely mechanism for the eversion of the suctorial organ is an increase in hemolymph pressure, similar to the increase in pressure that powers the leap of the jumping spider.

The suctorial organ obviously did not evolve in response to glass surfaces, although this exaptation is utilized while hunting at night on the glass windows of houses where abundant prey may be drawn to the light. Solifuges can be observed rapidly climbing "hand over hand" up vertical glass surfaces. The suctorial organ does assist in prey capture by allowing the solifuge to grasp its quarry and bring it within range of its chelicerae. This technique is versatile for use both on larger prey such as beetles and on tiny prey such as termites. In fact, the solifuge may "palm" a termite with the suctorial organ, demonstrating remarkable dexterity given the lack of an opposable thumb.

No other creature has a structure identical to the suctorial organ. The closest comparable structure is found in harvestmen in the family Grassatores. A fleshy structure called the arolium is located between the claws of the third and fourth leg tarsi. It functions much like the suctorial organ of solifuges, allowing harvestmen to hang even from glass. Found only in nymphal stages, it is probably used mostly during molting, allowing the harvestman to hang upside down. In this position gravity eases the process of extricating long, fragile legs from the old cuticle. Pseudoscorpions also have an arolium between the two claws on each tarsus that allows them to walk on vertical surfaces or the underside of an object. But these structures are fleshy pads compared with the balloonlike suctorial organ.

While hunting, solifuges utilize the palps and the first pair of legs in a sensory capacity as they constantly tap the surface of the ground, searching for chemical or vibrational traces of prey. In addition to the palps, solifuges have a unique set of structures called malleoli, or racquet organs, on the underside of the last pair of legs. Each of these organs (5 per hindmost leg) is fan-shaped and connected to the leg via a narrow stem forming a tennis racquet shape, hence the name "racquet organ." The edge of each racquet organ has an opening to a groove that has chemosensory dendrites, analogous to that seen in the pectines of a scorpion. Solifuges periodically "taste" the substrate by lowering their bodies so that the ventral edge of the malleoli touches the ground. Traces of prey, water, or potential mates may be detected with the help of the racquet organs.

Although solifuges lack trichobothria (the exquisitely sensitive hairs that spiders, pseudoscorpions, and scorpions possess), they do have an array of long hairs and an abundance of shorter bristles that are almost certainly sensory in function, much as a cat's whiskers are accessory sensory structures. They also have slit sensilla for detecting substrate vibrations. These narrow slits in the cuticle are covered by a thin, easily deformed membrane. Substrate vibrations deform the slit and trigger a nerve impulse, allowing the solifuge to locate moving prey. In fact, solifuges have been observed digging out subsurface prey and successfully capturing

it. Some species may also use visual cues when hunting prey, and there are accounts of solifuges stealthily stalking flies. Considering the active hunting strategy of solifuges versus passive (sit and wait) strategies, it is no surprise that their brains contain concentrations of the neurotransmitter acetylcholine comparable to concentrations found in actively hunting spiders (1.27 to 1.33 nmol/mg protein).

The slit sensilla and accessory sensory hairs may pinpoint the exact location of potential prey, but the attack is not fully launched until physical contact with the prey is made. Once the solifuge has contacted the potential prey, it can decide whether or not the quarry is edible. Desirable prey include crickets, grasshoppers, termites, beetles, cockroaches, true bugs, moths, scorpions, and centipedes. Even some small vertebrates such as lizards, small mammals, and baby birds may be eaten, although this may result in part from scavenging already dead prey. Species that are rejected include stinkbugs, velvet ants (a wasp with a powerful sting), blister beetles, ladybug larvae, velvet mites, and hairy caterpillars. Almost all the rejected arthropods possess effective defensive chemicals, irritants, or stings. The solifuge must be able to rapidly identify and sort potential prey from potential hazards. A slow solifuge may either lose an edible meal or, worse, become prey for another predator. Many a solifuge becomes a meal for a scorpion, giving fresh meaning to the expression "the quick and the dead."

Once a solifuge has located, identified, and captured its prey, the quarry is transferred to the chelicerae. The formidable chelicerae are powered by massive muscles, giving them a bulging appearance. Each chelicera has a fixed finger and a movable finger that opens and closes dorsoventrally (up and down) as the prey is crushed and masticated. Teeth along the edge of the chelicerae help in grinding the prey. Vertical rows of setae connected to the labrum (mouth) filter out particulate matter so that only liquefied food is ingested. Solifuges have been observed to eat truly prodigious meals relative to their own size. Because its flexible abdomen can accommodate a large volume and the diverticula of its gut can store this food internally, a solifuge can live off a large meal for some time.

Fecal material is stored in the stercoral pocket, where the last vestiges of nutrients and water are removed. Malpighian tubules also connect to the stercoral pocket and contribute to the excretion of

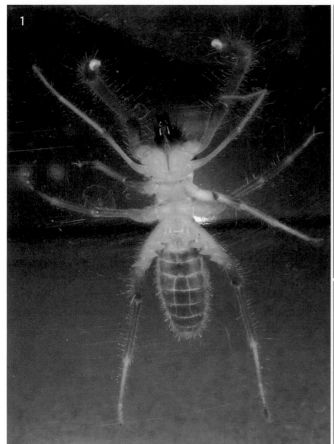

1. A solifuge climbs a smooth vertical glass surface using the suctorial organ at the end of each pedipalp. The suctorial organ is unique to solifuges and probably evolved as a prey-capture structure. A solifuge can grasp a large, smooth beetle or a small termite using the suctorial organ.

2. The tip of the pedipalp when closed. The lines across the surface indicate where the suctorial organ will emerge.

3. The everted suctorial organ resembles a small white pillow before it is pressed against an object.

4. Microscopic ridges permit the everted suctorial organ to "stick" to vertical glass.

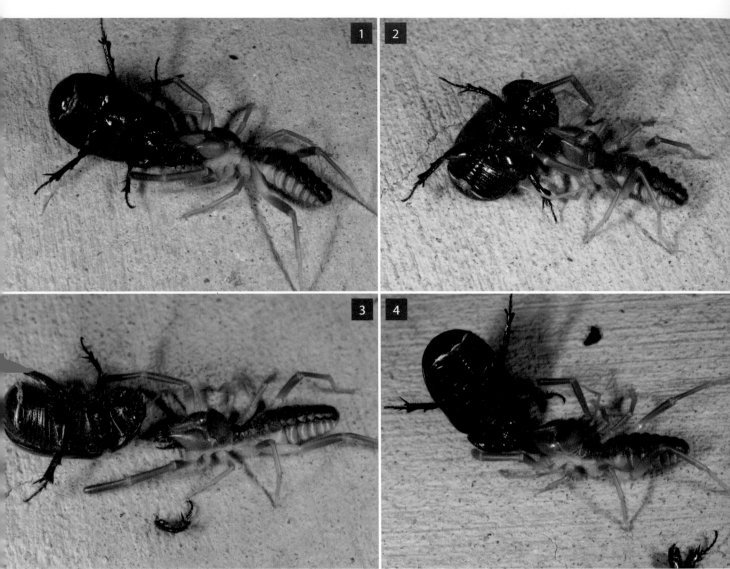

Above: Tearing off the crunchy bits:

1–4. Despite having no venom, solifuges are willing to tackle large, tough prey such as adult beetles. The first step in overpowering the beetle is to disable it by systematically removing its legs. The chelicerae have a bulgy appearance because they are highly muscular. As the determined solifuge attacks the still-struggling beetle, the chelicerae open and close dorsoventrally in order to remove each leg in turn.

Right: 5. A total of ten malleoli, also known as racquet organs, are located on the underside of the solifuge. These are most likely chemosensory in function.

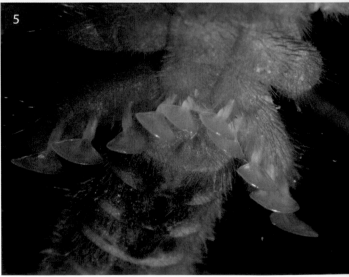

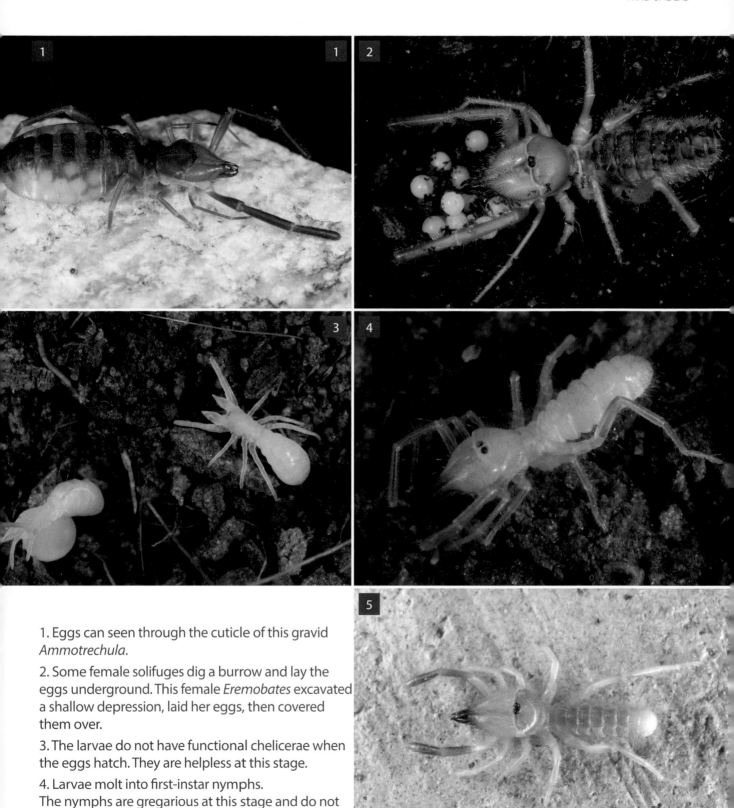

1. Eggs can seen through the cuticle of this gravid *Ammotrechula*.

2. Some female solifuges dig a burrow and lay the eggs underground. This female *Eremobates* excavated a shallow depression, laid her eggs, then covered them over.

3. The larvae do not have functional chelicerae when the eggs hatch. They are helpless at this stage.

4. Larvae molt into first-instar nymphs.
The nymphs are gregarious at this stage and do not yet hunt.

5. The second-instar nymphs are intolerant of their siblings and disperse to become solitary hunters.

nitrogenous waste. Cells lining the Malpighian tubules concentrate nitrogenous waste products consisting of guanine, hypoxanthine, adenine, and uric acid. These waste products precipitate out of solution to form insoluble crystals that are excreted. This permits the solifuge to reabsorb water, an important adaptation for living in arid climates.

In fact, solifuges are indicator species of desert biomes. Found primarily in the tropical and subtropical deserts of Africa, Asia, the Middle East, the Americas, and Southern Europe, they are restricted to regions that are both warm and dry with moderate or little surface vegetation. Most solifuges are nocturnal; however, diurnal species are found in some deserts in Africa and the Americas. Although the solifuges of the American Southwest are nocturnal, they are conspicuously active during the hottest nights in June, when the temperatures may hover over 90 °F (32.2 °C) well into the night, and the soil surface may remain warm for a considerable length of time.

During the heat of the day, they take refuge in an earthen burrow or under rocks or surface debris. Consequently, soil type is important in the construction of burrows, especially maternity chambers for the incubation of eggs. By digging a burrow in the correct microhabitat, the female solifuge gives eggs and young a degree of protection against the extreme temperatures and desiccation found on the surface. But first, the female must mate with a male; a risky proposition for these fiercely solitary creatures.

Although the courtship and mating of many species of solifuges has yet to be documented, it has been recorded for some species of North American eremobatids.

Male eremobatids probably employ their chemosensory malleolar organs in tracking down females, periodically touching and "tasting" the soil surface in search of chemical traces. Once a female has been located, the initial introduction consists of threat displays during which both solifuges rock back and forth with chelicerae gaping and palps held aloft. Once past these formalities, the male rushes at the female, grabs her between the prosoma and opisthosoma, and appears to palpate her with his pedipalps. If she is receptive, she adopts a position of submission. Her legs go limp, she drops her body to the ground, her chelicerae close, and she bends the front half of her body backward over her abdomen. The male starts to

nibble or chew just behind her head, and gradually works his way towards her genital opening. The male may adjust the female's position several times, lifting her or turning her on her side. He "chews" her genital operculum and the genital opening before depositing sperm directly from his genital opening to hers. He follows this up with inserting his chelicerae into her genital opening and once again chewing. At the end of mating, the male releases the female and escapes, or the couple may go into combat again. The male may be killed if he is not fast enough to escape.

In Africa, Asia, and Europe, the male solifuge may deposit the sperm in a little packet on the ground. The male then carries it with his chelicerae to the female's genital opening, unlike the North American male solifuges, which transfer sperm directly to the female.

Given the use of the male's chelicerae during mating, perhaps it is logical that solifuges demonstrate sexual dimorphism in the amount of dentition associated with their chelicerae. Mature males typically have fewer cheliceral teeth than do mature females.

In most families of solifuges, the mature male has a structure called a flagellum on the fixed, upper finger of each chelicera. This structure may be slender and whiplike, or may be flattened and transluscent, as seen in Ammotrechidae. It is not known how the flagellum functions, but it is probably used in courtship or in mating. Solifuges in the Eremobatidae lack flagella.

The male solifuge usually lives only a few weeks after mating even if he continues to hunt and feed. Meanwhile, the female becomes even more voracious, eating as much as she can until only a few days before she lays eggs, at which time she stops eating. The female may carry from 15 to 150 eggs, which can be seen through the semitransparent cuticle of her swollen opisthosoma. She digs a maternity chamber underground, and depending on the species, either she closes the chamber and abandons it after laying her eggs, or she may stay in the chamber and care for the eggs. After a few weeks of incubation, the larvae, also called postembryos, hatch. These helpless white larvae have legs with incomplete segmentation that lack tarsi and tarsal claws. The chelicerae are nonfunctional, and although the larvae have eye spots, the ocular tubercle has not yet developed. The larvae lie in the maternity chamber for 1 to 2 weeks while they undergo further development. At the end of this period, they molt into first-instar nymphs, the first of 8 nymphal stages. These first-instars more closely resemble an adult solifuge, but they are still pale in coloration and do not feed yet at this stage. Both the postembryos and first-instar nymphs are gregarious, clustering together. But as soon as the babies molt into second-instar nymphs, a dramatic transition occurs in their behavior. They become aggressive, far more physically active, and they no longer tolerate each other. This transition is attributed to changes in brain chemistry. Levels of serotonin and dopamine approximately double between the first and third instars. Thus, the docile, gregarious first-instar nymphs morph into aggressive hunters. Soon afterward, they disperse from the nest to live their solitary lives.

Because of their intimidating appearance and willingness to stand their ground, solifuges are summarily condemned and executed by most who encounter them. Even the naturalist R. W. G. Hingston wrote in *Nature at the Desert's Edge: Studies and Observations in the Bagdad Oasis*:

> Here is an odd creature, a monstrous apparition. How weirdly fashioned, how ill-proportioned, how evil and forbidding in its hideous shape and coat of bristling hairs! It might be some antidiluvial form, specially preserved for this unkindly soil.

Indeed, deserts have produced some of the most astonishing plants and animals in the world. The desert challenges the ordinary to become extraordinary. Plants evolve spines, Gila monsters have venom, and the fierce little solifuge can run as swiftly as the wind.

1. *Eremocosta* male (presumptive). The genus *Eremocosta* contains some of the largest of the North American solifuges.

2. *Eremocosta* male, threat display. Despite their fearsome appearance, solifuges have no venom. They can give a sharp nip with their chelicerae.

3. *Eremocosta* female. This impressive female was 2 inches (5 cm) in total body length; however, some species from north Africa have a leg span of 6 inches (15.2 cm), dwarfing the North American species.

4. *Eremobates* species (presumptive). From coastal southern California.

5 and 6. *Eremocosta striata* female, from Arizona. Note how all members of the Eremobatidae have a straight line marking the front of the propeltidium where it meets the chelicerae.

7 and 8. *Ammotrechula pilosa*. These small solifuges are abundant in southern Arizona. On hot summer nights, they can frequently be seen climbing on the outside of glass windows as they hunt for small insects. These are members of the Ammotrechidae family, the curve-faced solifuges. Note how the line of demarcation between the chelicerae and the propeltidium is curved.

9. Ammotrechid from southern Arizona.

CHAPTER 10 Ticks and Other Mites:
Parasitiformes and Acariformes

Two pairs of widely spaced ruby-red eyes easily distinguish this long-legged *Paraphanolophus* mite from harvestmen. This adult mite is a nocturnal predator of other small invertebrates.

The desert air shimmered with heat, creating the illusion of water in the distance. Over the baked surface of the ground a tiny predator darted, interspersing bursts of frenetic activity with brief periods of rest. This predator was so small, a mere one twenty-fifth of an inch (1.0 mm) in length, and moved so rapidly, that it was invisible to the casual human observer. But to the tiny prey it hunted, this predaceous mite was a fierce and dangerous adversary. Spinelike setae covering the legs and body of the hunter provided an impressive defense against possible retaliatory bites and kicks from its prey. Contraction of the muscles in the mite's body increased hemolymph (blood) pressure to the chelicerae, powering their action in stabbing into and subsequently sucking out the contents of the minuscule invertebrate prey. Wide-set beady red eyes punctuated the oval, tan body. A faint iridescence suffused the surface of the body, an effect created by the thin layer of waxlike lipid in the cuticle. With relative humidity below 10 percent and the ground temperature hot enough to melt and deform a plastic cup, this tiny hunter would desiccate almost instantly without this lipid protection. It would have become just one more speck of dust in an already dusty landscape.

The desert at midday in summer may seem like an unlikely place for any living thing, especially a delicate little mite. But these tiny arachnids have mastered some of the most extreme environments on Earth and, in doing so, have become truly ubiquitous. From the dusty Arizona desert to New England ponds, from the freezing Antarctic to volcanic steam vents in Hawaii, from several meters under the surface of the soil to thousands of feet up in the air, mites are found in virtually every environment on planet Earth. Mites may be found in our foodstuffs, in house dust, and even living in our eyebrow follicles. Mites may live as predators, parasites, or commensals on other animals, while other mites may feed on detritus, algae, fungi, or plants. From free-living predators to creatures that spend their lives in the ears of moths, mites have evolved to fill almost every imaginable niche. Even a single feather may have 4 different species of mites living on it, each species using a different part of the feather and making a living in a completely different manner from its neighboring mites. The evolutionary advantages of small size allow mites to "go boldly where no mite has gone before."

The first mites probably appeared on land from water about 400 million years ago, during the late

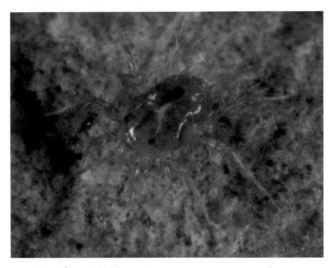

Tiny but formidable, a *Paratarsotomus* mite hunts for prey on the soil surface during the day, even when temperatures are over 100 °F (37.8 °C). A faint iridescence indicates the presence of lipid in the cuticle.

Silurian or early Devonian periods. These early mites all belonged to the Acariformes and, like some of their modern-day descendants, were particle feeders, consuming plants, algae, fungi, and detritus in soil and litter. Their massive chelate chelicerae were adapted for the "bite-and-swallow" strategy. These mites diversified to fill a variety of lifestyles. Among their descendants are the prostigmatid mites, many of which have piercing/sucking chelicerae adapted for the ingestion of only liquid food. The ancestors of the other group, Parasitiformes, may have appeared much later, anytime from the upper Triassic, 220 million years ago, to the late Cretaceous, around 65 million years ago, contemporary with the dinosaurs. These relatively large mites may have originally been predators, but over time, some became parasites.

During the late Mesozoic and early Cenozoic eras, 60 to 100 million years ago, a period of tremendous diversification of flowering angiosperm plants occurred. In turn, insects diversified, adapting to the increasing array of plants, which then selected for reciprocal adaptations by plants in response to herbivory and pollination by insects. Plants even evolved special structures that reward protective insects or mites that stay on that plant. These may take the form of food rewards or shelter rewards, called domatia. An example of a food reward is the beltian body, a nubbin of protein that some acacias produce at the tip of their leaves inducing ants to live on the acacia. In exchange, the ants protect the acacia from herbivorous insects. Shelter may take the form of myrmicodomatia, used by ants, or acarodomatia, used by mites. Acarodomatia consist of cottony tufts of hair, pits, or other structures on the undersurface of leaf vein axils that may provide shelter and refuge for predatory, fungivorous, and microbivorous mites, the predominant inhabitants of these shelters. The mites utilize the domatia during molting or for oviposition. Consequently, leaves with domatia have many more predaceous mites than leaves without domatia. These beneficial mites serve as protection against phytophagous mites, caterpillars, and other herbivores. As the coevolution of plants and insects produced an explosion of adaptive radiation, mites also diversified, coevolving with plants, insects, and vertebrates in a case of evolutionary synergism. Today, there may be 1 million species of mites, although only about 55,000 species have been formally described. Their complexity and diversity rival that of any other group of life forms.

J. B. S. Haldane once quipped that, judging from the great diversity of beetles, the creator must have had "an inordinate fondness for beetles." One can only speculate what he would have said had he known about the incredible variety of mating and reproductive strategies found in mites.

Mites demonstrate multiple strategies in regard to just about every aspect of reproduction. Sperm may be transferred from a male to a female, or there may be all-female parthenogenetic species with no need for fertilization of eggs by a male. Sperm may be transferred directly, with or without the aid of specialized organs, or it may be transferred indirectly via a spermatophore. In species in which sperm is transferred directly, it may be achieved venter-to-venter, or the male may have an intromittent organ, the penis or aedeagus. In some species, it is the female that possesses an intromittent organ, a spermaduct that is inserted into the male genital opening in order to facilitate transfer of sperm to the female. In some species, the male may use a special structure on his movable cheliceral digit, called a spermatodactyl, to transfer sperm to pores on the female's body. These sperm induction pores may be located in a variety of areas, including between the bases of the coxae, or even the trochanters of the leg femora. Once the sperm is introduced into the pores, it is stored internally in a sac. Eventually the ribbonlike sperm makes its way into the female's hemolymph (blood) and migrates to her eggs. The sperm induction pores may have evolved in response to traumatic insemination, where in the distant past the male may have stabbed into the female with his loaded chelicerae in order to accomplish the transfer of sperm.

Sperm may be transferred via a stalked sperm packet called a spermatophore. An almost endless variety of techniques for the transfer of spermatophores from males to females has evolved. In most species the spermatophore is transferred indirectly by being deposited on the substrate and afterward picked up by the female. The male may perform a "dance" before or after depositing the spermatophore in order to attract the female to the site, or he may deposit pheromone trails that lead her to it. In some species, the spermatophore may be transferred with the male's mouth parts directly to the female's genital aperture. Each species of mite has its own unique, distinctive spermatophore. In some species the spermatophore may be intricately sculptured, while in other species, a spermatophore "tree" may be constructed as each male attaches his own spermatophore onto the spermatophore stalk previously deposited by another male. Some of the most incredible spermatophores belong to ticks. Ticks produce a complicated double spermatophore that consists of an outer section, the ectospermatophore, and an inner section, the endospermatophore. Within the spermatophore is also a yeastlike fungus, *Adlerocystis*. First, the male tick places the spermatophore in the female's genital opening with his chelicerae. Shortly afterward, a carbon dioxide explosion occurs in the outer ectospermatophore that propels the endospermatophore as well as *Adlerocystis* into the female tick's reproductive tract. If there is a long delay between mating and laying the eggs, *Adlerocystis* attach to the spermatozoa and maintain the viability of the sperm. Argasid ticks may retain the viable sperm for years, eliminating the necessity of repeated mating.

Male mating behavior may be strongly influenced by the pheromones emitted by females. These pheromones are of three functional categories: arrestant pheromones that stimulate precopulatory guarding by adult males of soon-to-mature females, attractant pheromones, and finally, contact pheromones that stimulate actual copulation by the male.

The number of chromosomes that mites carry can be highly variable, depending on the species of the mite and its reproductive strategy. In the system most familiar to us, males and females each have two sets of chromosomes, one from each parent. This is known as diplodiploidy. Less common is haplodiploidy, where males have half the number of chromosomes of females. Haplodiploidy may occur as a result of males being produced parthenogenetically from unfertilized eggs while females are produced from fertilized eggs. This system, called arrhenotoky, is seen in roughly 20 percent of all animal species, the most familiar of which is the honey bee. Another scenario in which males have half the chromosomes may occur after an egg is fertilized, following which the paternal chromosomes are deactivated or eliminated, resulting in paternal gene loss and a haploid offspring. Finally, parthenogenetic, all-female species are found in a wide range of mites but most conspicuously among the plant-feeding mites. In this case, the female offspring are diploid, having a full complement of chromosomes, but are produced without a male fertilizing the eggs. This last system is called thelytoky.

Mites have a diversity of egg-laying strategies. Ticks may lay several thousand eggs at a time, whereas trombiculid mites may lay only 1 egg at a time. In these mites, the mother mite hides each precious egg under debris, placing it with the aid of an extrusible ovipositor close to a spot where a vertebrate host may walk by. Her larval offspring, known as chiggers, are external parasites of vertebrate hosts. While most mites do not show maternal care of their eggs or young after the eggs are laid, a few species may brood their eggs and actively defend their nests. In some cases, the mother mite makes the ultimate sacrifice for her offspring. She holds first the eggs, then the growing young, within her body. She dies as her fully mature offspring escape from her corpse. In some species, not only are her offspring fully mature when they emerge, but they may have already mated even before leaving their mother's body. In other species, the offspring stay in the body of the dead mother for several months before emerging. In a slightly more macabre variation, the eggs hatch within the mother and the young proceed to eat their way out as they feed on their mother's body.

In general, the life stages of a mite include the egg, prelarva, larva, protonymph, deutonymph, tritonymph, and adult. In some species, one or more of these stages is eliminated. The entire life cycle may require as little as 34 hours to more than 5 years, depending on the species.

The prelarvae are nonfeeding and most species are inactive, lying quietly within the egg chorion. However, in the Adamystidae and in the Anystidae, some prelarvae have 6 working legs. These elattostatic prelarvae are briefly mobile after hatching from their eggs, walking for a short distance before becoming inactive once again. The adaptive advantage of mobility may lie in the avoidance of cannibalism from siblings or possibly in finding a more humid spot for molting. Most species with elattostatic prelarvae do come from dry habitats. The next stage of life for the mite is the larva. This stage is typically 6-legged (hexapod), as compared with the 8-legged later life forms. In many species of mites, the larvae may be voracious predators or may be parasites. The larval morphology may be completely different from that of the later life stages, and, in fact, identification may be based on larval rather than adult characteristics in some species. Ideally, the larva is correlated with a later life stage, such as the deutonymph or the adult.

Most mites molt from a larva into a nymph (although in some fungivorous mites, the larva may molt directly into an adult). Most mites have two or three nymphal stages. Exceptions include ixodid ticks, which have only one nymphal stage, and argasid ticks, which may have as many as eight nymphal stages. The nymphs have eight legs and with some exceptions closely resemble the adults, though they are sexually immature. Many mites have only one active nymphal instar, the deutonymph, and two nymphal stages that are calyptostatic (or quiescent), the protonymph and the tritonymph. To the casual observer, the larva appears to molt directly into a deutonymph, and the deutonymph into an adult, because the intervening protonymph and tritonymph stages are cryptic, hidden inside the cuticle of the larva and the deutonymph, respectively.

In astigmatid mites (Oribatida), the deutonymph may facultatively take a form that is entirely different (heteromorphic) from either the larva or the adult. This special life stage, called the hypopus, resembles a mostly featureless sac except for ventral suckers or claspers with which it attaches itself to the smooth surface of an arthropod. Having a closed gut and lacking mouth parts, it does not feed during this stage, but travels on its arthropod ride until it is ready to molt into a tritonymph. This phoresy, or hitchhiking, may be a means to disperse to a more favorable habitat. Other hypopi are inert and may be dispersed by wind, or may simply wait until environmental conditions become favorable to activate.

Under some conditions, this deutonymph stage may be completely eliminated, in which case the protonymph molts directly into a tritonymph. The facultative nature of the hypopus suggests that this particular strategy is employed in response to environmental stresses. In the hypopus stage, the mite is highly resistant to adverse environmental conditions, and yet it is able to travel to other environments by means of phoresy or by air currents.

Mites have an array of sensory structures, weaponry, and defensive tools built into their bodies. First and foremost are the chelicerae and the palpi, which together constitute the primary tools for food acquisition. The chelicerae may be mandibulate, styliform, hooklike, or finely toothed. The type of chelicerae is specifically adapted for chewing, piercing, tearing, sucking, or scraping, depending on the lifestyle and food of the mite. Some mites are particulate feeders, while others ingest only liquid. The palpi are loaded with sensory receptors, analogous to insect antennae, but also may be modified for prey capture in predaceous mites, hold-fast structures in parasitic mites, or food filters in microbivorous species.

The legs may also serve a variety of functions. Locomotion is the most obvious purpose, but in some parasitic and commensal species, a suckerlike pulvillus on the terminal pretarsus allows the mite to hold onto a host. In many species, the first pair of legs immediately after the palpi are primarily sensory in function. Elongate and loaded with tactile and other sensory setae, these legs function like feelers rather than like walking legs, constantly and delicately tapping the substrate in front of the advancing mite. These sensory legs are similar in function to the antenniform legs of harvestmen, vinegaroons, and tailless whipscorpions.

Sensory setae are also found on the body, or idiosoma, of the mite. Among the array of sensory receptors are mechanoreceptors for tactile stimuli. Specialized cuplike sockets called bothridia and the trichobothrial setae associated with them are together referred to as bothridial organs. A number of mite families have bothridial organs on the dorsal surface of the body. These organs may detect movement from a distance as air currents disturb the delicate trichobothrial setae.

In addition to mechanoreceptors, there are chemoreceptors, thermoreceptors, hygroreceptors, and photoreceptors. Some structures may have one function in some mites and an entirely different

Hitchhiking on a giant crab spider, hypopus astigmatan mites cling to the smooth surfaces of the spider's chelicerae and eyes. The hypopus stage of this mite does not feed but is using the spider solely for transportation.

function in other species of mites. An example of this are eupathidia, the spinose, specialized setae found at the tip of the first two pairs of legs or on the tips of the palpi. A terminal pore on each eupathidium suggests that it functions as a contact chemoreceptor, allowing the mite to "taste" its environment; however, the hollow eupathidium on the tips of the palpi in the spider mites, Tetranychidae, has evolved to deliver silk instead of functioning primarily as a sensory structure. Some other groups of mites also have these specialized silk-delivery setae. Included among them are some immature whirligig mites (*Anystis* species) and some *Erythracarus* mites that produce silken cocoons in preparation for molting. (This is in contrast to the other major silk-delivery system found in mites, in which the silk glands are located in the main body cavity and the silk is delivered from the buccal, or mouth, cavity as modified salivary gland secretions.)

A variety of other sensory structures are found in mites. Porose, peglike structures may function for taste or olfaction (smell). Ticks (Ixodida) have a specialized sensory structure on the tarsus of the first pair of legs referred to as Haller's organ. This organ detects chemicals, heat, and humidity, assisting the tick in finding both hosts and mates. As a tick waits on vegetation, it stretches its first pair of legs out into the air, in preparation for detecting and grabbing hold of a passing host.

Mites have a noteworthy variety of sensory systems for detecting light. Many prostigmatids possess paired lateral ocelli (eyes), frequently ruby red in color. Many of these mites (bdellids, anystoids, and erythreaids) are predaceous, rapidly moving over the surface of the ground or on vegetation, actively searching for prey. In addition to the paired lateral ocelli, many prostigmatid and some oribatid mites also carry a median eye or pair of eyes on the underside of the naso, a "noselike" projection at the front of the mite. Elevated lenslike lenticuli are seen in many oribatid mites. Some mites that have no discernible eyes still react to a change in light intensity and may still have an optical nervous system. For ixodid ticks that possess eyes, the sensory cells for light detection are located on the dorsal surface along the edge of the shieldlike scutum. In argasid ticks that have eyes, those structures are located in the supracoxal folds. Perhaps the most novel method of light detection is found in a snake mite (family Macronyssidae). This mite has a pair of photosensitive spots on the pulvillar membrane of the first leg tarsus—essentially eyes on its "feet."

An important component of defense for mites is contained in the cuticle. Cuticular secretions may serve to protect against bacteria, fungi, or predators. Ixodid ticks secrete squalene, a defensive chemical effective against attack by ants. The squalene is produced as a derivative of host blood; the tick cannot synthesize squalene from scratch. The bright red coloration of many mites may advertise their unpalatability. Some mites, such as some *Scheloribates* (oribatids), produce toxic alkaloids, including pumiliotoxin, as protection against predators. In an interesting twist, dendrobatid poison dart frogs have evolved the ability not only to eat these toxic mites but also to sequester the toxin in their own skin, hijacking the mite's defense for their own. The frogs' bright aposematic colors warn potential predators of their toxicity. But there is a yet another twist at the end of the story. Humans in some tropical areas have learned to hijack the toxin from the frogs' skin for their own purposes, coating the tips of arrows or darts with the poison.

There are so many kinds of mites that it would take many volumes to describe all the major groups. Many texts are available regarding mites of medical or economic importance; therefore, they will only be touched upon here. Mites of particular interest because of their large size, abundance, or ecological significance

are included; however, because of the relative scarcity of aquatic environments in the arid southwestern United States, water mites are excluded from this volume. Therefore, the following profiles are not intended to be comprehensive, but instead comprise a sampler of groups selected for their interest and their significance, with emphasis on macroscopic mites found in the arid southwestern United States.

Parasitiformes: Ticks, Mesostigmatid Mites, and Opilioacarid Mites

Ticks: Ixodida

These relatively large mites are obligate blood-feeding parasites on vertebrate hosts. Consequently, ticks transmit a variety of significant diseases, including *Rickettsiae* and other bacteria as well as viruses, to humans and other animals. In the western United States, the Rocky Mountain wood tick (*Dermacentor andersoni*) transmits Rocky Mountain spotted fever, Q fever, Colorado tick fever, and tularemia. The American dog tick (*Dermacentor variabilis*) and the brown dog tick (*Rhipicephalus sanguineous*) also carry Rocky Mountain spotted fever. *Ixodes pacificus* in the western United States and *Ixodes scapularis* in the eastern United States transmit Lyme disease. *Ehrlichia*, a rickettsial bacterium, may be transmitted to dogs by the brown dog tick. Humans may also contract ehrlichiosis directly from the tick (not from the dog). Most of these hard ticks seem to favor wooded areas or habitats with tall grass. The exception is the brown dog tick, which is found in areas of human habitation since its main host is the domestic dog.

Ticks detect their hosts using Haller's organ, as well as with contact and olfactory sensory setae on their palpi and on the first pair of legs. They may be attracted to vibrations, dark objects, warmth, and carbon dioxide as a potential host animal passes near them. Once a host is contacted, the tick pierces the skin with its chelicerae and inserts the hypostome, an elongate structure lined with teeth that anchor the tick to the host. In hard ticks, salivary glands then produce a cement that helps glue the hypostome in place. Anticoagulants prevent clotting at the site of the wound, allowing the tick to continue to

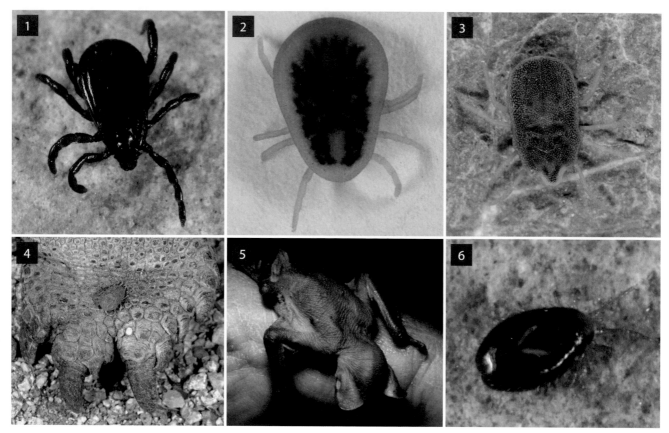

1. *Rhipicephalus sanguineus* male. The brown dog tick may carry a number of diseases that can infect humans. The arid southwestern United States is not favorable for most hard ticks; however, humans provide hosts (domestic dogs) that allow these ticks to survive almost everywhere.

2. *Argas cooleyi*. This argasid (soft) tick inhabits the nests of cliff swallows.

3. *Carios yumatensis*. Parasitic on bats, these soft ticks may be collected in caves.

4. *Ornithodoros turicata*. Found in the burrows of desert tortoises, these soft ticks may become numerous in a single burrow. This individual is feeding on the foot of a tortoise.

5. Clinging to a human finger, this baby pallid bat has a number of blood-feeding macronyssid mites on it (seen as brown specks).

6. *Steatonyssus antrozous*. This macronyssid mite feeds on the blood of bats.

feed for as long as it remains attached to the host. The salivary secretions of some ticks may induce a paralysis in the host, which is reversible if the tick is found and removed.

Most hard ticks (Ixodidae) are three-host ticks that require a different animal to feed on at each developmental stage of its life. The larvae and nymphs generally feed on small animals such as rodents or birds, while the adults seek out larger mammals. The hosts for a single species of tick may be as different as a shrew and a yak or a mouse and a deer. At each life stage (larva, nymph, and adult), ixodid ticks feed once on each host. After feeding, the larva and the nymph release the host, drop to the ground, and molt into the next stage. For the adult female, eggs are produced after a large feeding. On average, 2,000 to 8,000 eggs are laid, but as many as 20,000 may be produced at one time. Each egg is covered with antioxidants and a waxy material that makes it waterproof. The adult male tick may feed a bit on the vertebrate host or, in some species, may feed only on the engorged female tick. This is called homoparasitism, where the host is the same species as the parasite.

Soft ticks (Argasidae) have a slightly different pattern of feeding. Although the nymph may feed only once during each stage before molting to the next life stage, the adult argasid tick may feed multiple times on a host, releasing itself after each meal. In contrast to ixodid ticks, argasid ticks do not produce a cement in their salivary secretions. Therefore, argasid ticks can release their hold on the host without difficulty. Each feeding may last for only a short time (sometimes only minutes), after which the female deposits a few eggs. Because of their slow metabolism, these ticks can survive for extended periods of time between meals, the longest documented time being 11 years. Argasid ticks tend to be found in the permanent or semipermanent shelters of their host animals, such as birds, small mammals, and tortoises. Arid tropical and semitropical regions are especially favorable for argasid ticks. In the southwestern United States, argasid ticks may be abundant in cliff swallow nests, tortoise burrows, and bat roosts.

Mesostigmatid Mites: Mesostigmata

Mesostigmatid mites have an incredible diversity of lifestyles. Many of the 12,000 described species are free-living predators, while others are parasites or symbionts on a variety of vertebrates and arthropods. In Parasitinae many species of mites are phoretic on insects. The conspicuous presence of the deutonymph mites on their insect conveyance was initially thought to be parasitism; hence they were called parasitids. In hindsight, this was a misnomer. In many cases, the relationship of the mite to its associated insect is actually complex and mutualistic. For example, *Parasitellus* mites convey a clear advantage to their associated bumble bee hosts, since bumble bees with the mites have lower levels of parasitic nematodes than do those without mites. Somewhat more ambiguous is the relationship of *Poecilochirus* mites to their carrion beetle hosts. These phoretic mites are conspicuous passengers on burying beetles (*Nicrophorus*) in the southwestern United States. As soon as the beetle discovers some carrion and commences to feed, the mites scramble down to the carrion and dive in head first. They even use the beetle's head as a stepstool down to the food and may position themselves directly adjacent to the feeding beetle's mandibles. Curiously, the mites seem to know exactly when the beetle is finished feeding and is ready to move on. At that point, all the mites promptly scramble back onto their ride, with none

being left behind. If a mated pair of carrion beetles starts to bury a carcass, the mites dismount and molt into adults. If the beetle proceeds to lay its eggs in the carrion, the mites kill and feed on maggots and nematodes that may compete for the food with the beetle's own larvae. However, it is not known for certain whether or not the mites may also prey upon some of the carrion beetle larvae as well. Certainly, it is doubtful that the mites would destroy all the beetle larvae, since they and future generations of mites depend solely on these beetles to find a microhabitat (a dead animal) that is suitable for the rearing of their young. These microhabitats are both ephemeral in nature as well as spatially rare. Therefore, the mite is completely dependent on the beetle's powers to detect and then fly to the site of the carrion.

Phoretic mites are also found in a related group, the Dermanyssiae. Mites in this group may be parasites, free-living predators, or may have become specialized to feed on pollen, fungi, or nectar. The nectar and pollen-feeding mites in this cohort may use a variety of flower-visiting animals as transportation, including wasps, bees, butterflies, flies, beetles (especially scarab beetles), and birds. Some of these mites use honeyeater birds in much the same way that flower mites (Melicharidae) use hummingbirds, running in and out of the bird's nares as the bird feeds at a flower. Other phoretic mites may shelter in the space between the two palps of butterflies and moths or under the elytra (wing covers) of beetles as they ride from one location to another. Many of the phoretic flower-visiting mites may feed on nectar and pollen, while others may be predaceous, hunting tiny plant-feeding arthropods.

A related family, the phytoseiid mites, lives on vegetation, often in association with leaf domatia that provides shelter for the mites during egg laying or molting. In return, the phytoseiid mites feed on pollen, honeydew, fungi, and small arthropods, including spider mites. Consequently, some species of phytoseiid mites may be used as biological control agents.

The Dermanyssiae also include a diverse group of parasitic mites. Both vertebrates as well as invertebrates may be affected. One of the most economically damaging mites in this group is the *Varroa* mite (*Varroa destructor*), which parasitizes both adult and larval bees. *Varroa* mites not only cause bee mortality directly through feeding injury, but also indirectly kill bees by the transmission of viral pathogens. A lesser known but equally intriguing mite in the Dermanyssiae belongs

1 and 2. *Poecilochirus* mites. These mesostigmatid mites ride on carrion beetles (*Nicrophorus mexicanus*) in order to find food. The mites scramble down to the carrion and feed alongside the mandibles of the much-larger beetle. As soon as the beetle has finished feeding, the mites hastily climb back onto their ride. The mites at this stage are deutonymphs; they will not molt into mature mites until the *Nicrophorus* beetle has laid eggs in some carrion, thus ensuring future transportation for the next generation of *Poecilochirus* mites. The beetle may benefit from the mites, since the larval mites may kill and eat fly maggots and other insects that compete for food with the carrion beetle's offspring.

3 and 4. Digamasellid deutonymph mites utilize wood-boring cerambycid beetles for transportation. The beetles carry the mites over long distances to dead trees, where the mites can make a living under bark.

to the family Larvamimidae. Like the *Varroa* mite, larvamimid mites are specialists on social insects. In this case, they live in army ant colonies. Masters of deception, these mites mimic both the morphology as well as the signature chemicals characteristic of army ant larvae.

Also in this group are macronyssid mites, which parasitize birds, reptiles, and mammals. Like the *Varroa* mites, these may cause direct injury to the host, as well as indirect injury through allergic reactions or by the transmission of diseases. These bloodsucking mites are conspicuous in bat colonies, where the "sharing" of parasites is especially easy, given the nature of bats roosting in close proximity to each other.

Opilioacarida

Opilioacarid mites are relatively slow-moving, long-legged mites restricted to tropical and warmer temperate areas. Like harvestmen, opilioacarids eat a variety of food, including eupodid mites, oribatid mites, collembolans, pollen, and fungal spores. They eat particulate matter, cutting up larger items before swallowing chunks, or ingesting whole pollen grains. They have a mobile (elattostatic) prelarva stage followed by three octopod nymphal stages before molting into adults. Like some other mites, opilioacarids can molt even as adults.

Acariformes: Prostigmatid and Oribatid Mites

Prostigmatid Mites (Trombidiformes)

The prostigmatid mites are remarkably diverse, even by mite standards. From tiny eriophyoid gall and rust mites to giant desert velvet mites, the prostigmatids include some of the smallest as well as some of the largest of all mites. Prostigmatids live in terrestrial, aquatic, and even marine habitats. They may make their living as predators, parasites, fungivores, or plant feeders. Some may court and mate, while others are parthenogenetic, producing generation after generation of all-female mites. Many have stylet-like chelicerae for piercing and sucking, while others have chelate or even bladelike chelicerae.

However, there is one common characteristic shared by this diverse group of mites. Typical prostigmatid mites have stigmatal openings to the tracheal system located at the front of the body, either on the prodorsum or at the base of the mouth parts. These mites include some of the most ubiquitous inhabitants of the soil surface and of plants—the tiny herbivores, fungivores, and predators that make up a whole world every bit as complicated and dramatic as the Serengeti Plains with its myriad of predators, herbivores, and scavengers.

Family Bdellidae

Bdellid mites are just one of the many predators patrolling vegetation and leaf litter on the ground. Known as "snout mites" because of their elongate pointy chelicera and subcapitulum, these little hunters probe with their antennalike palps as they search for prey. Once a small arthropod such as a springtail is discovered, strands of sticky silk may be squirted to trap the prey, or the bdellid mite may run around the victim hobbling it with silk in

Neocarus texanus. Slow-moving but relatively tough, these opilioacarid mites are found even in low-elevation desert. They feed on a wide variety of small, soft-bodied invertebrates as well as on pollen and fungal spores.

much the same way that oecobiid spiders trap ants. A variety of small arthropods may be hunted, including springtails, woolly aphids, and soft-bodied mites. Silk is also employed in the construction of molting chambers. Bdellids mate by the indirect transfer of sperm via a spermatophore. After mating, eggs covered with spiny projections are laid. (Perhaps these spiny projections serve some protective function to deter egg predators.) These little mites remind one of tiny cats as they prowl the vegetation, suddenly leaping away from any danger.

Family Cunaxidae

Closely related to the bdellids, cunaxid mites stalk their prey on vegetation or on the soil surface. Some species of cunaxids hold their raptorial palps high, ever ready to strike. Soft-bodied arthropods such as Collembola are grasped with the palps and then impaled on the chelicerae. Cunaxids also use a few strands of silk to immobilize prey in much the same way bdellids do. However, some cunaxids take the use of silk a step further, constructing a silken snare that is analogous to a spider web. The mite crouches under the net, waiting for its victim to become entangled before it attacks.

1. *Neomolgus* species. The bdellid mites in this genus are commonly associated with seashore habitats. However, this species is found in the deserts of southern Arizona along streambanks in the spring.

2. *Bdella* species. Snout mites in the genus *Bdella* are characterized by two long setae projecting from each palp.

3. *Cyta* species. Heavy chelicerae enable this bdellid mite to kill armored beetle mites.

4. Bdellid mites are commonly found on vegetation and on the soil surface in a wide range of habitats. This bdellid is found in the coastal chaparral habitat of southern California.

5. *Armascirus* species. Holding its palps aloft, this cunaxid mite can stab down on its prey, impaling the victim on the pointed, needlelike chelicerae.

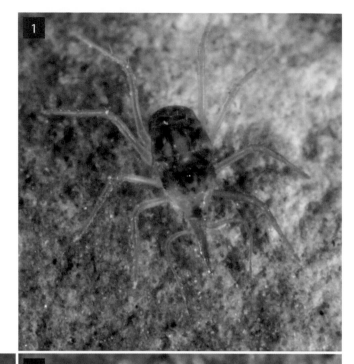

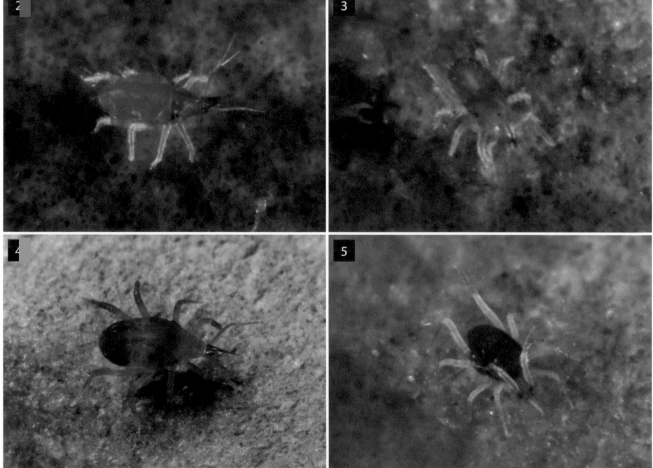

Eupodoidea: Families Eupodidae, Rhagidiidae, Penthalodidae, and Penthaleidae

Mites in the superfamily Eupodoidea may be found almost anywhere in the world, including some of the most extreme environments, from deserts to polar habitats. They are predators, fungus feeders, and plant feeders but are not known to be parasitic or symbiotic on other animals.

Mites in the family Eupodidae are typically found in moist microhabitats. Acceptable habitats may even include volcanic steam vents in Hawaii, which can reach temperatures of 106 °F (41 °C) and coastal intertidal zones. In arid climates, this usually means areas close to the ground, such as under rocks, under rotting logs, in leaf litter, and on low-growing vegetation such as lichens and mosses. The underside of rocks that are in contact with the soil is an especially important microhabitat. As water evaporates from the underlying soil, it condenses on the underside of the rock and provides critical moisture for more delicate creatures. Moisture is also a prerequisite for the algae and fungi on which many Eupodid mites feed. The enlarged femora of their hind legs give these mites the ability to jump. The eupodid mites in the genus *Linopodes* can not only jump but also move backward with alacrity. Their first pair of legs is incredibly long and delicate, being used almost exclusively as sensory feelers. They may be predators on fungivorous mites.

Mites in the family Rhagidiidae are found under rocks and other protected microhabitats during the cooler months of the year. These predaceous mites are commonly seen on the surface of moist soil as they search for prey. They can range in size from the tiny white *Poecilophysis* species (only just visible to the naked eye) to the aptly named *Robustocheles*, which can be several millimeters long. Jaguars of the mite world, these formidable predators overtake their prey in a short burst of speed. Several species are exclusively female, reproducing by thelytokous parthenogenesis. The larval instar is a mobile but nonfeeding elattostase. Prior to molting, some rhagidiids spin silk around themselves. Some species also utilize silk in the capture of prey, in the protection of eggs, and in the construction of communal shelters.

Mites from the family Penthaleidae are known as earth mites. These plant-feeding (phytophagous) mites are strikingly colored. Their cuticle is red to orange, but because of their dark internal color, these mites appear to be dark green or even black, with bright orange-red legs. Reproduction in many of these species is by all-female parthenogenesis.

Mite reproduction strategies highlight advantages of both all-female parthenogenesis as well as the more familiar system of two sexes mating. The odds of a single individual successfully colonizing a given area are at least doubled for a parthenogenetic species, since every single individual can produce offspring. (In contrast, a single male could never succeed in colonization by himself.) Also, the time it takes to populate a given habitat is shortened in parthenogenetic species, since approximately twice as many individuals can produce offspring in each generation compared with species that require males for reproduction. Therefore, parthenogenetic species are adapted for colonization of newly available and perhaps ephemeral environments, reproducing at a maximum rate with no "wasted" individuals.

In contrast, reproduction that requires mating between two sexes also has its advantages. Repair of DNA double-strand breaks may occur during meiotic recombination between nonsister chromosomes, or a damaged chromosome from one parent may be repaired by an intact homologous chromosome from the other parent. But more importantly, sexual reproduction reshuffles the genetic deck of cards with every individual produced. Like playing the lottery, there is always a chance that some new combination or mutation may be a winner that will be passed onto future generations through selection.

Parthenogenesis assists penthaleid mites in colonizing newly available habitat, but these mites have yet another adaptation for survival in ephemeral microhabitats. In climates that have seasonal variations, such as wet versus dry or warm versus cold, it may not be advantageous for eggs to hatch immediately after they are laid if the environment is too hostile for the survival of the young. Therefore, if eggs are laid under favorable conditions, they hatch after a short incubation period. But if conditions are unfavorable, such as at the start of a dry season, the eggs may shrivel and desiccate but not die. Instead, the calyptostatic prelarva in the egg remains dormant until conditions are favorable. With the proper moisture, the egg plumps up, and the active larva emerges. In some species of penthaleid mites, the eggs are retained in the mother's body even after she dies. They may remain there for months until the time is right for them to emerge.

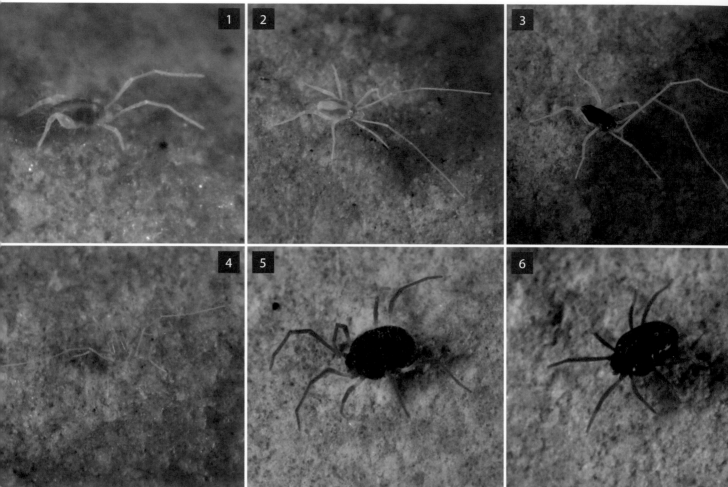

1. Eupodid mite, possibly *Eupodes* species. The enlarged femora of the hind legs enable these mites to jump.

2, 3, and 4. *Linopodes* species. Despite incredibly long, delicate front legs, *Linopodes* are still capable of jumping. This genus is found in moist areas, such as under rotting logs. The front legs are at least 3 times the length of the body.

5. Penthaleid mite near the genus *Chromotydeus*. These plant-feeding mites are most commonly found during the winter months along the edges of washes in the deserts of southern Arizona.

6. *Penthaleus* mite. Another plant-feeding mite, this genus has the anus visible as a small circle on the dorsal surface of its abdomen. It defecates as it hangs onto the underside of leaves.

7. *Robustocheles* species. This mite belongs to the Rhagidiidae, the jaguars of the mite world.

Two species within this family are economically significant crop pests: *Penthaleus major* (the winter grain mite) and *Halotydeus destructor*.

A related family is the Penthalodidae. These mites are found in association with mosses, upon which they are thought to feed. Like penthaleid mites, their greenish-black body contrasts with their brilliant orange-red legs and chelicerae.

Eriophyoidea

Another group of plant parasites belongs to the superfamily Eriophyoidea, also known as the rust and gall mites. Like many other prostigmatid mites, eriophyoids have stylet-type chelicerae evolved for piercing and sucking. In this case, they suck out the contents of individual plant cells. Each species of mite will live only on a specific species of plant or a closely related group of plants. Eriophyoid mites include rust, gall, erinose, blister, witch's broom, and big bud mites and may affect leaves, buds, flowers, and fruits as well as the tougher petioles and stems. Therefore, the diversity of plant species as well as the variety of different microhabitats within any one plant has provided this group of mites with a chance to undergo adaptive radiation and evolve into a tremendous number of species. The host plant is induced by the mite to form abnormal growths, and subsequently these structures provide a protected feeding spot for the mite. These abnormal growths may be galls, rusts, blisters, or the peculiar furry red patches called erinea. Eriophyoid mites may have evolved from mites that were initially fungivores or scavengers who opportunistically began to feed directly on the plant cells.

Eriophyoid mites use spermatophores for mating. Mites in this group are haplodiploid, in which females hatch from fertilized eggs and males are produced from unfertilized eggs (hence the males have half the chromosomes of the females). Therefore, a single unmated female may produce only males, which can then mate back with their mother, after which she can produce females from the now-fertilized eggs. This may be of great survival value for a species, since these mites disperse passively on the wind as aerial plankton, and therefore it may be a single female that randomly encounters the correct species of host plant. An entire colony could be founded by that single female. In fact, arrhenotokous parthenogenesis (the production of males from unfertilized eggs) is ancestral in the eriophyoid

mites, proving it to be a successful reproductive strategy from an evolutionary standpoint. Like the penthaleid mites with their thelytokous parthenogenesis, eriophyoid mites are able to colonize an ephemeral microhabitat with as little as one individual mite.

Tydeoidea

Because most prostigmatid mites ingest only liquefied food, they do not directly contribute much to the formation of soil, unlike mites that ingest particulate matter and excrete fecal pellets of solid matter; however, they may be indirectly critical to the formation of soil. It was discovered that in Chihuahuan Desert soils, prostigmatan tydeid mites were predators of nematodes. Without them, rhabditid nematodes exploded in number, overgrazing the soil bacteria. These bacteria were necessary for the process of organic decomposition, and without them, the rate of soil formation is reduced. Consequently, it was concluded that tydeid mites were in fact a keystone species for soil formation, at least in the arid region of the Chihuahuan Desert.

Other species of tydeids inhabit an incredible diversity of habitats, from feeding on citrus to living within the pulmonary structures of slugs.

Families Caeculidae, Adamystidae, Anystidae, Teneriffiidae, and Erythracidae: The "Anystina"

Mites from these families may be surprisingly abundant in the arid southwestern United States. They may be seen even on hot summer days as they run about on the surface of the desert or lie in wait for their prey. This group of predaceous mites seems especially well adapted to hot, arid conditions, and they have diversified to fill a variety of niches within this habitat. Perhaps their success is due partly to lack of competition; few other creatures of their size can withstand such hostile conditions as well as these mites do. They may be found on the ground, on exposed rock surfaces, and both on low herbaceous vegetation as well as up in taller trees.

Most of these mites have ocelli as well as a naso. The ocelli may be fairly conspicuous and ruby red in color in some species. They also have bothridial organs, the specialized mechanoreceptor hairs whose base sits in a cuplike socket. The bothridial organs may assist these predators in detecting and capturing their prey.

Mating is accomplished by indirect sperm transfer. Spermatophores are deposited on the substrate and left for the female to find and pick up.

Caeculid mites are distinctive with their tough body armor and their long, spiny front legs. Their heavily sclerotized cuticle affords them some protection from desiccation as they sit for hours, if not days, waiting to ambush prey such as springtails and booklice (psocids). They may be found baking in the sun on fully exposed rock surfaces as well as on the surface of the soil and even up in trees. Some caeculids become camouflaged with soil and debris that adheres to their cuticle. These mites are ambush hunters, grasping their prey with raptorial front legs.

Mites from the family Adamystidae are found primarily in drier environments. Like caeculid mites, adamystids can be found on the surface of the soil, on rocks, and up in trees. In some species, adamystids carry mate guarding to an unusual extreme. The male actually carries the female on his back until she is ready to mate, at which point he places her over his spermatophore. Like many other mites from this cohort, adamystid life stages consist of a prelarva, larva, three nymphal stages, and the adult. Active, nonfeeding elattostase prelarvae have been documented in some species of these mites. The temporary mobility of the prelarvae may be an adaptation for finding a humid niche in this dry habitat.

The Anystoidea include the families Anystidae, Erythracidae, and Teneriffiidae. Anystoid mites typically have a roundish or oval body and frequently have large setae on their legs. These fairly large mites also have conspicuous ocelli. They appear to be primarily diurnal. One of their most distinctive characteristics is their incredibly rapid running punctuated by brief periods of rest as they hunt for their tiny prey.

The most familiar of the anystoid mites may be the whirligig mites in the genus *Anystis*. These bright orange to red mites live up to their name by racing in circles as they hunt small arthropod prey on leaves or in leaf litter on the ground. Whirligig mites may use a paralytic toxin as they stab into their prey with their chelicerae. In arid desert regions, they are common along riparian areas and in cooler higher elevations but are rare in hot, dry environments.

Another frequently encountered anystoid mite is in the genus *Paratarsomatus*. These mites may be extremely abundant on the soil surface of low-elevation, hot, dry deserts as well as at higher, cooler elevations. As many as 30 or more individuals may be captured in a small pitfall trap within the space of a few hours. Females seem to be the only individuals found, which is suggestive

of parthenogenesis as a mode of reproduction in these mites. These mites hold a speed record in the *Guinness World Records* as the fastest animal (relative to body length).

Erythracarus species are so small and so fast that they are difficult to visually follow. They are found on the ground, frequently on or under rocks. They seem to be more abundant during the cooler months of the year. Some males from this group have an unusual behavior. They may "stack" their spermatophores, adding to others to make a spermatophore "tree." Perhaps this is a form of sperm competition. One species in this genus has elattostatic prelarvae.

Most mites in the family Teneriffiidae may be found in coastal habitats, but a couple of species are found in deserts. One species is found in the high deserts of Oregon, and one species is found in the low-elevation Sonoran Desert in Arizona. These Sonoran Desert mites are distinctive with their heavy chelicerae, deep yellow bodies, and red ocelli.

Parasitengonina: Erythraiae, Trombidioidea

This group of mites is extremely diverse and contains the largest, most spectacular mites in the world. Some of these mites are long-legged and earth-toned in color, while others have short legs and are brilliantly colored in velvety orange or red. Most species in this cohort show a dichotomy between their 6-legged larval stage and the 8-legged postlarval life stages. In most species, the larvae are parasites on either vertebrate or invertebrate hosts, while the deutonymphs and adults are predaceous on other arthropods. The larval morphology is so different from that of the deutonymph and the adult that only by rearing larvae to deutonymph stage or by hatching out eggs from known adults can correlations be made between the larvae and adults in order to characterize a species.

The parasitengonines are diplodiploid and reproduce sexually. Sperm transfer is achieved via spermatophore, which in terrestrial species involves depositing the spermatophore on the substrate for the female to find later. The egg hatches into a prelarva, which then develops into a larva, a protonymph, a deutonymph, a tritonymph, and finally the adult. The prelarva, protonymph, and tritonymph stages are all inactive calyptostases, and are cryptic because the reorganization of structure is hidden under the cuticle of the previous stage. Therefore, the three stages that

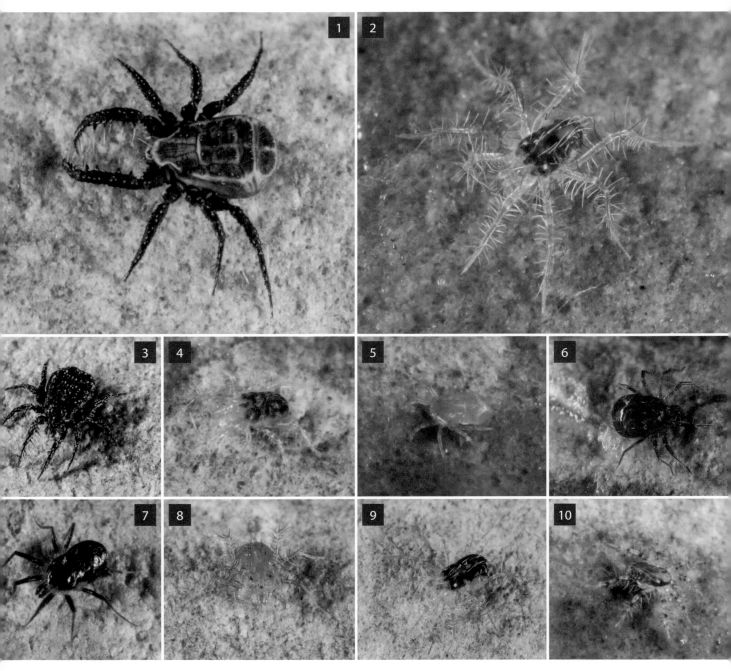

1. *Procaeculus* species, family Caeculidae. Raptorial front legs enable this ambush hunter to grasp its prey.

2. Many erythracids have heavy setae on their legs.

3. Caeculid mite, unknown species.

4. *Austroteneriffia* species. The family Teneriffiidae includes both desert and coastal species.

5. Adamystid mite. Large numbers of these tiny yellow mites may be found running about on rock surfaces.

6. *Anystis* species. Known as whirligig mites, these little predators run in circles.

7. Near *Tarsolarkus* species. Like other anystoid mites, *Tarsolarkus* are fast and nimble predators.

8. *Paratarsomatus* species or near *Paratarsomatus*. These diurnal predators may be extremely abundant on the hot desert soil surface.

9 and 10. *Erythracaurus* species. Moving at lightning-fast speeds, these tiny predators are easy to overlook.

are readily observed are the active larva, deutonymph, and adult stages.

Erythraiae: Smarididae and Erythraeidae

Smaridids are small predaceous mites found on the soil surface, in leaf litter, or in vegetation close to the ground.

Erythraeids are long-legged mites that are mostly parasitic as larvae and predaceous as adults. The larvae of most species are parasitic on other arthropods, including some other species of mites. Consequently, the life histories of some of the erythraeid mites can become interconnected and complicated. For example, *Paraphanolophus* larvae are parasitic on leafhoppers, but as deutonymphs and as adults, they in turn are parasitized by the larvae of another genus of erythraeid mite, *Lasioerythraeus*. Both the adult *Paraphanolophus* and adult *Lasioerythraeus* are predaceous on small invertebrates. One genus, *Balaustium*, has free-living predaceous larvae that also feed on pollen. Erythraeid mites may be frequently encountered on plants as they continuously cruise over vegetation, constantly tapping their long front legs in front of them in their search for prey. Some of these mites are significant predators of economically important agricultural pests such as the cotton bollworm and the tarnished plant bug. The mites generally feed on the eggs or immature nymphs of these plant parasites. In a few cases, erythraeids may be parasitic on vertebrates.

Trombidioidea: Trombiculidae, Trombidiidae, and Other "Velvet" Mites

Among the Trombiculidae are some of the most detested mites of the world—the chiggers. These mites are predaceous as adults and as deutonymphs, feeding on insect eggs, nymphs, and larvae. But trombiculid larvae are parasitic on vertebrate hosts, including birds, reptiles and mammals. The mother mite carefully places each egg where the larva may have a chance of encountering the correct host. Large, dark objects, such as a human casting a shadow, will attract the 6-legged larva. The chigger climbs up the legs of the unlucky vertebrate and finds a tight, protected spot to settle in and feed. On the correct host, such as a lizard, this spot would be the "mite pockets" at the base of the lizard's legs, somewhat analogous to "armpits" in humans. In a human, the snug spots tend to be where clothing fits tightly, such as the waistband of clothing or behind the knees. Once the

Balaustium mites. As they feed on pollen, *Balaustium* mites may form large aggregations, especially on flowers in the daisy family (Asteraceae).

chigger is settled, it tries to feed, inducing the formation of a feeding tube called a stylostome at the bite site. Unfortunately, most humans have a reaction to this that results in the formation of an itchy welt. The itching is so intense (and lasts for several days) that the welt may be scratched raw, and secondary bacterial infections may become a danger. If the chigger is on the correct host, it will eventually leave and molt into a predaceous deutonymph. If it is on the incorrect host, the chigger will detach from the host or possibly die. Unfortunately, it seems that the more "incorrect" the host, the more severe the reaction to the bite. Lizards are the correct host for many of the chiggers of the southwestern United States, so humans do tend to develop intensely itchy welts from these chigger bites. Harvest mites in Europe also cause itchy lesions in humans. But perhaps the itching caused by North American and European chiggers is relatively innocuous compared with the danger from the notorious chiggers from eastern Asia. These chiggers (*Leptotrombicula*) transmit scrub typhus to humans, a rickettsial disease found in rodents.

In contrast to the trombiculids, some trombidiid deutonymphs and adults could be the poster children for "most charismatic mites." Many species of these mites are quite large and have an attractive coat of red setae. The larvae of these mites parasitize arthropods. Depending on the species of velvet mite, the host may be as small as an aphid or as large as a grasshopper. The smallest velvet mites may not be much larger than a grain of sand, while the giant desert velvet mites of the Sahara may be 0.6 inches (1.5 cm) in body length.

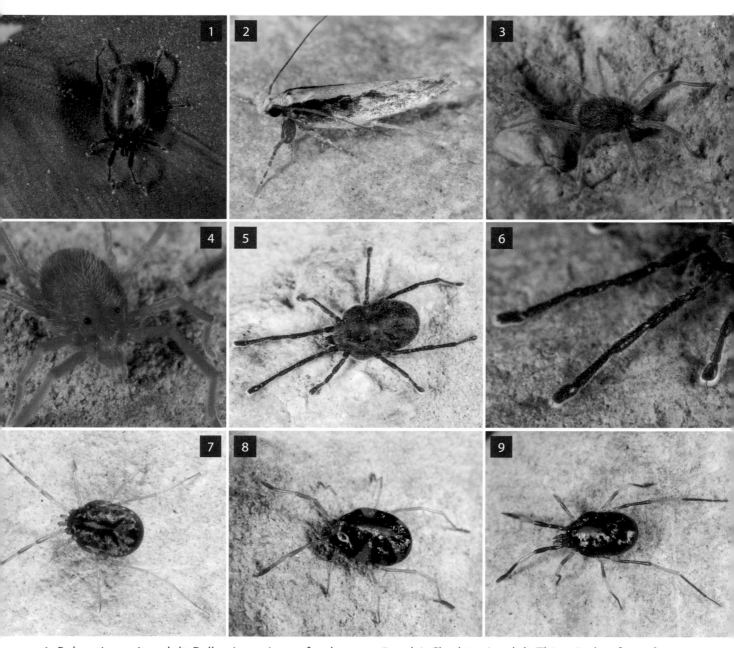

1. *Balaustium* mite adult. Pollen is a primary food source for adult, deutonymph, and larval *Balaustium* mites. These mites can be found on a variety of flowering plants, including trees. This mite had been feeding on pollen from a flower in the morning glory family (*Evolvulus arizonicus*).

2. *Callidosoma* mite larvae. The hexapod larvae are parasitic on moths.

3 and 4. *Callidosoma* deutonymph and adult: as with many other erythraeid mites, the predaceous deutonymph and adult mites have entirely different morphology compared with the hexapod larva.

5 and 6. *Charletonia* adult. This mite has fuzzy feet (tarsi), enabling it to easily grip vegetation as it hunts for its prey. This individual was found in an oak tree.

7, 8, and 9. *Erythraeus* species. A number of different species of these large, colorful predaceous mites can be found from low-elevation desert (7 and 8) to cool mountain forests (9).

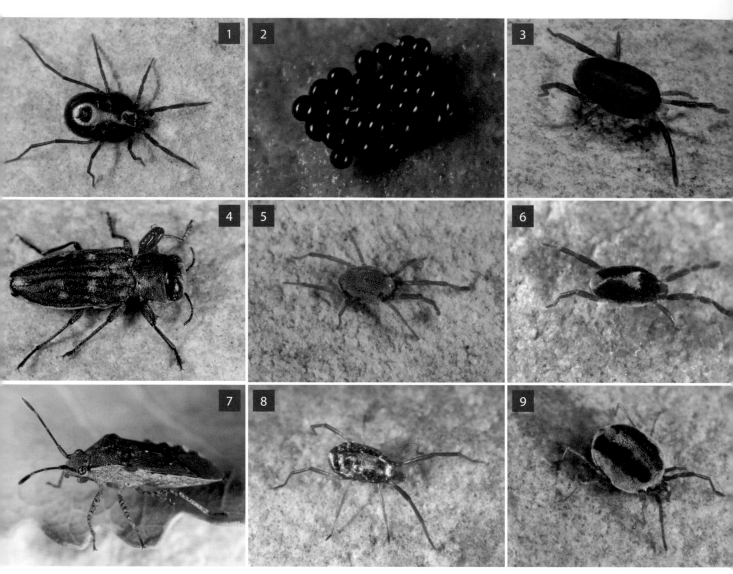

1. *Erythraeus* mite adult. Huge numbers of these iridescent mites are seen in early June at high elevations (8,000 ft. or 2,440 m) in the mountains of southern Arizona. A number of mite species are highly seasonal in their occurrence.

2. *Erythraeus* mite eggs.

3. *Leptus* species adult.

4 and 5. *Leptus* species larva and deutonymph. This particular larva was parasitizing a wood-boring beetle (Buprestidae). Most erythraeid mite larvae are parasitic on other arthropods before molting into predaceous deutonymphs.

6. *Leptus* species adult. Many *Leptus* mites are boldly patterned with contrasting white and dark markings.

7. *Leptus* larvae on plant-feeding bug. Only about 0.2 inches (5 mm) in length, this small bug was host to 9 *Leptus* larvae belonging to two different species of mites. This is an example of multiparasitism.

8. *Curteria* species. Mites in this genus have intensely iridescent cuticle. They may be seen running about on low vegetation and on the soil surface even at high temperatures.

9. *Augustsonella* species. This group of mites is found only in the western United States, usually in dry habitat. Like *Curteria*, these mites move rapidly on the substrate and on low vegetation.

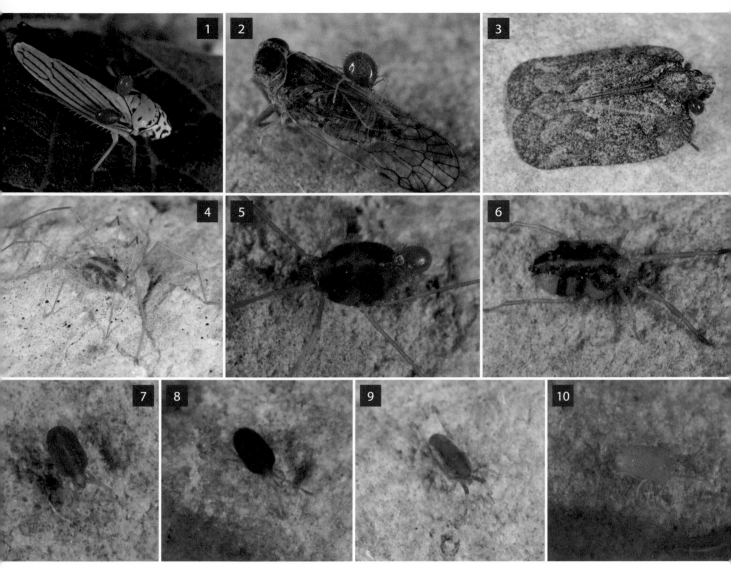

1. As larvae, some erythraeid mites parasitize plant-sucking homopterans, including leafhoppers.

2. *Paraphanolophus* larva on the plant-feeding cixiid *Oecleus*.

3. *Paraphanolophus* larvae on *Flatoidinus fuscus*, another plant-feeding homopteran. These hexapod larvae bear no resemblance to the adult mites.

4. *Paraphanolophus* adult. The long-legged adult is a nocturnal predator of other small arthropods.

5. *Lasioerythraeus* larva parasitizing *Paraphanolophus* adult. The genus *Lasioerythraeus* commonly parasitizes other erythraeid mites, especially *Paraphanolophus*.

6. *Lasioerythraeus* adult. The adult mites are nocturnal predators.

7. *Fessonia* species, family Smarididae.

8 and 9. Smaridid mites.

10. *Trichosmaris* species, family Smarididae. Smaridid mites are tiny "velvet" mites usually collected in pitfall traps set in the ground; however, some are capable of climbing. The mite shown in box 9 was collected on a windowsill at night, and the *Trichosmaris* was collected on a juniper tree under bark. Much still remains to be discovered regarding the natural history of many mite species.

Deutonymph and adult velvet mites are predaceous on arthropods.

One of the most conspicuous of these mites is the giant desert velvet mite of the southwestern United States, appropriately named *Dinothrombium*. These magnificent mites (reaching a length of about 0.4 inches, or 1.0 cm) are covered in a pelt of red "fur" that has a silky luster. Beady ruby-red eyes are in evidence as the mite trundles along on its short legs. The adults feed on winged termites. Therefore, these mites emerge in large numbers coinciding with the emergence of the winged virgin termite queens that are mating and then dispersing from the natal nest after heavy summer monsoon rains. As many as 50 mites per square meter may be on the surface of the ground in ideal habitat. In a few hours, the velvet mite can eat enough to sustain it for another year. (This feeding behavior parallels that of some other desert inhabitants, such as the Gila monster and the spadefoot toad, which have also adapted to a "feast or famine" environment. A Gila monster may eat one or two very large meals in a year. Calories are stored as fat in its tail, sustaining it through long periods of fasting.) *Dinothrombium* buries itself after feeding and goes into a state of torpor until it is time to emerge the following year. The workers of the encrusting termites (*Gnathamitermes perplexus*) react to the giant desert velvet mites as enemies, attacking and biting any mite immediately upon contact. The larvae of these mites are parasitic on grasshoppers, and so they are found in fairly undisturbed desert habitat in which both grasshoppers and termites are found.

Among the Microtrombidiidae, Johnstonianidae, and Neothrombidiidae are some very small velvet mites. Some microtrombidiids may continue to molt upon reaching maturity, in one case molting 7 times as a mature mite. Hosts for microtrombidiid larvae include small flies such as mosquitoes (among other types of arthropods such as grasshoppers), while johnstonianid larvae parasitize only flies associated with water, especially crane flies. The adults, barely larger than a grain of sand, may be found in large numbers along the edges of moist streams, where their larvae will be more likely to encounter the correct host.

Parasitism and phoresy may be closely related strategies for some mites. Sometime in the distant past, a phoretic mite may have gotten a little snack from its ride and thus may have started on the slippery slope of a parasitic life. There are risks involved in a parasitic lifestyle. There is a high probability that the correct host will not be encountered, and then the parasite will die. In fact, in one study, 98.9 percent of mite larvae that parasitize brine flies died before finding a host. But the alternative may be even worse. Habitats such as small bodies of water may be extremely ephemeral. For a mite dependent of these habitats, the only way to find a new home after the loss of the previous home would be via a larger arthropod.

Certainly, there must be a selective advantage to parasitism, since so many groups of mites have convergently evolved this strategy. Some of these parasites have evolved specific behaviors that maximize the survival of their host as well as their own survival. This is logical, since the continuing existence of the host directly benefits the parasite. An example of this is seen in mites that parasitize moth ears. *Dicrocheles* (a mesostigmatid mite in the family Laelapidae) live in the ears of moths. These ears consist of paired tympanic chambers on the body of the moth and are critical to the moth's ability to detect the clicks of a hunting bat and therefore to take evasive action to avoid being eaten. The mites that live in the ears of moths use one of two strategies. Some species live in only one ear, feeding on the tympanic membrane and destroying the function of that ear, but they leave the other ear fully intact and functional. Other species of mites do not destroy the tympanic membrane or the function of the ear, and therefore can parasitize both ears of the moth. In either case, the moth can still hear the hunting bat, and so it and its parasitic mites can survive.

Pterygosomatoidea

Many mites in this superfamily are parasitic on lizards and tortoises. Because they may settle in the mite pockets of lizards, they are frequently misidentified as chiggers. One genus, *Pimeliaphilus*, is an ectoparasite on kissing bugs (*Triatoma*), which is somewhat ironic given that kissing bugs are themselves bloodsucking ectoparasites of vertebrates.

Tetranychoidea: Families Tetranychidae (Spider Mites) and Tenuipalpidae (False Spider Mites)

All the mites in this superfamily are obligate plant parasites. Like many other prostigmatids, their chelicerae are adapted for piercing and sucking. In this case, the movable digits form a long, recurved, hollow stylet specifically evolved for piercing individual plant cells and sucking out their contents. Tetranychoid mites are able

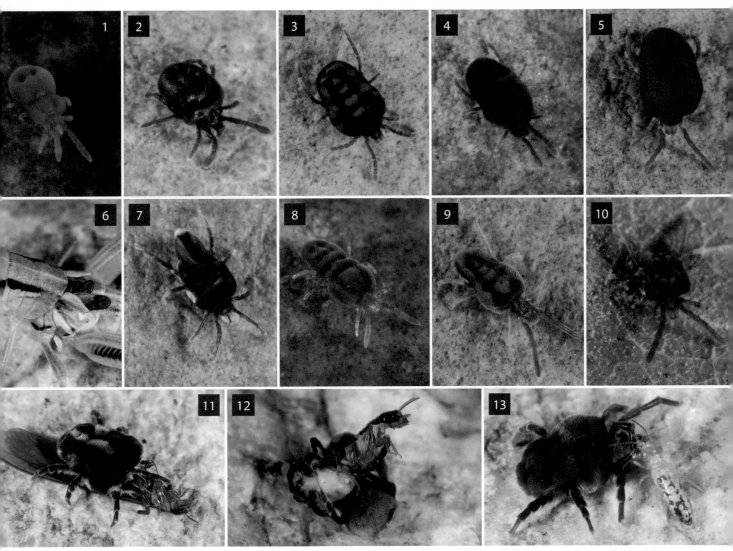

1. An adult chigger, possibly genus *Eutrombicula*, family Trombiculidae.

2. *Centrothrombidium* species, family Johnstonianidae.

3. *Empitrombium* species, family Microtrombidiidae.

4. *Valgothrombium* species, family Microtrombidiidae.

Centrothrombidium, *Empitrombium*, and *Valgothrombium* are all found in moist areas such as along streambanks. The larvae may be parasitic on dipterans such as mosquitoes. The adults are roughly the size of a grain of sand.

5. *Sphairothrombium* species, family Microtrombidiidae.

6. *Eutrombidium*, family Microtrombidiidae. The larvae are parasitic on grasshoppers.

7. *Eutrombidium* adult.

8. *Dasitrombium* species, family Neothrombiidae.

9. *Allothrombium*, family Trombidiidae. The larvae are parasitic on leafhoppers and other homopterans.

10. *Parathrombidium*, family Trombidiidae. A medium-sized velvet mite found along streambanks.

11 and 12. *Dinothrombium* species feeding on alate termites. In two hours, the mite can ingest enough nutrients to sustain itself for more than a year.

13. Worker termites immediately attack and bite any *Dinothrombium* mite.

to rapidly colonize a plant, in no small part because their reproduction is based largely on either a haplodiploid arrhenotokous system (where males are produced from unfertilized eggs and females are produced from fertilized eggs) or on a diploid thelytokous system (where parthenogenetic females can reproduce without any males). In either case, a single female could potentially start an entire colony. Both spider mites and gall mites (Eriophyoidea) disperse as aerial plankton, and both groups frequently require specific plants as hosts. It is probably not a coincidence that these two unrelated groups of mites both utilize haplodiploid arrhenotokous reproduction. The likelihood of a single mite randomly encountering a correct host plant may be slim, but still possible. The likelihood of two mites of the same species but opposite sex randomly encountering a correct host plant may be almost impossible. But by virtue of needing only a single female to start a colony, these mites have become amazingly successful. Tetranychidae now contains 1,200 described species worldwide.

Tetranychid mites are able to spin silk from modified hollow setae (eupathidia) on their pedipalps that are connected to large silk glands. The silken webs associated with these mites give them the common name of spider mite. Silk is used both in dispersal and in the construction of shelters. When a spider mite is ready to ride the wind, it doesn't balloon quite the same way that spiderlings do. Instead of pointing its abdomen upward and releasing silk from spinnerets as spiderlings do, spider mites hang from a silken thread until a breeze breaks the thread and carries the mite away. Sheets of silk shelter the colony against the elements, protecting both adult mites and their eggs from desiccation and from rain. Silk also serves a protective function against predaceous mites. In response to herbivore damage, plants release volatile chemicals that attract predators. Predaceous phytoseiid mites can enter the colony only through small, easily defended openings in the thick sheet of silk. In some species of social spider mites, the adult mites cooperate in the defense of the home nest. Males and females use teamwork to chase and kill predators.

Finally, silk may be used as a weapon in fights between males. In some social species of spider mites, males fight over females, "spitting" silk and attempting to stab each other with their chelicerae. In small colonies, a dominant male may kill his rivals and therefore obtain exclusive breeding rights. But in larger colonies, an aggressive male may be so busy fighting other males

that a less aggressive male can then mate with a female. This is a bit analogous to the situation seen in other species of animals, such as fence lizards (*Sceloporus*), where the "sneaky," less dominant male slips into an alpha male's territory and mates with a female while the dominant male is distracted with territorial disputes. Consequently, there is selection for both aggressive alpha-type males as well as less aggressive males.

Although Tetranychid mites are not generally vectors of plant viruses, they are of economic concern. They can damage virtually every major plant crop, and because favorable mutations are rapidly fixed in the haploid males, these mites are especially quick to develop resistance to pesticides. The use of pesticides negatively impacts populations of beneficial predators, while plant-feeding spider mites evolve resistance and survive.

Oribatida (Sarcoptiformes)

More than 9,000 species of oribatid mites make up a diverse group of particle-feeders found in soil, leaf litter, and plants such as mosses that grow close to the ground. They are known as beetle mites, armored mites, or moss mites. Many do have a beetlelike appearance, due to their heavily sclerotized, hard, dark, and sometimes even shiny cuticle. These mites may be extremely abundant in favorable habitats such as moist, temperate forest soils with high organic content. There may be as many as 100 to 150 different species collectively totaling as many as a staggering 100,000 mites per square meter in the upper layer of soil. In very dry desert soils with low organic content, they are found in much lower numbers. These mites are generally saprophages and fungus feeders but may also opportunistically feed on algae, nematodes, and other tiny animals such as springtails.

Oribatid mites are critical to the process of soil formation. Dead organic matter, fungi, and other material is bitten off and swallowed in particulate form. As the food undergoes digestion, it forms first a bolus and then a pellet, which is then excreted. Many microorganisms in the pellet not only are still viable at the time they are excreted but may also be "awakened" and stimulated to grow after passing through digestion. Adding to the positive effect of the process is the fact that the microorganisms are excreted in a nice little nutrient packet, the fecal pellet of the oribatid mite. Finally, because the mite can move from one area to another and is fairly long-lived, both nutrients and microorganisms may be transported from one area to another, thereby enabling saprophytes to gain access to fresh supplies of organic matter. This is analogous to the role of cattle in the dissemination and selective enhancement of trees and shrubs in grassland communities. Cattle feed on mesquite pods, and the seeds pass through the digestive system of the cattle still viable and ready to germinate. The seeds are then excreted in the nutrient-rich manure, which gives the seedlings a head start. Oribatid mites move saprophytic fungus spores and bacteria in very much the same way that cattle move mesquite seeds, and are almost certainly as significant to the ecology of the ecosystem, albeit on a microscopic scale.

A large number of oribatid mites are partheno-genetic, reproducing generation after generation with only females. One of the puzzles is how a nonsexual reproductive system managed to diversify into so many species. An even bigger puzzle is how an entirely thelytokous lineage of oribatid mites produced the sexually reproducing astigmatan mites. This is just one of many mite mysteries waiting to be solved.

Mites possess an almost infinite array of lifestyles and strategies for reproduction and survival. Their diversity rivals that of any other group of complex organisms. Because their small size frequently puts them under the radar of our awareness, mites are poorly understood. But they richly reward those who look into their Lilliputian world.

1 and 2. *Tetranychus* species, family Tetranychidae. Spider mites in this genus produce sheets of silk that protect the mites from rain, desiccation, and predators. Their numbers can increase explosively, causing the death of plants.

3. *Bryobia* species, family Tetranychidae. This species of plant-feeding mite does not build a silk web. Many mites in this genus are all female, hatching from unfertilized eggs.

4. The larger mite is an oribatid mite in the family Damaeidae, a fungus-feeder. The smaller mite is in the family Camerobiidae, the stilt-legged mites, and is predaceous on other small arthropods.

5. Oribatid mites are also called beetle mites. These detritivore mites are important components of soil ecosystems.

6. Oribatid mite, family Galumnidae. Photo by Scott Justis.

7. *Pimeliaphilus* species parasitizing *Triatoma rubida*. This mite is an ectoparasite on kissing bugs in the genus *Triatoma*. Ironically, *Triatoma* kissing bugs are themselves blood-feeding parasites (on vertebrate hosts). Therefore, the mite would be considered a hyperparasite (a parasite of a parasite).

CHAPTER 11 Spiders:
Araneae, the Eight-Legged Puzzle

A mother spider (*Anelosimus arizona*) has brought a captured fly to her hungry babies. As the young spiders grow larger, they will cooperatively kill and share prey. Just before they reach maturity, these subsocial spiders will disperse and each female will form her own colony.

The stereotypical spider is solitary, uses an orb web for prey capture, eats its mate, and its behavior is determined purely by rigid instinct. Like many stereotypes this has a kernel of truth to it, and like many stereotypes it sometimes could not be further from the truth. The problem with spiders is that they are so incredibly diverse that almost any generality about them has conspicuous exceptions.

Despite this, a few generalities can be made with some degree of confidence. Spiders have two main body parts: the prosoma, also known as the cephalothorax, and the opisthosoma, also known as the abdomen. These are connected via a narrow "waist" called the pedicel. Spiders, like other arachnids, have 12 appendages: 8 legs, 2 pedipalps, and 2 chelicerae. In spiders, each chelicera includes a fang instead of a fixed and movable finger as is seen in most other arachnids. Another distinctive feature in spiders is that they have silk-producing spinnerets, usually at the end of their abdomens. Some other arachnids produce silk, but not from abdominal spinnerets. Finally, spiders have a unique method of transferring sperm while mating. The pedipalp of the adult male is the actual copulatory organ. The presence of fangs, spinnerets, and copulatory pedipalps not only help to define spiders from a morphological viewpoint but also help in understanding how a living spider functions. Adding another dimension to the challenge of understanding spiders, these arachnids demonstrate a startling degree of behavioral complexity regarding sociality and intelligence. A closer examination of silk, sex, sociality, and intelligence may better elucidate this 8-legged puzzle, the spider.

Sociality

In the popular view, spiders are the very antithesis of sociality. It is true that the vast majority of species are indeed solitary; however, this group of predators has independently evolved some degree of sociality in a wide range of different families. To piece together the puzzle of spider sociality, it must be examined in the context of the underlying conditions required for the evolution of sociality, as well as the short- and long-term consequences of sociality.

The evolution of sociality in spiders has three prerequisites. First, sociality requires sufficient resources to support a relatively dense population of individuals in a small area. This is true in human societies as well;

the appearance of towns did not occur until agriculture provided enough food to support large, permanent aggregations of humans. Likewise, a permanent colony of spiders containing many individuals must occur where prey is so abundant that even a relatively small area can support many individuals. These areas tend to occur in the tropics.

Second, the spiders must have preexisting behaviors that can serve as a foundation for more advanced sociality. These preexisting behaviors include maternal care of the young and tolerance between conspecifics. Numerous species of spiders show some rudimentary degree of sociality. Many mother spiders actively guard their egg sacs and their earliest instar offspring, and young spiders tend to tolerate their siblings for at least a short period of time. The transition (through natural selection) to more fully social spiders requires only degrees of modification to this elementary sociality. Consequently, a gradation in sociality can be observed in spiders. For example, subsocial (periodic-social) spiders may live together and cooperatively capture prey until they are close to maturity; at that point, they disperse. Fully social (permanent-social) spiders tolerate each other even as adults, communally maintain a living area, cooperatively capture and share prey, and communally raise their young. The transition from a subsocial species to a species in which a colony contains multiple generations of permanent residents is merely a difference in degree of sociality.

Third, social spiders must be attracted to one another and to the group as a whole. Nonterritorial social spiders could, in theory, live on their own—but they don't. In at least some species, lone individuals from a social species will group together if given the opportunity to do so. Tolerance of other adults and abundant food does not in itself produce groups of spiders that live together; there must be a glue, or attraction, that holds each spider to the group. This interattraction between individuals might occur because of a pheromone linked to silk, or perhaps a chemical change occurs in the spider when it lives as part of a group, as opposed to when it is living by itself. In one experiment, social spiders (*Agelena consociata*) that were raised singly grew more slowly and had a higher mortality rate than did their siblings that were allowed to live as a group. Despite all other conditions being equal, the single spiders failed to thrive compared with their group-living siblings. In any case, this intangible "glue" makes a difference in the physiology of social spiders and presumably is one of the prerequisites to sociality.

That sociality in spiders is advantageous in the short run is evident in several ways. In web-based spiders, a communally maintained web requires less effort and silk on the part of each spider than individually maintained webs. Therefore, a greater efficiency is achieved in respect to the construction of the web. Perhaps more important is the ability of a group of spiders to overcome prey too large for a single spider to tackle. This enables a colony to add large prey as a food resource. Permanent-social species can easily find mates within a colony, thereby avoiding the mortality associated with dispersal and wandering in search of a mate. Once young are produced, if the mother dies, others in a colony will communally raise the babies. Agonistic encounters are reduced or eliminated, thereby reducing injury and mortality, as well as reducing energy wasted during disputes between conspecifics.

The benefits of sociality can be measured in part by the size of some colonies. A single *Anelosimus eximius* (Theridiidae) colony may include more than 50,000 individual spiders. The social sheet web weaver *Agelena consociata* (Agelenidae) has colonies that may contain more than 1,000 adult spiders.

Another measure of the success of sociality (and subsociality) in spiders is that it has evolved independently in a wide range of spider families. Among these families are dwarf tarantulas (Theraphosidae), cobweb weavers (Theridiidae), lynx spiders (Oxyopidae), spitting spiders (Scytodidae), mesh web spiders (Dictynidae), giant crab spiders (Sparassidae), crab spiders (Thomisidae), sheet web weavers (Agelenidae), and velvet spiders (Eresidae). Many other families display lesser degrees of sociality, including extensive maternal care of the young and tolerance between individuals. Nonetheless, fully social spiders are the exception rather than the rule. The vast majority of species are solitary, or at most subsocial. Given the advantages and rewards of sociality, why is the solitary lifestyle the norm for spiders?

First, as was mentioned previously, most social spider species are limited to the tropics or close to the tropics. A dense aggregation of predators can survive only where prey is abundant; for spiders, that condition is found principally in warm, tropical climates. However, a more serious limiting factor may

be genetic. Examining the phylogenetic distribution of social spider species, it is noteworthy that although sociality has evolved independently in many families of spiders, any one fully social species does not seem to have diversified into an array of other species. It would appear that each social species is a "dead end," evolutionarily speaking.

The underlying problem with web-based social spiders is their inherent lack of outcrossing. A study comparing fully social *Anelosimus* spider species with subsocial *Anelosimus* species demonstrated a significant loss in genetic variability in the fully social species. The subsocial spiders disperse just before reaching sexual maturity, and so encounter unrelated individuals to mate with. In contrast, the fully social spiders live in one colony for generation after generation, mating with close relatives. It was found that the fully social *Anelosimus* had a 50 percent reduced mitochondrial sequence divergence and less than 10 percent of the nuclear genetic variability compared with the subsocial *Anelosimus*.

So, although the short-term benefits seem to favor sociality in spiders, the long-term cost in genetic variation ultimately takes a toll on the success of a species. Genetic variation is the raw material upon which natural selection can work. Species must adapt to changes and new challenges in their environment. Without genetic variation, not only does a species have less potential to diversify into new species, but the inbred species may become extinct in the face of new stresses such as climate change, disease, or parasitism. With less raw material (genetic variation) to work with, speciation cannot outrun extinction rates. Therefore, it is not surprising that extant fully social species are not very old from an evolutionary perspective. Fully social *Anelosimus* species are estimated to be only 2 million years old at most, whereas subsocial *Anelosimus* are estimated to have originated 10 to 15 million years ago. In the dynamic interplay between adaptation and extinction, web-based social spiders die out before they can adapt.

An interesting case is the subsocial huntsman spider from Australia, *Delena cancerides* (Sparassidae). Up to 300 individual spiders may belong to a single colony. These colony members share a refuge (under the bark of a tree), share some food, and completely tolerate each other. Strangers are not accepted into the colony unless they are young-instar spiders. Mature and older-instar

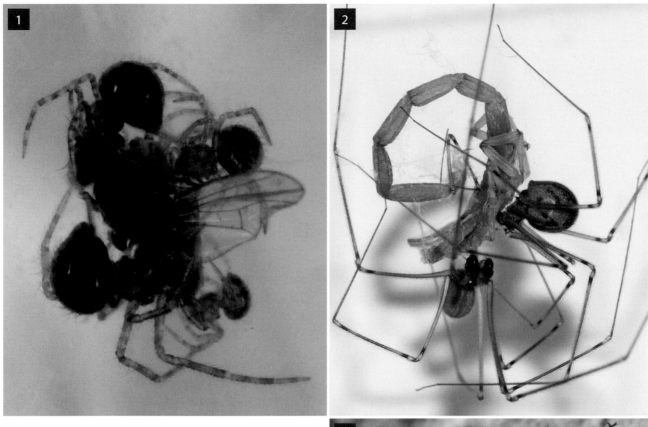

1. *Theridion cochise* siblings share a fruit fly they killed. A number of theridiid species show some degree of sociality.

2. Chivalrous pholcids. The male *Physocyclus* (facing the camera) had succeeded in killing the bark scorpion (*Centruoides sculpturatus*) and was preparing to feed on it when the female *Physocyclus* approached. After considerable leg waving and other interactions, they both settled down to the feast, which lasted 24 hours.

3. *Dictyna calcarata* male and female. Dictynid spiders sometimes cohabit after mating. An unrelated mated pair that is living together may cooperatively kill and share prey, clearly an advantage for such a tiny spider. In this case, a male (behind the fly) and the female (in front of the fly) are sharing the prey that they killed together.

spiders from other colonies are not accepted into new colonies and are treated with aggression. Unlike web-based social spiders, these subsocial huntsman spiders have a lot of genetic variability within each colony. Perhaps because sparassids must leave the confines of their refuge in order to hunt and capture prey, some younger spiders may wander from the home colony and join a neighboring colony, thereby contributing to the genetic variability of the colony, and to the species as a whole.

These subsocial huntsmen spiders also demonstrate a quality not normally associated with spiders. In a disturbing experiment, triplets of *Delena cancerides* were set up together. Two out of the three were from the same colony, but the third was from a different colony. The spiders were then starved. Under starvation conditions, the spiders attacked and killed the "different" spider. But 100 percent of the remaining spiders actually starved to death rather than kill and eat a member of their own colony. *Delena cancerides* and social spiders challenge our most deeply held beliefs and biases regarding spiders in particular and sociality in a broader context.

Silk

With only one known exception, spiders are primarily carnivorous predators. (That one exception is a jumping spider, *Bagheera kiplingi*, which lives principally on nubbins of protein produced by acacia trees.) Spiders capture their prey by using a wide range of strategies, including ambush, stalking, chasing, and snares. Orb webs are among the most conspicuous and aesthetically pleasing snares, but other snares may consist of sheet webs, irregular tangle webs, gumfoot webs, or lacy crevice webs. Among the most unusual snares is the mixture of silk and glue that the spitting spider shoots out over its prey. Another snare consists of a single blob of glue at the end of a short thread of silk which the bolas spider swings and aims at flying moths. In yet another example, the ogre-faced spider *Deinopis* holds a stretchy "cat's cradle" snare with its front legs and scoops up its prey. The one constant tying together the huge diversity of spider snares is silk.

Silk is an integral part of most spiders' lives. It is the essential element of snares and may also be used for wrapping prey after it has been captured. It is used in the construction of shelters, from retreats connected to aerial webs, to lining underground burrows. Almost

every spider releases silk behind it wherever it goes, and this ubiquitous dragline may function both as a safety line and as a means of communication via pheromones incorporated into the silk. Pheromones also advertise gender, maturity, and species in web silk. Silk is used in other spider communications, relaying messages between individuals through transmitted vibrations. These messages may be between two courting spiders, or between members of a colony, or it may even be the deceptive vibrations mimicking struggling prey used by a pirate spider to lure another spider to its death. Silk is used in several capacities related to reproduction. Sperm is deposited on a silken sperm web preparatory to the male spider loading his pedipalps. Some male spiders wrap the female in a "bridal veil" of silk before mating. Silk is also used for protecting vulnerable eggs against desiccation and predation. Finally, silk is even utilized as a means of transportation, allowing ballooning spiders to disperse and populate new areas. A spider was the very first animal documented on the volcanic island of Krakatoa only a year after its eruption (in 1883). Hawaii, which is 2,000 miles (3,218 km) from the nearest continent, has a number of endemic species of spiders. The linyphiids of Hawaii may be descendants from one species of spider that ballooned to Kauai, the oldest island of the archipelago.

Given the diverse use of silks, it is not surprising that a single spider may have up to 7 different glands for the production of several distinct silks. Even the construction of a single classic orb web (of an araneid spider) may involve up to 6 different types of silk. The outer frame and radial spokes are composed of strong silk from the major ampullate glands. Temporary scaffolding silk is produced by the minor ampullate glands. Viscid silk from the flagelliform glands forms the sticky capture spiral. As this silk is extruded, it is coated with aqueous glue from the aggregate glands. Piriform glands produce the silk that cements intersecting threads to each other. As a finishing touch, the thicker silk forming the conspicuous stabilamenta is generated from the aciniform glands. Those same aciniform glands produce the swathes of silk used in wrapping prey. Dragline silk is furnished by the major ampullate glands. Egg cocoon silk comes from the tubuliform glands. Finally, cribellate silk is produced from the platelike cribellum, with its hundreds or thousands of microscopic silk spigots. This silk is a combination of pseudoflagelliform silk surrounded by the dry cribellate threads.

The epitome of elegance, the delicate beauty of an orb web belies its strength and elasticity. Strong framing silk combined with a spiral of sticky catch silk create a perfect aerial trap. The geometric design transmits vibrations to the waiting spider and helps to dissipate the energy from an insect's impact.

Spiders re-engineer protein from ingested prey in order to synthesize this extremely strong, versatile material we call silk. In an effort to synthesize man-made silk that duplicates spider silk, researchers have discovered the basic ingredients of silk as well as its structure. The most completely characterized silk is the major ampullate silk produced by araneid orb weaver spiders. This is composed of 5 layers. First,

the outermost coat is an extremely thin lipid coat that does not contribute strength to the silk but acts as a carrier for pheromones used in species and gender recognition. The second coat is a layer of glycoproteins. These glycoproteins may provide some protection against bacterial or fungal degradation, but more importantly, they regulate the water balance of the silk. The hygroscopic property of the glycoproteins causes the absorption of water from the atmosphere, maintaining the elasticity of the fiber. The third layer is the skin. The skin consists of proteins that provide mechanical support to the silk by confining the core layers. It is also important in protecting the silk from chemical and microbial attack. The fourth and fifth layers make up the core, which is the operative part of the silk. Composed of proteins called spidroins, the core confers both strength and elasticity to silk. The outer layer of the core has a greater percentage of beta-sheet crystals, giving it strength. The inner layer has fewer crystals and is more elastic.

Dragline silk and the framing silk for orb webs must be simultaneously strong and flexible, and major ampullate silk fulfills both these conditions admirably. Therefore, unlocking the secret of the core components of major ampullate silk has been the holy grail of much research. Discovering the primary structure, which is the amino-acid sequence, is only part of the solution. The secondary and tertiary structure of a protein give the molecule its three-dimensional configuration, which in turn confers its physical properties. This structure is determined largely by the hydrogen bonds within the molecule, which provide the scaffold underlying the folded, three-dimensional shape of the molecule. This configuration has an enormous influence on the physical properties of the protein. For example, the liquid silk "dope" from a silkworm caterpillar can be dried. The dried protein has the proper amino acid sequence, but because the three-dimensional configuration of the protein assumes an alpha-helical conformation instead of a beta-sheet, it does not remotely resemble the native silk produced by the caterpillar. Spider silk presents much the same challenge. The major ampullate spidroin proteins consist of sequences of beta-sheet crystals alternating with "amorphous" sequences of helical and beta-turn structures. The beta-sheet crystals are the primary source of the silk's strength, while the "amorphous" sections give silk its elasticity. The crystals are composed primarily of stacked poly-alanine beta-sheets forming blocks. Glycine pairs are also part of the beta-sheet regions; alternating glycine with either serine or alanine causes the spontaneous formation of the beta-pleated sheet crystals. These crystals make up about 25 percent of the silk protein. The other part of the spidroin protein is the "amorphous" component (which alternates with the crystalline component). These glycine-rich residues form well-defined three-

fold elongated helical and beta-turn structures that are aligned along the fiber direction, giving the silk its elasticity.

The conversion of the liquid silk "dope" to the solid silk fiber is not dependent on drying. (In fact, one species of spider, the European water spider *Argyroneta aquatica*, can even produce underwater silk structures.) On the contrary, the conversion requires elongation, conformational changes, alignment, and polymerization of the spidroin protein. This is achieved both within the silk-producing gland and by the action of reeling out the silk. This almost magical transformation of liquid dope to silk fiber begins in the main body of the gland with the formation of some beta-sheets. These are probably initiated by shear forces generated by the movement of liquid against the inner walls of the gland, and these may serve as nucleation sites for subsequent beta-sheet aggregation. An alteration in pH, along with some water elimination, induces polymerization of the spidroin. As the liquid continues to traverse the gland and moves through a draw-down taper to the valve, the majority of the beta-sheets are formed. Conformational changes and alignment of the polypeptide chains occur as the liquid approaches the valve. Skin proteins are probably added along the inner part of the duct walls, where high shear forces enhance the orientation of these proteins. The glycoprotein and lipid layers are added just before the silk exits the gland. A draw-down of the fiber occurs as it is reeled from the spinneret, inducing final crystallization. The ultimate product is silk fiber: strong, elastic, resistant to microbial attack, and versatile.

The rate at which silk is reeled from the spinnerets modifies the tensile strength of the silk. The faster the reeling speed, the stiffer and stronger the silk. This is because at higher reeling speeds the crystals within the silk protein are smaller and more uniformly aligned. At slower reeling speeds, the silk from the same gland is more elastic. Hence, the silk from the major ampullate gland produces both dragline silk that is stiff and can support the dropped weight of the spider, as well as the more elastic frame silk for an orb web, able to dissipate energy from an insect's impact. Another example of a spider producing two different silks from a single gland is seen in the theridiid spider *Achaearanea tepidariorum*, which produces gumfoot snare silk as well as scaffolding silk from a single gland. These two silks are structurally different from each other and

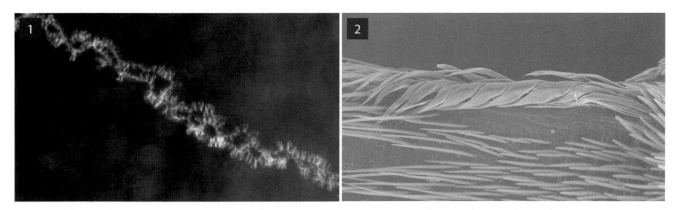

1. A cribellate silk strand in the web of *Kukulcania arizonica*. This strand is made up of a multitude of microscopic silk threads, which are combed out with the calamistrum (right).

2. The calamistrum of the *Kukulcania* consists of a row of thick, curved setae on the metatarsus of the fourth leg.

therefore function in separate capacities. In addition to reeling speed, the spider can control the diameter of the silk by adjusting muscular tension on the valve.

In contrast to the major ampullate silk, viscid silk produced by the flagelliform glands is composed almost entirely of glycine and proline and lacks the poly-alanine beta-sheet crystals. Consequently, this sticky silk is more elastic but not as strong as the major ampullate silk. The three-dimensional structure, so important to protein function, consists of beta spirals that act as molecular nanosprings. Additionally, the protein molecules in this elastic capture-silk contain sacrificial chemical bonds that can be broken and reformed, imparting a "self-healing" property to the silk. Finally, as this silk is extruded, a coating of watery glue is applied from the aggregate glands. The glue, made up of water, sticky glycoproteins, and hygroscopic molecules, coalesces into tiny droplets all along the capture-silk fiber. The presence of water is critical to the capture-silk elasticity; moist silk can stretch up to three times its length, whereas dry silk may stretch only one-third of its length before breaking. Anchored by the stiffer, stronger radial silk (from the major ampullate glands), this viscid silk makes up the spiral capture threads of the classic orb web.

The delicate beauty of a spider web belies the toughness of its silk. This toughness can be demonstrated experimentally. On a graph, plotting stress (force) versus strain (change in fiber length) generates a curve. The point on the curve at which the material fails determines its strength and extensibility.

The area under the stress vs. strain curve is the value for toughness. High tensile steel is in fact stronger than major ampullate silk, but because the silk has a much higher value for extensibility, the silk is technically tougher than steel. Viscid silk has different ratios of strength and extensibility, but it is still almost as tough as the major ampullate silk. In fact, it takes more energy to break a fiber of spider silk than to break high tensile steel, collagen, bone, synthetic rubber, nylon, Kevlar fiber, or carbon fiber. But major ampullate silk has another remarkable property. It converts more than half the kinetic energy of impact into heat, and the rest into elastic deformation. The dissipation of impact energy into heat is critical in the mechanics of insect capture. A vertical web intercepts primarily fast-flying insects which strike at right angles to the fiber axis. If the kinetic energy of the impact was not dissipated, the insects would simply be catapulted off the web. By absorbing the energy, the silk is better able to bring the insect to a halt. Not only does the silk possess remarkable mechanical properties, but the design of the web itself is also tremendously important. The geometric design of the vertical orb web is ideal for spreading the energy of an insect impact out over the web, as well as transmitting any vibrations to a spider waiting at the hub.

Silk may not have always been used as a snare. It may well have first been used to line underground burrows and extended outward from the entrance more by accident than by design. However, silk radiating outward from a burrow transmitted vibrations of insects

1. The tarsus (foot) of a western black widow (*Latrodectus hesperus*) has a variety of specialized setae. A central hook and several toothed setae allow the spider to ascend or descend the silk with perfect control.

2. The envy of anyone who has climbed rope, an orb weaver tarsus has all the tools necessary for maneuvering on silken lines. This *Neoscona oaxacensis* tarsus bears some resemblance to the black widow tarsus (left). It is thought that theridiids (such as the black widow) are fairly closely related to the araneids (orb weavers).

walking on the surface, permitting early detection of prey at a greater distance for the resident spider. This very likely gave a selective advantage to spiders who capitalized and enlarged on this early detection network of silk. Over 400 million years, spiders have been continuing to evolve different capture strategies using silken snares. As far back as the Triassic, spiders were probably starting to build orb webs. By the Cretaceous, the coevolution of plants and animals resulted in an explosion of adaptive radiation with the appearance of a multitude of new species. The proliferation of flying insects provided a newly profitable niche for spiders who could build aerial webs.

Orb weavers diversified into two main superfamilies: the Araneoidea and the Deinopoidea. Among the Araneoidea are the "garden orb weavers" with their large vertical orb webs. Among the Deinopoidea are the cribellate orb weavers such as *Uloborus*, with its relatively small, horizontal orb web. The vertical orientation of araneid webs selected for sticky, strong, elastic silk that could dissipate the energy of a flying insect's impact. In contrast, the hackled cribellate silk of the Deinopoidea lacks glue; instead, the microscopic silk threads form a woolly silk that catches the hairs on insects' legs. These small, horizontal orb webs are better at catching small insects such as dipterans (flies) and leafhoppers. Compared with the small prey

captured by cribellate orb webs, larger insects are more easily captured in the vertical webs of araneids. Also, the initial cost of producing viscid silk is lower than the cost of producing a comparably sized cribellate silk web, because the viscid web uses less silk in its manufacture. However, araneid webs require some humidity in order to maintain their elasticity and must frequently be replaced. In this regard, cribellate silk has its own advantages. Cribellate webs can be utilized even in very dry conditions, and a single web may last for an extended period of time without requiring replacement. Ultimately, both these designs are successful from an evolutionary perspective.

The success of the Araneoidea and their aerial webs catalyzed the diversification of spiders into a variety of specialized niches. Currently, Araneoidea includes 15 families of spiders, including Araneidae (garden orb weavers), Linyphiidae (sheet-web weavers and dwarf spiders), Nesticidae (cave cobweb spiders), Tetragnathidae (long-jawed orb weavers), and Theridiidae (combfooted spiders).

The irregular, three-dimensional webs of theridiids may appear untidy but are actually well structured. One advantage of a three-dimensional web over the two-dimensional orb web is protection against hunting wasps, which may have greater difficulty executing a surprise attack through an obstacle course of silk lines. As spiders expanded into new niches, concurrent selection for silks with new mechanical properties also occurred. The silk of the gumfoot web trap differs significantly from silks making up a vertical orb web. Instead of dissipating energy (as a vertical orb web does), gumfoot silk relies on storing energy for its success. The silk of a gumfoot trap is attached to the substrate and is under tension. When a struggling insect breaks the silk's connection to the substrate, the stored potential energy causes the elastic silk to spring upward, lifting the captured insect with it. This allows theridiids to exploit terrestrial prey, capturing abundant insects such as ants.

Given the multitude of different spider webs, humans have only just begun to discover the mysteries of silk.

Sex

Spiders have a unique method of mating. Sperm is transferred from the male to the female with specialized copulatory organs, the male spider's pedipalps. The pedipalp functionally consists of three parts: a modified tarsus, called the cymbium, a reservoir for the sperm, and a narrow structure through which the sperm passes as it is transferred to the female, called the embolus. The cymbium of mature male spiders serves as support and protection for the more delicate structures more directly involved with the storage and transfer of sperm. These structures may be relatively simple or incredibly intricate. The female's reproductive anatomy corresponds to the male palpal structures. Consequently, the reproductive structures of both male and female spiders fall into two general categories: simple, as seen in mygalomorph and haplogyne spiders, and complex, as seen in entelegyne spiders.

Simple reproductive structures probably represent the ancestral model. Among the spiders with more simple reproductive structures are mygalomorphs such as tarantulas, as well as haplogynes such as cellar spiders (Pholcidae), desert shrub spiders (Diguetidae), sowbug killers (Dysderidae), crevice weavers (Filistatidae), brown spiders (Sicariidae), and spitting spiders (Scytodidae). The bulb (reservoir) and embolus of these spiders tends to be fairly simple, frequently resembling a rubber squeeze bulb (used for flushing out ears) with a pointed end. In the female haplogyne spider, the passage for receiving sperm (the copulatory duct) is the

same passage through which eggs pass for oviposition. The male spider's bulb is inserted into the female's genital opening (gonopore) and sperm is transferred and stored in the female's seminal receptacle, to be used later to fertilize the eggs. The eggs pass through the gonopore as they are laid. Recent research indicates that the female may be able to exercise choice in whose sperm she uses, if more than one male has mated with her.

Entelegyne reproductive structures are probably more recently evolved. The majority of spiders fall into this category, including jumping spiders (Salticidae), wolf spiders (Lycosidae), crab spiders (Thomisidae), garden orb weavers (Araneidae), black widows (Theridiidae), funnel web spiders (Agelenidae), and many others. The male entelegyne palp is highly differentiated, containing complex, hardened structures (the sclerites) as well as a membranous structure (the hematodochae) that can be inflated with hemolymph, almost resembling a water balloon. Entelegyne palps also include a cymbium and an embolus. The embolus of the entelegyne palp may be tremendously long or complicated and in some species may be coiled up in the palp when not in use. During mating, the membranous hematodochae fills with hemolymph and the hardened sclerites become erect. The embolus is inserted into the female's specialized copulatory structure, the epigynum. The cuticle of the epigynum is sclerotized (hardened) and may include complex convolutions and infoldings; these consist of the seminal receptacles for storing sperm as well as the sperm ducts. The morphology of the embolus must exactly match the convolutions of the internal sperm ducts within the female's epigynum. Sperm is stored in the female's seminal receptacles. At the time the female is ready to lay her eggs, the sperm migrate through special fertilization ducts to the eggs. The eggs pass through a separate external genital opening as they are laid.

In haplogyne spiders, the male inserts the entire bulb into the female's genital opening while mating. Many haplogyne males use a "two-fisted" approach, inserting both bulbs simultaneously. This is in contrast to entelegyne spiders, which insert only the embolus (tip) of one pedipalp at a time. In haplogyne spiders, the last male to mate with a female has priority in the fertilization of her eggs, whereas in entelegyne spiders, the first male to mate with a female has sperm priority.

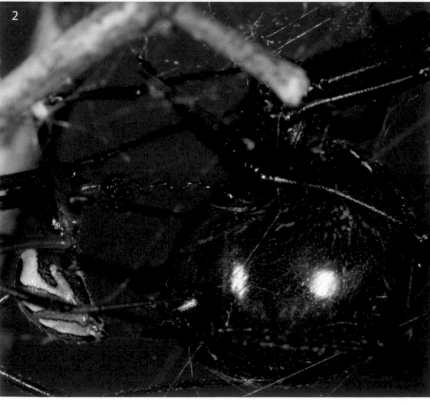

1. The brown and white male western black widow (*Latrodectus hesperus*) mates with the large female. The membranous hematodocha has become inflated with hemolymph and almost resembles an irregular, clear water balloon. This structure assists in positioning the sclerites correctly. Each pedipalp is used in turn by entelegyne spiders.

2. Having completed transferring sperm from that pedipalp, the hematodocha deflates and the embolus is withdrawn. In the black widow, the embolus is extremely long and slender. The embolus of each species of spider must correspond exactly with the female spider's internal anatomy, thus avoiding accidental hybrids.

3. A ventral view of the male black widow shows the pedipalps, each with its neatly coiled embolus.

This difference occurs because of the different internal arrangements of the fertilization ducts. In haplogyne females, one duct does double duty for both fertilization and oviposition. In entelegyne females, there are separate passages for fertilization and oviposition.

In many species of spiders, males will "camp out" near a penultimate female in anticipation of being the first to mate with her after her final molt. In web spiders, the male may take up residence in the female's web. In other spiders, such as jumping spiders (Salticidae) or ground spiders (Gnaphosidae), the male may construct a silken nest for himself next to the female's nest. He then mates with the female shortly after her final molt. In this way, he not only gets the chance to be the first to mate with that female, but he may also survive the encounter if her fangs have not had time to completely harden after molting. In some species, the males remain with the female after mating as a form of mate guarding, in an effort to prevent competing males from attaining access to the female.

Prior to mating, the male must load his pedipalps with sperm in a process called sperm induction. This can be accomplished only following the male spider's final molt, at which point the structure of the pedipalp is functional. The mature male spider must first build a special web upon which he deposits a drop of sperm. The sperm web is constructed with epiandrous silk, originating from silk glands found on the male spider's epigastric furrow. After depositing sperm on the web, each pedipalp can be loaded in turn.

The requirement for an exact fit between the epigynum and the embolus has been likened to a "lock and key," preventing accidental cross-species mating. Complex reproductive structures are just one of several barriers to accidental hybridization. Complicated courtship displays and signature pheromones also provide some insurance against hybridization. Many male spiders must actively court the female, using acoustic, vibrational, and/or visual displays. The elaborate displays used in courtship may provide a more effective barrier to hybridization than morphological differences alone do. For example, two species of wolf spiders, *Schizocosa ocreata* and *Schizocosa rovneri*, have mutually compatible reproductive structures and may be found in the same habitat. But hybridization normally does not occur, because the acoustic and vibrational sequence ("song") performed by each species of courting male is significantly different from

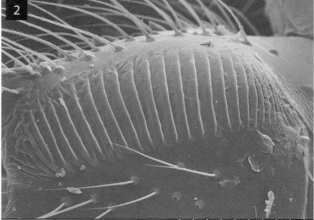

1. The male pholcid spider (*Physocyclus*) has tiny ridges on his chelicerae. During courtship, he rubs these ridges with a hardened area of his pedipalp, thus stridulating and so signaling to the female spider.

2. Micrograph of the cheliceral ridges.

any other species' song. Conspicuous decorations and highly specific dances also contribute to species recognition and prevent cross-breeding. Finally, pheromones cannot be underestimated in importance. Male wolf spiders initiate courtship displays as soon as they encounter the dragline silk of a female wolf spider of their own species. The male black widow does not start to court the female until he physically encounters the pheromone-laden silk of the female's web.

During courtship and mating, spiders demonstrate a surprising variety of behaviors, from simple to incredibly elaborate. Some species of crab spiders (Thomisidae) have a minimalist approach in which the male simply climbs onto the female and, with virtually no preliminaries, proceeds to mate with her. Others, such as the black widow, send vibrational signals between the courting couple via the web silk. The male first plucks the female's web. Then both the male and female vibrate their abdomens in turn, each answering the other until the male approaches the female and mates with her. The male desert blond tarantula, *Aphonopelma chalcodes*, initiates courtship by drumming his pedipalps on the ground near the female's burrow entrance. If the female is receptive, she replies, also drumming her pedipalps on the substrate before climbing up out of her burrow to meet the male. The male must hold her body upright by hooking the tibial spurs of his first pair of legs under the female's fangs in order to mate. In another example of the male holding the female's fangs, long-jawed orb weavers (Tetragnathidae) interlock their enormously long chelicerae and fangs against each other while they mate. The male must have a long reach with his pedipalps in order to successfully mate while in this position. The visually oriented wandering hunters, such as the jumping spiders (Salticidae) and wolf spiders (Lycosidae), incorporate complex visual displays into their courtship. In many of these species, especially the jumping spiders, the males have conspicuous tufts of hair or bright colors and must perform a dance with specific leg-waving as well as drumming and vibrating the abdomen. The male nursery web spider *Tinus* (Pisauridae) presents the female with a gift of wrapped prey before mating. This may buy some time for the male to mate while the female feeds on the nuptial gift. In another version of buying time, the male sand spider *(Homalonychus)* wraps the female in a "bridal veil" of silk before mating. Although she can quickly

Pholcid spiders mating. The male (on the upper right) is using a "two-fisted" technique, inserting both bulbs simultaneously.

extract herself from her bonds, it may give an alert male a chance to escape should she become aggressive. In some species, such as the redback spider of Australia, the male sacrifices himself during or immediately after mating; he actually does a back flip right into the female's chelicerae. By providing nutrition to the female, he may increase the number of eggs she lays and therefore help to ensure that his own genes will be passed on. In other species, the male endeavors to escape immediately after mating, and is killed and eaten only if he is too slow. If he is quick enough to escape with his life, he may mate with other females. Finally, in some species, the males cohabit with the female after mating and even steal food from her web. In some species of mesh weavers (Dictynidae), the male and female not only live together but they also cooperatively kill and share prey.

The range of courtship and mating behaviors in spiders therefore runs from one extreme of the males being killed and eaten to the opposite extreme of couples living and working cooperatively in the same web.

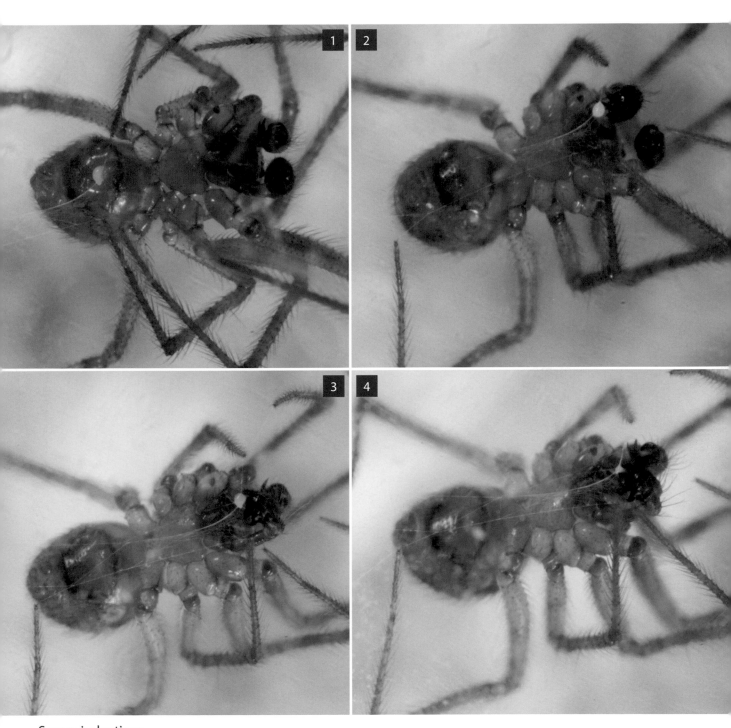

Sperm induction:

1. Preparing to load his pedipalps, a male *Theridion cochise* is in the process of depositing sperm onto a small sperm web specifically constructed for this purpose.

2. Positioning himself under the sperm web, the male spider can now reach the drop of sperm with his right pedipalp. He draws sperm up into the bulb of the pedipalp.

3. The left pedipalp is then loaded with sperm from the same drop.

4. The last of the drop is drawn up into the pedipalp. The male spider is now ready to mate.

Intelligence

It is an almost universal assumption that intelligence requires a large brain. The arthropod brain is by necessity small; therefore, it is only logical that arthropods have little or no mental capacity. Consequently, insects and spiders have been viewed as little more than diminutive robots, slaves to rigidly preprogrammed instinct. Certainly, there are many examples of instinctive behaviors in insects and spiders. But some behaviors simply refuse to fit neatly into the category of instinct, and some behaviors have every appearance of involving intelligence. Spiders demonstrate several capabilities that are normally associated with intelligence. They appear to have a working memory, they show plasticity and adaptability in their routine activities, they are able to learn, and at least a few appear capable of assessing a problem, devising a solution, and carrying out the solution.

Many spiders have a good working memory, especially in regard to spatial memory and remembering numbers of objects. An orb weaver may capture two prey items in close succession. If she wraps one insect and has to leave that part of the web to deal with a second insect, she remembers to return to the first capture afterward. If either of the insects is removed from the web, the spider actively searches the specific area from which the insect was removed. Other spiders appear to have a good spatial memory for relocating their refuge. Some species have been documented as wandering hundreds of feet from their burrow, meandering in an erratic, nonlinear path, but returning to their burrow by the shortest possible route.

Spider behaviors are a complex mixture of both instinct and adaptive responses to specific conditions. For example, building an orb web is an instinctive behavior; spiders do not have to be taught how to do it. However, orb weavers build larger webs if they are in areas with less prey, and some build different webs depending on what kind of prey is available. One species, *Parawixia bistriata*, normally constructs a fine mesh orb web that captures small dipterans; however, during the season in which winged termites are dispersing, the same species may construct a wider-mesh web for termite capture. Even the circadian rhythm of the spider may change in response to prey availability. Not only does *P. bistriata* modify the web mesh size during termite season, it also builds its web during the day instead of at dusk, as is its normal pattern. Spiders also show plasticity in regard to the process of building their webs. If a portion of a spider's web is destroyed while she is still building it, she will replace only the missing area instead of starting over completely. One of the most remarkable examples of adaptation in orb weavers was observed during experiments in outer space. Spiders normally use gravity while constructing the frame and other portions of an orb web; they simply add the vertical sections of the frame by using their own weight to drop down while releasing silk behind them. But in conditions of zero gravity, this instinctive behavior is no longer an option. After a couple of days of struggling to build their webs, orb spiders did adapt to weightlessness and succeeded in constructing reasonably good orb webs. The ability to meet this challenge was even more impressive, given that silk becomes solid only while under the tension of being reeled from the spinnerets. Therefore, under conditions of zero gravity, an orb weaver must reinvent the entire process for constructing the web, even including actively reeling out the silk for the frame instead of being able to passively use her own weight to exert tension on the silk as it is being extruded.

If learning is defined as altering behavior based on experience, then spiders demonstrate the capacity to learn. Naive young jumping spiders initially attack ants but quickly learn to avoid them because of the ants' ability to bite, sting, or spray formic acid. Sheet web spiders are able to capture more dangerous prey as they gain more experience. Finally, although jumping spiders are frequently thought of as intelligent by spider standards, an example that defies all our preconceptions is the incredible jumping spider *Portia*.

The genus *Portia* includes 17 described species of jumping spiders (Salticidae) native to Australia, Africa, and Asia. These jumping spiders specialize on preying on other spiders. Any one species of *Portia* is capable of hunting and killing a variety of spider species, including web-based spiders as well as other wandering hunters.

When *Portia* first sees a target spider in its web, it assesses the situation and decides on a specific plan of attack. During this phase, *Portia* integrates visual information regarding the type of spider and whether or not the spider is carrying an egg sac. A spider carrying an egg sac in its chelicerae is reluctant to release the egg sac, making it more vulnerable to attack. A spider without an egg sac must be drawn into striking range with care. *Portia* uses trial and error in order to discover what signal will draw the spider closer without the target spider actually attacking the jumping spider. *Portia* starts by plucking at the web, running through an array of different signals until one signal elicits the desired response on the part of the target spider. If the target spider responds too rapidly or aggressively, *Portia* stops and retreats off the web. If the target spider stops responding to the signal, *Portia* starts the process over, running through the array of signals again. If the target spider still refuses to approach close enough, *Portia* stops signaling. Approaching on the web is tricky; unless there is a breeze disturbing the web enough to mask *Portia*'s approach, the target spider may attack and kill *Portia*. So, the jumping spider uses an entirely different tactic. *Portia* scans its surroundings and decides to approach the target spider from above, without using the web. In order to achieve this, *Portia* must maneuver through foliage and over rocks, losing sight of its quarry in the process. The detour may take an hour or more, but *Portia* retains the location of the target spider in her working memory and is able to position herself for the attack correctly. Once she is above the target spider, *Portia* descends on her drag-line silk. Finally, she lunges and grabs at the unsuspecting victim and kills her target before it can react.

Some species of *Portia* have been observed to use aggressive mimicry while hunting other species of jumping spiders. With one species, *Portia* mimics a courting male, using the same signals that the male jumping spider would use to entice the female to come out of her protective nest. Another species of *Portia* even hunts a species of spitting spider (Scytodidae) whose own specialty is hunting jumping spiders. In this case, because the spitting spider has the advantage of being able to spit a mixture of glue and silk over a distance, *Portia* deliberately attacks the spitting spider from behind. However, if the spitting spider is a female carrying an egg sac in her chelicerae, then *Portia* approaches from the front, knowing that the spitting spider mother is reluctant to release the egg sac in order to spit the glue mixture.

In all the above cases, *Portia* shows a truly astonishing degree of intelligence. Robert Jackson, author of "A cognitive perspective on aggressive mimicry" in the *Journal of Zoology*, writes:

> Portia identifies a problem (how to reach the prey), derives a solution, makes a plan and then acts on that plan, with the problem's solution not being derived by actual trial-and-error in the physical environment, but instead by neural processing that can be likened to running a simulation in a virtual, or mental space… Portia's success as a raider in other spiders' webs depends on active decision-making, planning and flexibility… A more rigid routine might often be fatal.

Whether it is the ability of an orb weaver to construct her web in zero gravity, or a little jumping spider's ability to plan and carry out an attack that rivals a Special Forces military operation, spiders truly challenge our preconceived ideas regarding their intelligence. For such a small package, spiders hold a lot of surprises.

CHAPTER 12 Tarantulas, Trapdoor Spiders, and Their Kin: Mygalomorphs

An iconic symbol of the Arizona desert, the desert blond tarantula (*Aphonopelma chalcodes*) is an imposing spider with a leg span of about 4 inches (10 cm). The southwestern United States has a rich diversity of tarantula species.

On a hot night in late July, a tiny tarantula emerges from an underground burrow. As it walks away from the maternal burrow, it is followed by another almost identical baby, and that one by yet another. Soon, a long line of spiderlings purposefully march in single file as they leave their natal home. Dozens if not hundreds of small, fuzzy spiders form a living chain that stretches across the desert floor. Eventually the chain fragments as each little spider strikes out on its own to start its solitary life. This is a night of dispersal for the desert blond tarantula, a remarkable event that few humans have been lucky enough to witness.

It may seem incongruous that the epitome of a huge, hairy spider starts out as a tiny pinkish spiderling. In fact, by the time the spiderling leaves the maternal burrow, it has already undergone as many as three molts. It all begins when the baby first hatches from the egg, at which point in time it is considered a postembryo. The postembryo is essentially helpless. Neither its legs nor its chelicerae are functional yet. It remains in the egg sac, absorbing the yolk stored in its abdomen and undergoing further development. When this development has progressed sufficiently, the postembryo molts into a first-instar spiderling. Both of these earliest life stages are referred to as "eggs with legs" because the abdomen is so full of yolk, the baby spider looks like a sphere with legs and a tiny head attached to it.

The first-instar spiderling still lives in the egg sac. As it continues to grow and develop, it molts into the second-instar spiderling. This life stage has functional chelicerae and legs and may have emerged from the egg sac, but it is still confined to the maternal burrow. In some species, second-instar spiderlings may cannibalize their litter mates even while still in the egg sac. Finally, after yet another molt, the third-instar spiderling is ready to make its own way in the world.

In the desert blond tarantula (*Aphonopelma chalcodes*), the third-instar spiderlings are pink and white, with long, feathery setae and a conspicuous solid black circle on the dorsal surface of the opisthosoma. This solid black area owes its color to the presence of microscopic urticating hairs. These barbed hairs readily break free to become suspended in the air if kicked loose by the tarantula. They can become lodged in the skin and mucous membranes of the eyes and respiratory system of birds and mammals. If a predator attempts to kill or eat a tarantula, it may encounter

these urticating hairs and suffer extreme discomfort. Even breathing the air near a hair-flicking tarantula may be dangerous, especially to those individuals who have become sensitized to these hairs. Most tarantulas from the Americas depend on kicking urticating hairs as a primary defense and are somewhat less inclined to bite than the tarantulas of Africa, Asia, and Australia. The Old World tarantulas actually lack urticating hairs, and consequently their primary defense consists of relatively powerful venom and a willingness to use it.

A bewildering array of more than 900 species of tarantulas is found throughout the warmer regions of the Americas, Africa, Asia, and Australia. Tarantulas come in a stunning range of colors, including purple, flaming orange, iridescent green, velvety black, and even cobalt blue with bright yellow highlights (*Poecilotheria metallica*). Their coats may be short and snug, or they may be long and "fuzzy." Some of the most spectacularly beautiful species, such as *Poecilotheria* and *Avicularia*, are arboreal, hunting prey and building their silken retreats up in trees. For the pink-toed tarantula (in the genus *Avicularia*) found in the tropical forests of South America, the fuzziness of the tarantula is a useful adaptation for living in forests that periodically flood. The weight of the tarantula distributed via many hairs is not sufficient to break the surface tension of water, and so the spider can walk across the water as it moves from one tree to another in a flooded forest. Other spiders live in burrows in the ground or build silken refuges in some other protected spot. Some may live in deserts, and others may live in tropical rain forests, even such moist environments as riverbank burrows.

Tarantulas range in size from the largest spider in the world, *Theraphosa blondi* of South America, with a leg span of up to about 10 inches (25 cm), to *Aphonopelma paloma*, with a leg span of only 0.75 inches (2 cm).

Tarantulas also have a variety of lifestyles and behaviors, from the stereotypically solitary burrow dweller to the subsocial behavior of some communal species, such as the dwarf tarantula species *Holothele* (from South America) and *Heterothele* (from Africa). These communal spiders may cooperatively kill prey and young spiders share the kill. Circumstantial evidence suggests that even some species that live in underground burrows may have extended maternal care of young. In a number of instances, young tarantulas well beyond the third instar have been found

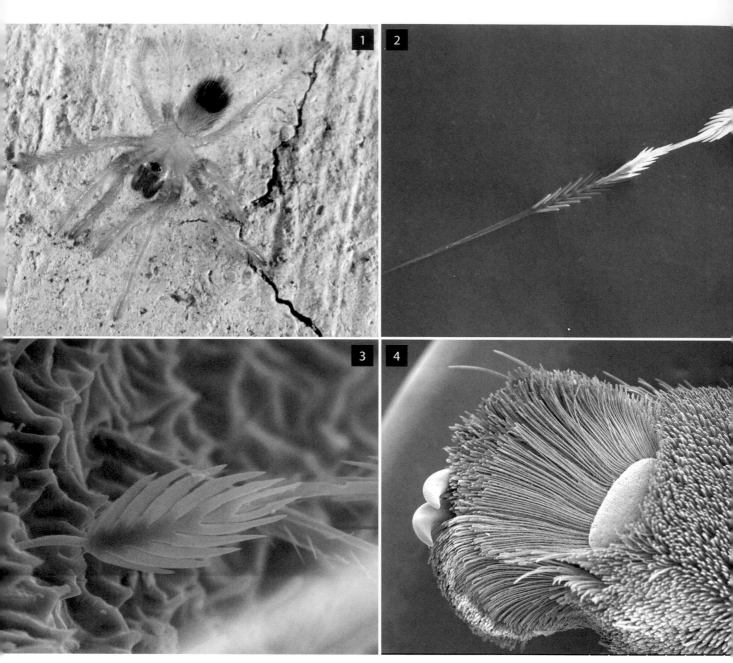

1. Only a few millimeters long, this tiny desert blond tarantula gives no hint of the 4-inch (10 cm) leg span it will have when it reaches maturity. The black coloration on its abdomen is due to microscopic urticating hairs (setae).

2. An electron micrograph reveals the structure of one urticating hair from the molted cuticle of a tarantula. These specialized setae are barbed in two directions.

3. Urticating hairs have an easy break point near the base of the hair. When a tarantula (from North America) feels threatened, its first line of defense is to kick hairs. This may superficially look as though the tarantula is scratching an itchy abdomen, but the tarantula is actually kicking the hairs to break them free of the cuticle and get them airborne.

4. *Aphonopelma* means "foot without sound." Scopula hairs allow the tarantula to walk and climb silently.

sharing an adult female's burrow, leading to speculation regarding whether the mother shares food with her offspring.

The tarantulas of the southwestern United States belong to the genus *Aphonopelma*. These range in size from fairly large species such as *Aphonopelma chalcodes*, with a leg span of about 4 to 5 inches (10–12.7 cm), to the tiny *Aphonopelma paloma*. A number of *Aphonopelma* are intermediate in size and are restricted to the mountains of southern Arizona. These tarantulas have a leg span of only about 2 inches (5 cm). The males mature in late fall or winter and may be seen as they wander in search of females even when there is snow on the ground. Because these mountain ranges are separated by barriers of low desert, many of the "sky island" populations have been geographically separated long enough that they are separate species.

One of the most common and conspicuous species is *Aphonopelma chalcodes*, also known as the desert blond tarantula. This handsome spider lives in an underground burrow in the low-elevation deserts of Arizona and may take about 10 years to reach maturity. The male looks markedly thinner and leggier than the female and, in addition, acquires a tibial spur on his front legs with his final molt. The males leave their burrows upon reaching maturity and go wandering in search of females. They are a familiar sight in the southern Arizona desert during the summer monsoon season, cruising at night or during the late afternoon, especially after a summer rain storm.

The mature male tarantula has prepared himself for mating prior to his wanderings. Shortly after his final molt, he constructs a special sperm web, shaped somewhat like a small tent. The male tarantula lies belly-up under the tent and deposits a drop of sperm on the sperm web. He then climbs on top of the tent and reaches down and around with his pedipalps in order to touch the drop of sperm, which is then drawn into the bulb of each pedipalp. This process is called sperm induction. At this point, he is ready to go in search of a female.

When the male encounters the silk near a burrow identifying it as belonging to a female *Aphonopelma* of his own species, he starts to drum on the surface of the ground. If the female is receptive, she answers him with her own rapid drumming down in the burrow. The vibrations are carried by the earth, conveying the messages back and forth between the courting couple. If the female ignores him and fails to answer

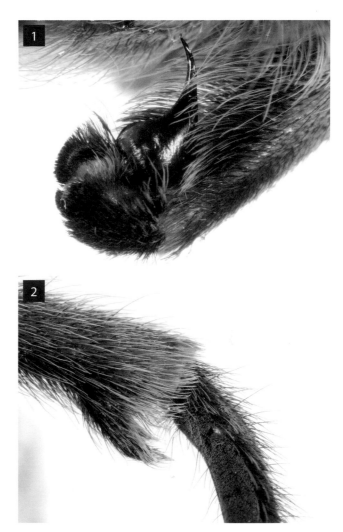

1. As with all other spiders, the pedipalp of the mature male tarantula has a specialized structure, the bulb, for transferring sperm to the female. The bulb of the male tarantula is fairly simple compared with the highly complex structures found in some araneomorph spiders. The male loads the bulb with sperm prior to searching for a receptive female.

2. The mature male tarantula has a pair of tibial spurs on each of his front legs. A receptive female presents her fangs to the courting male so he can hook his tibial spurs under her fangs. In this way, he can lift up the front of her body in order to mate with her.

his drumming, he may crawl head-first down into her burrow. He lowers his cephalothorax to the ground, tilts his abdomen upward, and stretches his first pair of legs as well as his pedipalps out in front of him, flat to the ground. He then slowly backs out of the burrow in the same posture with the female following.

If the female is ready to mate, she comes out of her burrow with her fangs extended. This is not a threat display; instead, she is presenting her fangs in order for the male to hook his tibial spurs under them and lift her upright. He may also support her midsection with his second pair of legs. By locking his tibial spurs under the female's fangs, he is not only able to position her for mating, but he may also be simultaneously protecting himself from her fangs. After some preliminary stroking of her abdomen with his pedipalps, he inserts the bulb of one pedipalp, then the other, into the female's gonopore and transfers sperm to her spermathecae. Once the male has finished mating, he suddenly pushes the female away (sometimes pushing her so hard that she falls over backward) and beats a hasty retreat. This is done with good reason. Many a male tarantula has paid with his life for the privilege of mating. From a strictly practical view, it makes sense for the female to kill and eat the male after mating. Since his days are numbered anyway, his body would at least benefit the female, and hence his offspring, if she is well nourished at this point in time. However, the male tarantula does not seem to share this viewpoint, and unlike the redback spider of Australia, he attempts to escape with his life if he can.

The female stores the sperm in her spermathecae until she is ready to produce an egg sac. The eggs are fertilized at the time they are laid. The female constructs a silken sheet upon which she deposits the eggs as well as a considerable quantity of fluid. She then bundles the eggs up in the silk sheet, thus making the egg sac. She rolls and kneads the egg sac, moving the eggs and fluid. Soon, the fluid disappears, probably absorbed by the eggs. This rolling and kneading is absolutely critical, or the eggs will stick together and die. It is also necessary to continue to periodically shift the eggs during the incubation period, or the embryos will die.

The mother tarantula guards the egg sac with fierce devotion, defending it against potential predators. She also practices active thermoregulation of the incubating eggs, bringing the egg sac to the mouth of the burrow in order to benefit from the warmth and radiant heat of the sunlight. After about three months (in the case of *Aphonopelma chalcodes*), the eggs hatch into helpless postembryos. This stage is followed by the first-instar spiderlings, then the second-instar spiderlings, and finally third-instar spiderlings. At this point, they are ready to disperse.

Like all other arthropods, tarantulas must molt in order to accommodate growth. Prior to molting, a new cuticle is generated beneath the old cuticle. In a process called apolysis, a layer of the old cuticle (called the endocuticle) is broken down by enzymes, thus separating the old exoskeleton from the new one. During the period of time leading up to a molt, spiders become lethargic and usually refuse to eat. The molt renews many of the internal surfaces of the spider as well as the external ones. Consequently, the cuticular lining of the mouth, pharynx, and stomach are replaced as well as the more obvious external cuticle. The abdomen may appear to be somewhat swollen and dark, due to the fluid between the layer of old and new cuticle.

When they are actually ready to molt, many tarantulas construct a silk sheet called a molting mat. The tarantula lies on its back on this molting mat, and after the old exoskeleton splits open, the spider very carefully extracts itself from the old exuvia and pushes it up and off to the side. It may take several days for the new exoskeleton to harden, during which time the spider does not feed. The hardening and darkening of the cuticle occurs as the molecules of arthropodin (making up sclerotin) become crosslinked. The cuticle is composed of both sclerotin and chitin (nitric polysaccharide) covered by a thin layer of waxlike lipid that reduces water loss.

Molting is a necessary risk for arthropods. It not only permits the spider to grow in size but also allows it to repair damaged cuticle and regenerate lost limbs; however, it comes at a cost. During both the actual

Facing page:

In order to mate, the male tarantula (in the foreground) must hook his tibial spurs under the female's fangs, thus lifting the front of her body. In this position, he can reach her genital opening with his pedipalps.

molt as well as the time immediately before and after molting, a spider may be unable to run from danger or defend itself. Any injury during this time may result in death. And for male spiders, molting holds special risks. Freeing the fully developed pedipalps from the old exuvia can be a challenge that not all individuals succeed in accomplishing. Some spiders become trapped in the old exuvia and may die or have to amputate the pedipalp. Even if they survive the amputation, they would be unable to mate with a female. Male tarantulas have fairly simple pedipalp structures, so most make it through their final molt successfully, but many other species of spiders with more complex structures suffer some mortality during the final molt. If a male tarantula survives to molt yet again after reaching sexual maturity, the palps are frequently so damaged in the process that he cannot successfully mate again. Female tarantulas may live for many years after reaching sexual maturity (the longevity record being 49 years) and may molt once every year or two during this time. When the mature female tarantula molts, the lining of her gonopore, including the spermathecae (where sperm is stored), is shed with the old cuticle. Consequently, she is once again a "virgin" and must mate again before she can produce viable eggs.

Molting allows a tarantula to regenerate a lost or damaged leg as long as it has time to grow a new leg before the next molt. Tarantulas, like many other spiders, have an easy break point near the base of each leg. If a predator grabs a leg, the tarantula can autotomize that leg. As the leg breaks free, the tarantula escapes with its life at the cost of its leg, a small price considering that the leg may be regenerated. However, the tarantula must be conscious for this to occur correctly. If a leg is damaged and the tarantula is anesthetized before the leg is removed, the break will not be a clean break and the tarantula may bleed to death.

Molting also gives an opportunity to renew the exterior cuticle and its complex structures. The exterior cuticle has a marvelous array of different setae. Some are used as sensory hairs, some are used for defense, and some are used during locomotion. In fact, the scientific name for *Aphonopelma* means "foot without sound"—an appropriate name since their tarsi, or feet, have hundreds of scopula hairs, giving them a soft, silent tread. Each scopula hair in turn has hundreds of microscopic cuticular extensions; hence the resemblance of the scopula hair to a brush (*scopa* means broom in Italian).

Armed with a powerful sting, this tarantula hawk (*Pepsis* wasp) has paralyzed a large tarantula and is carrying it to the burrow before depositing an egg on the spider. The wasp larva will feed on the still-living tarantula, finally killing it just before pupating. Photo by Timothy A. Cota.

Each cuticular extension is rounded and flattened at the tip. The scopula hairs give the tarantula foot an extraordinary ability to cling to smooth surfaces. In fact, some tarantulas can walk up a vertical pane of glass.

Tarantulas may have two kinds of setae employed in defense: the previously mentioned urticating hairs, and plumose bristles used for stridulation. Plumose bristles may be found on the chelicerae, pedipalps, or forelegs, depending on the species. If a tarantula feels threatened, it may rub the plumose bristles against each other, producing an audible hissing sound. Stridulation is a common defense strategy among a wide range of arthropods. Presumably, the sudden hissing sound may startle a potential predator and cause it to hesitate. In conjunction with stridulation, the tarantula may exhibit a threat display. By raising the front part of its body, giving a clear view of the fangs, and holding the first pair of legs in an elevated position ready for a rapid strike, the tarantula positions itself for attack. The fangs of tarantulas aim downward like the teeth of saber-toothed cats (instead of the pincer-type arrangement of non-mygalomorph spiders). Consequently, the raised prosoma of a tarantula would give it a greater range for the downward thrust of its fangs.

Tarantulas may be hunted by a variety of predators. Coatimundis have powerful front legs and heavy, bearlike claws for digging in the ground and tearing open rotten logs. This member of the raccoon family reportedly digs up tarantulas and rolls them on the ground in order to dislodge the urticating hairs prior to eating them. Grasshopper mice are tiny but formidable carnivorous mice capable of tackling a variety of arthropod prey, including scorpions and many other invertebrates. In fact, grasshopper mice can withstand the sting of even the bark scorpion *Centruroides* with no ill effects. An interesting hypothesis proposes that the reason tarantulas in the Americas have urticating hairs is that these hairs were selected for primarily as a defense against predaceous rodents. Another possibility is that the hairs were selected as a defense against predaceous marsupials such as opossums. This theory is plausible because the tarantulas with urticating hairs occur in geographic areas that also include opossums, whereas the tarantulas of Africa and Asia have no urticating hairs, and no marsupials are native to Africa or Asia. But perhaps the most notorious predator of tarantulas is the tarantula hawk, a very large wasp in the genus *Pepsis* that specifically hunts tarantulas. When it locates a tarantula, the wasp gives the spider a paralyzing sting, then drags the spider down its own burrow and lays an egg on it. There, the wasp larva will feed on the helpless, still-living spider until the larva is close to pupation, at which time it finally kills and finishes eating the spider. Tarantula hawks have one of the most painful stings of any North American invertebrate. Other parasites of tarantulas include small-headed flies (Acrocerid flies) and nematode worms.

In addition, a significant cause of mortality in tarantulas may be bad weather. During violent monsoon storms, they may be flooded out of their burrows, and little is known of the effect of extremely low temperatures or extended drought on populations of tarantulas in the wild. Out of the hundreds or even thousands of offspring produced by a single female, only a rare spiderling will reach maturity.

All tarantulas are superb predators. Most employ an ambush strategy, staying immobile near their refuge until their trichobothria or other sensory structures tell them that prey is nearby. Then, with lightning speed and pinpoint accuracy, they pounce on the prey and capture it. Once the prey has been captured, the tarantula may perform what appears to be a victory dance, gracefully turning in a circle as it sways and dips its abdomen, repeatedly touching the ground with its spinnerets. The purpose of this dance is to construct a silk feeding mat. The tarantula then masticates its food, mixing digestive fluids with the prey before ingesting only the liquid and leaving behind the bolus of fragmented, indigestible cuticle.

Tarantulas have an extremely low metabolism, consistent with their dependence on the metabolically conservative book lungs for respiration. Consequently, tarantulas can fast for an extended period of time—months or even years in extreme cases; however, they also have little stamina, and so are limited to short bursts of speed in order to capture prey. They can sprint surprisingly fast for short distances, an obviously useful adaptation for a sit-and-wait predator. They can also capture prey that is relatively large in relation to their own body size; in fact, they can even capture small vertebrates, including mice. The ability to capture both large prey as well as smaller arthropods would be another useful survival skill for a predator that may have to wait a long time between meals. It can take advantage of every opportunity that comes its way.

Ever since Maria Sibylla Merian published an engraving of a South American tarantula eating a bird in the 1700s, these spiders have captured our imaginations. The mere idea of an invertebrate hunting and killing a vertebrate such as a bird or a mouse seems contrary to the natural order, even akin to sacrilege. When the invertebrate predator not only belongs to one of the most despised and feared groups of animals, the spiders, but also is large and hairy, the horror becomes multiplied severalfold. Hollywood has effectively exploited these fears, conjuring up giant tarantulas, deadly venomous tarantulas, and tarantulas that wipe out entire towns. But these animals have so much more to offer than cheap thrills. Their beauty and their diversity in both appearance and lifestyle defy the imagination and far surpass Hollywood's wildest dreams. Certainly, they compel our respect, as does any predator capable of self-defense, but they also deserve our appreciation and protection.

1. *Aphonopelma saguaro*. With a leg span of only about 1.5 inches (3.8 cm), this miniature tarantula can be found in the same habitat as *Aphonopelma chalcodes* and *Aphonopelma vorhiesi* in southern Arizona.

2. *Aphonopelma chalcodes* adult male. These large males have a leg span of about 4 inches (10 cm) and are a common sight as they wander during the summer monsoon season (July and August) in southern Arizona.

3. *Aphonopelma mojave*. Found in the Mojave Desert of southern California, this miniature species is syntopic with the larger *Aphonopelma iodius*.

4. *Aphonopelma vorhiesi* adult male. This species is found in the same area as *A. chalcodes* but is smaller, with a 3-inch (7.6 cm) leg span, and its breeding season starts in September.

1. *Aphonopelma paloma* mature female. Almost an oxymoron, this tiny tarantula is syntopic with the much larger *Aphonopelma chalcodes*. Her burrow has a characteristic crescent-shaped pile of excavated soil near the entrance.

2. *Aphonopelma catalina* male. Restricted to the Catalina Mountains in southern Arizona, this species typifies the diversification of species in the unique biodiversity hotspot of the sky islands of southern Arizona.

3. *Aphonopelma madera* mature male. Like *A. catalina*, the adult male *A. madera* has a leg span of only about 2 inches (5 cm) and is found in the mountains of southern Arizona (the Huachuca and Santa Rita mountains). Both *A. catalina* and *A. madera* belong to the Marxi species group, and the males of these two montane species mature in the cold winter months.

4. *Aphonopelma superstitionense* mature male. Part of the *paloma* species group, *A. superstitionense* is another miniature, found in the Superstition Mountains of Arizona.

Ctenizidae and Antrodiaetidae: The Trapdoor Spiders

Invisibility has its benefits. Whereas artistic masterpieces made by humans stand out and demand attention, the quality of the trapdoor spider's workmanship is measured by its inconspicuousness. Trapdoor spiders have perfected the art of invisibility over millions of years, becoming masters in the art of deception. The spider's survival depends on it.

The trapdoor spider lives in a silk-lined underground burrow. As the name of the spider would imply, the entrance to the burrow is concealed by a hinged door. Both *Ummidia* and *Bothriocyrtum* belong to the family Ctenizidae, the "cork-lid" trapdoor spiders; however, the thickness of the trapdoor varies significantly between these two different genera of spiders. *Ummidia* constructs a fairly thin, flexible trapdoor, whereas *Bothriocyrtum* constructs a thick, "cork-bark" door with a slanted edge. The upper edge of *Bothriocyrtum*'s burrow is also slightly beveled so the door can close snugly. The resident spider can hold the door shut from inside, as seen by two fang marks on the underside of the door. Bits of leaves and other debris may be attached to the upper surface of the door, camouflaging it.

Spiders in the family Antrodiaetidae also live in underground burrows, but this family has a variety of structures relating to their burrow entrances. Within this family are species that build a turret-like collar that projects above the ground and other species that construct a flexible folding door (*Antrodiaetus* species). Others, like *Aliatypus*, build a wafer-type trapdoor lid covering the burrow entrance. *Aliatypus* can be found in a surprisingly diverse array of habitats, from the low, hot deserts of Death Valley in southern California to the cool mountain forests of Northern Arizona. The trapdoor itself may be the secret of its success. With the burrow entrance sealed, precious moisture can be conserved, thus allowing these spiders to survive during extended dry periods. The deserts of southern California typically have only one rainy season per year, which occurs during the winter months. The rest of the year usually has little or no precipitation. In drought years, even the winter rains may not come. Trapdoor spiders must endure these long dry periods by sealing their burrows with silk, a soil plug, or a combination of the two. This reduces evaporation within the silk-lined burrow. The burrow also protects spiders against brush fires, a common occurrence in the chaparral of the southern California coast. In Arizona, *Aliatypus* is found at higher elevations, where it can become extremely cold, well below the freezing point. The protection of a closed, underground burrow permits survival in these challenging environments.

Bothriocyrtum also can be found in a variety of habitats, including mild coastal areas of southern California as well as the low, hot creosote desert flats near Florence, Arizona. Once again, the trapdoor may be the key to this spider's success. In the creosote flat habitat, the trapdoor may serve double duty in conserving critical moisture during dry periods while sealing the burrow against flooding during heavy rains. In the Arizona desert, summer monsoon storms can drop a lot of water in a very short period of time; in fact, enough water accumulates that spadefoot toads breed in the area. The trapdoor appears to seal the burrow against flooding. Whether a spider is surviving heat, cold, drought, fires, or floods, a snug burrow may be critical. Most of these spiders require years to reach sexual maturity, and so the ability to endure these periodic stresses would be necessary for the survival of the species.

The burrow is constructed in stages. For *Aliatypus*, after a shallow starter burrow is excavated, the lid is constructed. The spider can then proceed to deepen the burrow beneath the protection of the closed lid. With *Aliatypus*, the dirt is simply pushed out of the burrow, whereas with *Bothriocyrtum*, the excavated soil is formed into pellets that are catapulted some distance from the burrow's entrance. Soil is molded and packed down around the rim of the entrance and the lid is repeatedly tested for fit until there is a perfect seal. A "plaster" of soil is applied to the walls of the burrow, and then the walls are lined with silk. *Aliatypus* builds its burrow on steep slopes along ravines and along road cuts. Burrows are situated among plant roots or under the partial shelter of a rock overhang, thereby reducing damage from erosion.

Trapdoor spiders in the family Ctenizidae are especially well equipped for digging burrows. A series of large spines on the margins of the chelicerae form a structure called the rastellum that is used to dislodge soil as the spider digs the burrow. The trapdoor spider then molds the soil into a ball and carries it out. The sides of the burrow are first smoothed, waterproofed, and stabilized

In an imposing threat display, an Arizona trapdoor spider in the genus *Bothriocyrtum* raises its front legs and remains motionless. This not only makes it appear larger but also enables the spider to better strike downward if it becomes necessary.

1. Camouflaged by soil and bits of dead vegetation, this trapdoor lid betrays its presence by being slightly ajar. This particular trapdoor is the lid to an *Ummidia* burrow.

2. A closer examination of the lid reveals a silk trip line extending out from the trapdoor. This silk line might assist the waiting spider in detecting nearby prey.

3. Oblivious to nearby danger, a cricket (*Gryllus* species) has wandered close to the *Ummidia* burrow. The tarsi (feet) of the waiting spider can be seen clinging to the edge of the burrow. The patient spider must wait until the prey is close enough to grab without the spider's hind feet losing contact with the burrow.

4. The hungry *Ummidia* drags the cricket into the burrow, where it will feed at its leisure.

Facing page:

In a single explosive lunge, the trapdoor spider *Ummidia* has captured its quarry.

Photos by Timothy A. Cota.

with a mixture of saliva and soil, and then lined with silk. The burrows built by this family are simple, single tunnels, but trapdoor spiders from other families may construct more complicated branched burrows.

With the door ever so slightly ajar, the hungry spider waits until it detects the presence of prey moving nearby. In a flash, the trapdoor spider pops out of its burrow and grabs the hapless prey. Some species such as *Ummidia* do not entirely leave the safety of the burrow even while capturing prey. These spiders maintain contact with the burrow using their hind legs while grabbing at prey with their front legs. They immediately disappear back into the burrow with their quarry to feed in safety.

After feeding, the spider may need to defecate. The excretory products are white, insoluble nitrogenous wastes consisting of guanine, uric acid, adenine, and hypoxanthine. The trapdoor spider shoots the excretory products a considerable distance from the burrow in a kind of projectile defecation. This projectile defecation is seen in species of trapdoor spiders from many different families and probably serves a protective function for the spider. An accumulation of conspicuous white excrement at the burrow entrance would advertise to predators the location of the spider. Shooting the wastes a considerable distance and in random directions from the entrance makes it harder for a predator to pinpoint the location of the burrow.

1. *Ummidia* species. This mature female was found wandering on the surface of the ground during a heavy monsoon rain in the Santa Rita Mountains in southern Arizona. She might have been flooded out of her burrow.

2. *Ummidia*, immature. This individual was found in creosote flats in the low-elevation desert of southern Arizona. Note the saddlelike depression on the third-leg tibia, a helpful feature for identification of this genus.

3. *Ummidia* species, mature male. This male was found in a low-elevation rocky canyon in southern Arizona. The males of this genus mature during the summer monsoon season. The extremely long pedipalps of male trapdoor spiders might assist them in mating with a female that is still in the burrow.

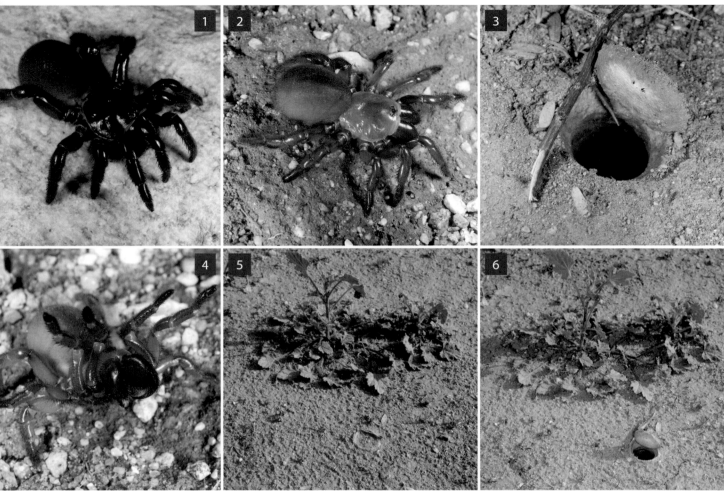

1. *Bothriocyrtum californicum.* Found in southern California and into Baja California, this species lives in a variety of habitats, including desert and oak scrub.

2. *Bothriocyrtum* species, Arizona. This species has been documented only in creosote flats in southern Arizona.

3. *Bothriocyrtum* lid, Arizona species. The lid is thick and beveled at the edges for a snug fit. The two holes on the underside of the lid allow the spider to hold the lid from below with her fangs.

4. *Bothriocyrtum* species, Arizona. Short, heavy setae on the legs and the front edge of the chelicerae are used to excavate soil from the burrow.

5 and 6. *Bothriocyrtum* species, Arizona. The burrow is well camouflaged. Excavated dirt is thrown a considerable distance from the entrance.

7. *Aliatypus isolatus*: Two isolated populations of this species occur high in the mountains of Arizona.

Unfortunately for the spider, one predator can find it despite all the spider's precautions at concealment. Wasps in the family Pompilidae are spider hunters by trade. These wasps walk rapidly across the surface of the soil, antennae quivering, as they investigate all possible sites for a spider burrow. Once a burrow is found, the wasp attempts to enter. The trapdoor spider holds the lid shut from inside, hooking its fangs into the underside of the door and bracing its legs against the walls of the burrow. But the wasp chews its way through the protection of the door, and once inside the burrow, the agile wasp quickly stings the spider, paralyzing it. It then lays an egg on the helpless spider. The wasp larva feeds on the paralyzed spider, not killing it until the larva is almost ready to pupate. The snug home of the spider becomes its grave.

Small-headed flies in the family Acroceridae also contribute to burrowing spider mortality. This parasitoid maggot grows enormously large inside the spider's abdomen, finally killing the host when the fly is ready to pupate.

Trapdoor spiders are syntopic (living side by side) with other species of burrowing mygalomorph spiders and may sometimes form dense populations in especially favorable habitats. In California, *Aliatypus* is just one of five species of mygalomorph spiders belonging to four different families in one densely populated area. In southern Arizona, four species of burrowing mygalomorphs are found in a single location in bleak, hot creosote desert flats. These four species are the large desert blond tarantula (*Aphonopelma chalcodes*), the tiny tarantula *Aphonopelma paloma*, *Bothriocyrtum*, and *Ummidia*. Perhaps the particular soil type in this location is especially conducive to burrowing. In addition, perhaps the lack of vegetative or topographic complexity of this site, as well as the lack of sheltering rocks, provides little opportunity for predators above the soil surface to survive, reducing some of the competition for prey in this bleak habitat.

During breeding season, the task of finding a mate falls to the male while the female stays in her burrow. Consequently, males are found on the surface as they wander in search of females. *Ummidia* males mature in the summer, and *Bothriocyrtum* and *Aliatypus* males mature in late fall or in winter (although it is not known for certain when *Aliatypus isolatus* matures in the mountains of Arizona). The male trapdoor spiders have extremely long pedipalps and, unlike tarantulas, have no spurs for lifting up the female. It is thought that the female might stay in her burrow even during courtship and mating, and therefore the male needs long pedipalps in order to reach the female's genital opening. After mating, the female uses her burrow as a maternity chamber, incubating her egg sac within its protected confines. The female *Aliatypus* hangs the pendulous egg sac from the walls of the burrow using silk suspension lines. The egg sac is positioned close to the bottom of the burrow, where the temperature and humidity are more constant. The mother spider probably assists the young to escape from the egg sac.

After emerging from the egg sac, the babies may remain in the maternal burrow for some time. *Aliatypus* young in California emerge from the maternal burrow at the start of the winter rainy season. This is advantageous timing, since they are less likely to desiccate and the moist soil is easier for the tiny spiderlings to dig as they excavate their own burrows. Many juvenile *Aliatypus* are found in close proximity to the maternal burrow, not surprising since *Aliatypus* is not known to balloon. In the genus *Ummidia* (found throughout the southern United States and into Central America), the young disperse by ballooning, an unusual behavior for a mygalomorph. Young *Ummidia* leave their mother's burrow in single file and climb up onto trees or other vertically projecting objects. Unlike most spiders, which balloon by tipping their abdomens upward as they produce aerial silk, *Ummidia* drop and hang by dragline silk from their perch until a breeze breaks the connection with the substrate and the spider is swept away. This requires stronger air currents than the gentle updrafts preferred by most ballooning spiders. In contrast to nonballooning trapdoor spiders, *Ummidia* subadult burrows are only rarely found in close proximity to an adult female's burrow. Ballooning may permit them to disperse a considerable distance from the natal burrow, and they may even be able to disperse over water in this way.

Dipluridae

The name Dipluridae means "with a double tail" in Greek. It is easy to see why this name was chosen for this family of spiders. Two long, fingerlike spinnerets extend well beyond the end of the abdomen, indeed resembling two tails. As the diplurid walks, each long spinneret bends and waves independently of the other, giving the spider a whimsical appearance. The terminal segment of each elongated spinneret is equipped with multiple silk spigots which enable the spider to produce a swath of silk from each spinneret. The wide spacing of the two long spinnerets and their maneuverability allow the diplurid to construct a sheet web as well as a tubular retreat. Interestingly, another group of funnel web spiders, the agelenids, also have long spinnerets. Since these two groups of spiders are not at all closely related to each other, this represents a case of convergent evolution, in which long spinnerets evolved independently in species that build both sheet webs and funnel retreats.

The funnel part of the web serves both as a retreat and as a strategic area to lie in wait for prey. It may have hidden side tunnels that open out onto a primary capture site comprising irregular, interconnected sheet webs in layers. The whole web somewhat resembles a chunk of Swiss cheese, with silken tunnels between the multiple layers of sheet web. The sheet web transmits the vibrations of any insect on the silk to the waiting diplurid. As soon as an unlucky cricket or other insect disturbs the surface of the web, the spider rushes out from its hiding place. Lifting up its fangs, it stabs down on the prey in a style reminiscent of a saber-toothed cat. Then it carries its victim into the safety and privacy of the funnel retreat to dine at its leisure.

The three species of diplurid found in the arid southwestern United States belong to the genus *Euagrus*. This half-inch-long (12 mm) brown spider lives an inconspicuous life in woodland habitats at about 6,000 ft. (1,828 m) elevation. Its irregular sheet web and funnel retreat are tucked under the shelter of a rock or a piece of dead wood, where ground-traveling insects may stumble onto the web. Another funnel web spider, *Agelenopsis*, may live in the same habitat. It is not known whether these two unrelated spiders with similar hunting strategies directly compete with each other, or whether they differ in some way that allows them to coexist without direct competition.

Like other mygalomorphs, diplurids have two pairs of book lungs, and their relatively small eyes are clustered together on top of their carapace. They do not appear to use eyesight in the detection of prey, being able to hunt equally well at night as in daylight. Instead, other sensory systems assist the spider in pinpointing the location of prey on the sheet web. One system consists of trichobothria, the specialized hairs that are extremely sensitive to vibration or air movement.

Other sensory structures are the tarsal sensilla, which consist of slits covered with a thin membrane on the feet of the spider. Finally, the dorsal surface of the tarsus has distinctive club-shaped sensilla that are also used in the detection of substrate vibrations.

The mature male *Euagrus* has a pair of large tibial spines on each of his second pair of legs. These spines are probably used in holding the female upright during mating, analogous to how mature male tarantulas use their tibial spurs.

Female diplurids continue to molt after reaching maturity, unlike males, who die not long after becoming adults. For the females, the entire cuticle lining their reproductive tract is shed during the molt, so they are once again "virgins" and must mate again in order to produce fertile eggs.

The family Dipluridae is famous as a result of the unenviable position of having a species included on the endangered species list. *Microhexura montivaga* has an extremely restricted range on the peaks of some of the southern Appalachian Mountains. This tiny spider, only 0.1–0.24 inches (2.5–6.0 mm) in length, is found only on moss mats in fir forests. It probably hunts springtails. One of the factors in its disappearance is the loss of the Fraser firs due to an introduced European insect pest, the balsam woolly adelgid (*Adelges piceae*). Without the shade of the fir trees, the mossy rock outcrops dry out too much for this little spider, and it is now almost extinct. Fortunately, the other species in the genus, *Microhexura idahoana*, has a large range in the northwestern United States and is reasonably common. It lives under fallen logs and other debris.

Another noteworthy species in this family lives in Panama. In the genus *Diplura*, this large spider (about 1.6 inches or 4 cm in length) is host to a tiny parasitic spider, *Curimagua bayano*. Unlike other kleptoparasitic spiders, *Curimagua* doesn't steal small prey from the web of its host. Instead, it rides around on the cephalothorax of the *Diplura* and waits for the big spider to kill and start to feed on an insect. In the process of feeding, the *Diplura* masticates the prey and regurgitates digestive enzymes onto it. At this point, the tiny parasitic spider crawls down the chelicerae of the *Diplura* and sucks up some of the semidigested liquid from the captured insect. The chelicerae of the minuscule, 1-millimeter-long *Curimagua* are so small that they are probably unable to capture their own prey, relying entirely on the tolerant host *Diplura* for their sustenance.

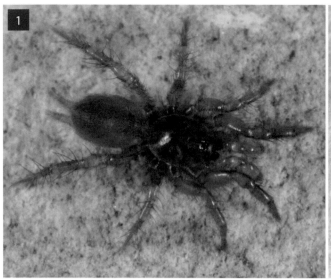

Both *Microhexura montivaga* and *Curimagua bayano* demonstrate the importance of the "web of life." Without the fir trees, the *Microhexura* is at the brink of extinction, and without the *Diplura*, *Curimagua* would not survive. The use of the word "web" implies a far-reaching interconnection between the organisms that share this planet. The loss of each species leaves a hole in the web that may never be filled again.

Mecicobothriidae

Spiders in this family resemble diplurids, with the exception that mecicobothriids possess two patches of sclerotized cuticle located at the front of the dorsal surface of the abdomen. These thickened, sclerotized patches are fused in some species, thus appearing as a single patch. In other respects, these spiders are very similar to diplurids. They have long, flexible spinnerets adapted to building sheet webs, which they usually construct under rocks, dead wood, or other debris on the ground. The web is composed of horizontal layers for prey capture connected by tunnels of silk.

The genus *Megahexura* includes the largest species within this family. These spiders are approximately the same size as the diplurid species *Euagrus*, reaching a length of 0.4 to 0.7 inches (10–18 mm). Found only in California, they construct their webs in sheltered nooks and crannies along the banks of shady ravines in oak or conifer habitat.

The smallest mecicobothriids belong to the genus *Hexurella*. Rivaling *Microhexura* as the most diminutive mygalomrph in North America, a mature male *Hexurella* may be only 0.1 inches (2.5 mm) in body length. These spiders build tiny sheet webs under rocks or fallen logs or even in leaf litter. In contrast to the thousands of eggs produced by the larger species of tarantulas, *Hexurella* deposits only 4 to 7 eggs in each little egg sac. Its tiny abdomen is unable to accommodate more than a few eggs at a time.

Hexurella apachea occurs at higher elevations in the mountains of southeastern Arizona. It may be found in oak, juniper, and conifer habitats.

Hexurella, though tiny compared with most mygalomorphs, is still robust compared with the smallest mygalomorph in the world. *Micromygale diblemma* (in the family Microstigmatidae) occurs in Panama on the forest floor. These minute spiders are distinctive in several respects. They have only 2 eyes, a thickened scutum covers the dorsal surface of the abdomen, and they have no lungs. Unlike other mygalomorphs that have 2 pairs of book lungs, at only 0.03 inches (0.75 mm) in body length, these spiders are so tiny that apparently they can obtain oxygen through their cuticle. This is probably possible because they have a high ratio of surface area to their mass.

Certainly, *Microhexura*, *Hexurella*, and *Micromygale* challenge the perception that all mygalomorphs are large and intimidating. These diminutive spiders have evolved to fill a tiny but fascinating niche.

Facing page:

1. *Hexurella apachea*, subadult. Tiny and secretive, these mecicobothriids live under rocks and in debris in the mountains of southern Arizona. This male was only 0.1 inches (2.5 mm) long at maturity. Characteristic of the genus, this spider has two faintly sclerotized patches which can be seen on the dorsal surface of the abdomen just behind where it connects to the cephalothorax.

2. *Megahexura fulva*. The largest of the family Mecicobothriidae, this species reaches a body length of only 0.5–0.7 inches (13–18 mm), a modest size compared with most mygalomorphs. *M. fulva* builds its web and retreat among rocks or tree roots in forested areas, especially in shaded ravines. It occurs in California. Although it has two sclerites (dark patches) on the abdomen, only one is readily visible.

3. *Euagrus* species, subadult. Two long spinnerets were the inspiration for the family name Dipluridae, meaning "two tails." Each spinneret moves independently of the other, giving this spider a whimsical appearance as it walks.

Euagrus lives in cooler forested areas in Arizona, New Mexico, and Texas. It builds a sheet web under rocks or debris, with silken tunnels giving it rapid access to the web. Sometimes more than one sheet is built on different levels, interconnected by the silk tunnels. As a medium-sized spider, *Euagrus* has an average body length of 0.26–0.67 inches (6.5–17.0 mm). In many respects, *Megahexura fulva* and *Euagru*s are comparable in size, habitat preference, and web construction. The long spinnerets of the mecicobothriids and the diplurids are an adaptation for sheet web construction, giving the spider tremendous control as it adds a swath of silk to the sheet web. Convergent evolution has selected for long spinnerets in other, unrelated families of funnel web spiders, such as Agelenidae.

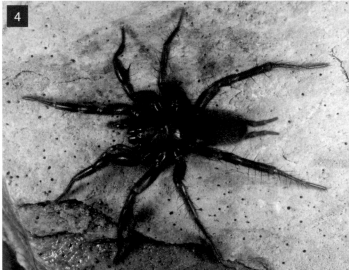

4 and 5. *Euagrus* species, mature male. After the final molt, the male *Euagrus* has a pair of tibial spines on each of the second pair of legs. Presumably, these spines are used in courtship, perhaps in much the same way that tibial spurs are used by male tarantulas; however, courtship and mating have yet to be documented in *Euagrus*.

CHAPTER 13 Orb Weavers:
Araneidae, Tetragnathidae, and Uloboridae

Surreal in color and form, the spiny orb weaver, *Gasteracantha cancriformis*, builds its web in trees and other tall vegetation. This genus occurs primarily in the tropics; however, this particular species is also found across the southernmost states in North America.

Araneidae

A delight to the eye and an engineering marvel, the orb web epitomizes the stereotypical spider web. It is built in a vertical plane, with strong, nonsticky silk radiating out from a central hub like the spokes of a wheel, supporting a spiral of evenly spaced sticky silk threads. A gap in the sticky silk near the hub allows the orb weaver to rapidly climb from one side of the web to the other, depending on which side of the web a flying insect has blundered into. Some orb weavers wait in the center of the web, legs stretched out in contact with the radiating silk lines that convey the vibrations of a struggling insect. Others build a little retreat at one side of the web, maintaining contact with the radiating lines via a signal thread leading to the hub. Lying in wait in the retreat, the spider rests with one leg touching the signal line. At the first indication that an insect has been caught, the spider moves into the web and tugs at the radial lines, testing to see the general location of the prey. It then uses the nonsticky radial lines as a quick pathway leading to the insect. Once the prey is reached, the spider uses large amounts of silk to wrap and immobilize it prior to settling in for the meal.

Many orb weavers build a fresh web every night and eat the silk by the next morning. Experiments with radioactive labeling have shown that spiders are the ultimate recyclers; up to 90 percent of the old silk is recycled into the new web, and such ingestion and reuse of the silk protein can occur in as little as 30 minutes. The spiral silk of the orb weavers owes its stickiness to the addition of little beads of viscous glue along its length, like the beads of a necklace. Neither the radial threads nor the hub threads have this glue, allowing the spider easy and rapid access to all parts of its web.

Some orb weavers build a web that remains in place for more than one day. Among these diurnal spiders are some that incorporate a special structure into the web, called the stabilamentum. The stabilamentum is composed of a thicker kind of silk, frequently appearing as a conspicuous white area in the web. It may look like a lace doily, or like one or more heavy zigzags in the web. Another type of stabilamentum consists of a line of silk above and below the resting spot in the hub of the web. The empty husks of insect prey are attached to this line, forming irregular clumps of detritus. Sitting motionless in the open spot in the middle of this detritus, the orb weaver *Cyclosa* appears to be just one more clump of

debris in the stabilamentum. Camouflage protects the spider against predation by birds. Yet a different type of protection from birds may be derived from the presence of stabilamenta. In *For Love of Insects*, Thomas Eisner describes one experiment in which 60 percent of webs marked by stabilamenta survived by noon, but 92 percent of webs without stabilamenta were destroyed in the same period of time. Birds were found to have flown into webs, but it appeared they avoided flying into the webs marked conspicuously with stabilamenta. In experiments by other researchers, webs with stabilamenta attracted more insects than webs without them. The stabilamenta silk of *Argiope* species reflects ultraviolet light. This may attract pollinators, which are drawn to the reflected ultraviolet light found in the nectar guides of flowers. Even the webs with the detritus stabilamenta of *Cyclosa* spiders were discovered to attract almost 150 percent more insects than webs without it. *Cyclosa* spiders make use of this by building smaller orb webs as long as they include the stabilamenta. This greater efficiency allows them to capture more insects with a small web than if they built a large web without the stabilamentum. They even conserve the stabilamentum, moving the detritus clumps from old webs to new ones.

Orb weavers are more flexible in their ability to react to different circumstances than one might imagine. They build larger webs when they are hungry or if they are in areas of low prey availability than when they are well fed or in areas of high prey availability. The Brazilian colonial orb weaver *Parawixia bistriata* (formerly *Eriophora bistriata*) builds two kinds of webs, modifying the size of the mesh depending on the availability of the prey at that time. Small, fine mesh webs are spun at sundown on a daily basis that catch small dipterans (flies). But during termite season, large, wide mesh webs are constructed during the day when diurnal termite swarms are flying. Both web design and the timing of its construction are synchronized with the type of prey and its availability, requiring the adjustment of the spider's circadian rhythm. In addition, orb weavers modify their approach to different types of prey in the web depending on whether the prey is potentially dangerous or not. They seem to know what kind of prey has been captured (perhaps based on the vibrations transmitted from its struggles) even before the spider physically makes contact with the prey. Some undesirable prey, such as stinging insects, are deliberately cut loose and released from the web. Other prey, like stink bugs, may be carefully wrapped so as to avoid eliciting a release of defensive chemicals until the killing bite can be administered in safety.

For many spiders such as *Cyclosa*, camouflage is a good defense against predation; however, spiny orb weavers of the genus *Micrathena* (meaning "little spine") have a completely different and surprising defense strategy. They produce an audible buzz if molested, startling a predator so it might drop the spider. This sound is produced by stridulation, the file being on the cuticle covering the book lungs and the scraper being on the femur of the last pair of legs. The book lungs may actually amplify the sound, acting as a resonating chamber. Many other arthropods (including kissing bugs, weevils, and velvet ants) use stridulation as a defense against predation in the same way.

The bolas spider is a special kind of araneid that no longer builds an orb web. Hers is a different and remarkable strategy. The name *bolas* refers to the apparatus used by the South American cowboy, or gaucho, to capture cattle or game animals. The gaucho swings several braided leather cords (connected together) to which weights or balls (bolas) have been attached. When the gaucho releases the rope, the momentum of the attached weights wraps the rope around the target animal's legs, thereby capturing the animal. At night, the bolas spider *Mastophora cornigera* uses a similar weapon. As she hangs from a single line of silk suspended between two twigs, she produces a short strand of silk with a sticky gob of glue at the end, forming the "bola." As a moth flies in close, the spider swings the bola and hits the moth with the sticky gob. The moth sticks to the glue, permitting the spider to reel it in and wrap it with silk. Researchers were puzzled by the success of this hunting method. It seemed improbable that enough moths would fly close enough to be captured by the spider; however, the solution to the mystery became apparent when it was discovered that *all* the captured moths were males. It turns out that the bolas spider baits her trap by releasing chemicals that mimic female moth pheromones. Male moths detect as little as 1 molecule of pheromone per cubic meter of air and may fly as far as almost 7 miles (11 kilometers) following the pheromone trail in order to find and mate with a female. Instead of becoming lucky in love, these male moths become dinner.

The bolas spider can produce more than one fake pheromone and can adjust the ratio of pheromone

Camouflaged within a line of detritus, *Cyclosa turbinata* conceals itself at the center of the trashline stabilamentum. This species of orb weaver saves the empty exoskeletons of its prey, adding them to the stabilamentum. The "trashline" is conserved each time a new web is built. Not only does the detritus provide a hiding place for the spider, it also attracts more prey than does a web without the trashline.

1. The bolas spider gets its name from the sticky blob of glue and silk with which it captures its prey. The female bolas spider prepares the bolas only after detecting potential prey nearby, male moths lured in by the fake pheromones released by the spider. Photo by Matt Coors.

2. The pea-sized egg sac is extremely hard. The tiny spiderlings inside the egg sac must chew a hole in it in order to escape.

3. Only 0.06 inches (1.5 mm) in body length, this tiny bolas spider emerged from the egg sac as a fully mature male. The lines of the finger whorl give scale to this minute individual.

4. When at rest, the female bolas spider *Mastophora cornigera* resembles a bird dropping. Predation by birds has selected for mimicry of inedible bird droppings in a number of unrelated invertebrates, including spiders and caterpillars.

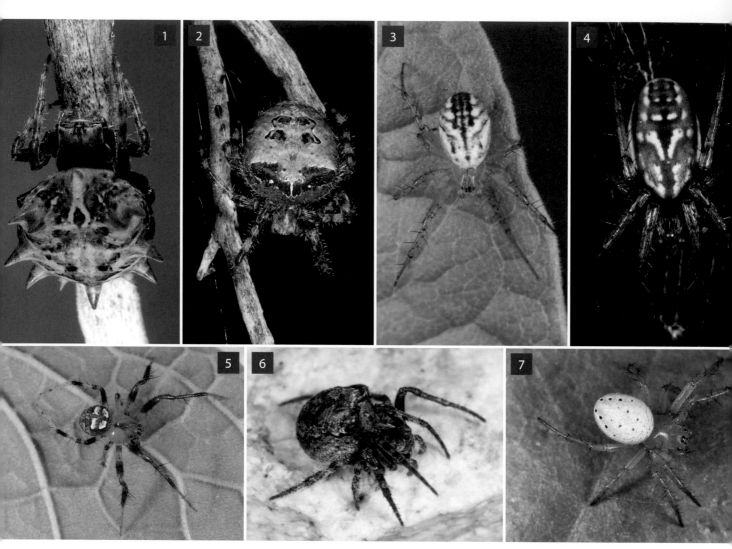

1. *Acanthepeira stellata*, the starbellied orb weaver. In the arid southwestern United States, this spider can be found in the slightly cooler oak zone. The spiny abdomen breaks up the outline of the body, making the spider harder to see.

2. *Araneus illaudatus*: The female of this large species can reach almost an inch (2.5 cm) in body length, and the orb web is correspondingly large and strong. This individual had a web that stretched across the entrance of a cave.

3. *Mangora fascialata*: This species is primarily a Central American species but has been found in southern Texas and rarely in Arizona.

4. *Mangora passiva*: This species has only rarely been documented and is found in Arizona.

5. *Araneus pegnia* male: The name butterfly orb weaver refers to the distinctive markings on the dorsal surface of the abdomen.

6. *Ocrepeira globosa* (presumptive): This genus does not use any sticky capture silk in its orb web. Radial silk lines connect to twigs, forming an asterisk web. Any insect disturbing the network is fastened to a twig with silk.

7. *Araniella displicata*: The sixspotted orb weaver has several different color morphs. In the hot southwestern United States, it occurs at higher, cooler elevations among pine forests.

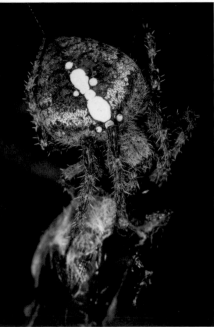

Many species of orb weavers have a bewildering array of color morphs.

All the individuals on this page belong to one species, *Eriophora edax*. This medium-sized orb weaver constructs a new web each night and hides in foliage during the day. Presumably, the variability in appearance may provide some protection from diurnal predators, such as sharp-eyed birds, since it is more difficult to form a search image for so many different color morphs.

This species may be found in areas where moisture is nearby, such as desert canyons as well as urban gardens.

Eriophora is primarily a tropical genus but has been documented in California, Arizona, and Texas.

1. *Argiope trifasciata*: The banded garden spider maintains a large orb web even during the day. A heavier zigzag line of silk forms the stabilamentum. This large species occurs throughout the Americas.

2. *Hypsosinga funebris*: Tiny and shiny, this orb weaver reaches a maximum body length of only about 0.2 inches (5 mm). It builds a fine-meshed orb web.

3. *Metapeira grinnelli*: Several similar species of *Metapeira* occur in the United States. Their web is easy to recognize, consisting of both an orb web as well as an irregular tangle web that might serve as a deterrent to hunting wasps. There is frequently a cluster of debris or egg sacs in the web. Some species, such as *Metapeira spinipes* in the southwestern United States, are facultatively communal; each spider builds and defends its own web, but webs may be interconnected by scaffold silk. Aggregations may reach several hundred individuals in areas with abundant prey. In less productive areas, the same species is solitary.

4. *Argiope aurantia*: The black and yellow garden spider mimics the warning colors of stinging insects. Note the zigzag stabilamentum.

1. *Eustala rosae*: Some morphotypes of this species are more reddish in color.

2. *Neoscona crucifera* has several color morphs. On this individual, the pattern on the abdomen forms a cross, but not all individuals have such distinctive markings. This species hides in vegetation by day and builds a fresh web each night.

3. *Neoscona oaxacensis*. The western spotted orb weaver occurs from Texas to the west coast. In the arid southwestern Sonoran Desert, this species is extremely abundant during the summer monsoon season, July through September. The substantial moisture during the monsoons results in a dramatic increase in invertebrate populations. Beetles, flies, and other flying insects are captured in the large web of this species.

4. *Larinia directa*: Although the color pattern can be variable, individuals of *Larinia directa* all have long, relatively thin abdomens. This species rests on narrow stems or grass blades during the day and builds its web at night.

5. *Micrathena funebris*: The genus *Micrathena* is primarily tropical, but a few species occur in the southern states of North America. This species is found in Arizona.

components in a blend relative to the flight time of the target species of moth. Adjusting the pheromone cocktail gives the bolas spider a greater chance of hunting success, since some species of moths fly early in the evening while other species fly later during the night.

The production of these fake pheromones is an example of aggressive mimicry. In this case, a predator actively mimics the prey's potential mate in order to lure it to within striking distance. Other examples of aggressive mimicry are seen in the light signals from "femme fatale" fireflies, the caudal (tail) lures used by some venomous snakes that imitate invertebrate prey, and the lighted lures used by some deep-sea species of fish. Humans have adopted a form of aggressive mimicry when they release massive quantities of synthetic pheromones in a deliberate attempt to disrupt the reproduction of a pest insect.

After an adult female bolas spider has succeeded in capturing a number of moths and feeding well, she may produce 1 or more egg sacs. Each of these sacs is a hard sphere about the size of a small pea. Perhaps the strength and rigidity helps to protect the contents from desiccation as well as from parasitic wasps and other predators. Several may festoon a bush or tree, almost resembling tiny tan Christmas tree ornaments suspended in the twigs. When the spiderlings are ready to emerge from the egg sac, they chew a single exit hole. The male bolas spiders emerge as penultimate (almost mature) or as fully mature spiders. These males require only 2 molts from the time they hatch from the egg to the time they are mature, and both of these molts may take place while the spider is still living within the egg sac. Consequently, the mature male bolas spider is quite tiny, only about 1.5 millimeters in body length. No other spiders are known to mature in so few molts. In contrast, the female bolas spider eventually grows to the size of a pea, and as she sits still among vegetation, she closely resembles a bird dropping.

Bolas spiders in the arid southwestern states are extremely difficult to find, although they do occur. One factor limiting their abundance is that the adult males and the juvenile females feed exclusively on moth flies (family Psychodidae). Each species of bolas spider preys on a particular species of moth fly, attracting the fly with pheromones. The larvae of these flies require wet, mucky areas. The scarcity of these wet areas in the arid southwest makes this specialized spider an even greater rarity.

The risk of specialization is clear if populations of bolas spiders are compared with populations of a generalist orb weaver such as *Neoscona oaxacensis*. *Neoscona* are highly successful, judging by their abundance. Building an orb web requires a greater initial investment, but a generalist can then exploit a wide range of potential prey species. In contrast, the bolas spider requires a smaller initial investment in silk, but at a cost of reducing variety in its available prey. Ultimately, the highly specialized bolas spider may disappear, unable to adapt to changes in the environment.

Tetragnathidae: Long-jawed Orb Weavers

The orb web that tetragnathids build is similar in structure and form to araneid webs but is usually on a plane that is between vertical and horizontal (instead of the vertical orb webs of araneids). Long-jawed orb weavers favor grassy meadows and moist habitats, and are common along streams in desert canyons. Frequently their webs actually overhang a small stream, where gnats and mosquitoes are abundant. Perhaps the angle of the web facilitates the capture of these small insects as they fly over a stream. Unlike other orb weavers, tetragnathids can easily walk on the surface of the water. In fact, they move more rapidly on water than they do on land.

In situations in which there is an extreme abundance of prey, this group of spiders may be the foundation for the formation of a giant, communal web. These webs are truly amazing in their dimensions, sometimes hundreds of feet in length and covering several adjacent trees. Thousands if not millions of spiders may inhabit this canopy of silk, each spider contributing its own small web to the composite whole. It is noteworthy that in such high densities, the long-jawed orb weavers abandon their usual orb design, instead making irregular webs. This modification suggests an unusual degree of plasticity of behavior which may be unexpected in arachnids. Despite the smaller individual webs, each spider succeeds in obtaining enough food and is therefore tolerant of its neighbors and of the crowded conditions. Other species of spiders may also form a part of the giant web.

In 2007, a giant web of this type formed in Texas near Lake Tawakoni after heavy rains had produced a bumper crop of mosquitoes. More than a dozen species of spiders inhabited the web, 60 percent of which were

1. *Tetragnatha versicolor*.

2. *Tetragnatha laboriosa*, presumptive.

3. *Tetragnatha laboriosa*. In a contest between the sexes, the male and female long-jawed orb weavers lock their chelicerae together while mating. This probably reduces sexual cannibalism, at least for as long as the fangs are held immobile. Once mating is completed, the male releases the female's chelicerae and makes a hasty escape.

4. The male tetragnathid has long chelicerae and huge fangs. Each chelicera has a prominent spine, which assists the male in locking his "jaws" against the female's chelicerae. Long pedipalps are required to reach the female's genital opening in order to transfer the sperm.

long-jawed orb weavers, followed by jumping spiders (18.4 percent) and araneid orb weavers (7.6 percent). In addition, funnel-web spiders (*Agelenopsis*) and cobweb spiders (Theridiidae), including kleptoparasitic dewdrop spiders (*Argyrodes*) and the social spider *Anelosimus studiosus*, were living in the giant web. The giant web may persist as long as the food supply remains abundant and as long as a storm does not destroy the canopy of silk. One can only guess at the millions of mosquitoes and gnats consumed by such a legion of spiders, to the benefit of their human neighbors. Such webs are rare occurrences but have been documented in Texas, California, and other parts of the world where sudden blooms of mosquitoes follow heavy rains.

The name "long-jawed orb weaver" is well chosen, owing to the fact that these spiders possess very large chelicerae and fangs. During courtship and mating, each spider pushes its chelicerae against the chelicerae of its mate, and the fangs interlock over the ends of the chelicerae in a sort of "spider kiss." The male also has pointed, small "spurs" on his chelicerae that assist him in holding the female's jaws open. As long as this locked position is maintained, the safety of the male continues. Once mating is completed and he releases his hold on the female's chelicerae, he makes a quick getaway from his mate. Perhaps this practice of a "spider kiss" has selected for the very large chelicerae and fangs seen in this genus of spiders, creating a sort of arms race between the sexes. Only a strong male with relatively large chelicerae and fangs can hold his own against the larger female and succeed in mating with her.

Uloboridae

Although the uloborids utilize a web design similar to that of the araneids and tetragnathids, the uloborid's web differs in several important respects. First, it is generally oriented on a horizontal plane instead of the vertical or intermediate orientation of the araneid and tetragnathid webs. Second, instead of being chemically sticky (via tiny drops of glue), it owes its "stickiness" to the physical structure of the silk, called cribellate silk. The cribellum is a platelike structure just anterior to the other silk spinnerets. It contains hundreds, if not thousands, of microscopic silk spigots. Together they produce a silk very like a strand of yarn, made up of many finer threads. The uloborid spider roughs up this woolly silk using a comblike structure on the hind

Uloborus diversus. Spinning its prey rotisserie-style, an uloborid tightly wraps a fruit fly in silk. Spiders in the family Uloboridae are unique in that they completely lack venom; therefore, they must immobilize their prey with silk

leg called the calamistrum. When a fly stumbles into the web, the tiny hairs on its leg become caught in the tangles of this Velcro-like silk, and the spider can then capture the fly. An advantage of the cribellate silk over the chemically sticky araneid silk is that it does not require humidity and moisture to remain effective. Therefore, a single web can last for days or even weeks without the necessity of daily reconstruction.

Uloborid spiders are unique in that they completely lack venom. The ability to produce venom was lost by this family somewhere along the evolutionary path; perhaps the cost of producing the complex cocktail of proteins that make up venom outweighed the cost of simply wrapping prey in silk. Consequently, the only way that an uloborid spider can subdue prey is to wrap it in silk, using a combination of thin and thick silk. The spider first throws silk over the prey while still a short distance from it, and then proceeds to rapidly spin the prey rotisserie-style while wrapping more silk around it, until the captured insect is properly and completely swaddled. In fact, the spider wraps its prey so tightly that the cuticle of the prey is disrupted, essentially squashing the prey. The uloborid spider then regurgitates digestive enzymes directly onto the entire package. Both the swaddled prey as well as the thinner wrapping silk are liquefied and ingested by the spider. The thicker wrapping silk is not affected by these enzymes and remains wrapped around the food package, acting as a filter. Because the uloborid spider has no venom, a single spider is limited to relatively small prey. Perhaps this may in part explain why members of this group of spiders are sometimes communal. The collective efforts of several spiders may be able to subdue larger prey than one spider alone could handle.

The genus *Philoponella* (in the family Uloboridae) includes several communal and facultatively communal species. Facultatively communal species can either live alone or may live in a colony of interconnected webs. Each spider in the colony constructs its own orb web, which shares boundaries with other orb webs. There may be multiple "stories" of these webs, resulting in a colony of dozens of individuals. The individual webs are defended by their occupants, but in some species cooperative killing of prey does occur. *Philoponella oweni* is a facultatively communal spider found in the mountains of southern Arizona. So far, cooperative killing of prey has not been observed in this species. However, since progressively more species of *Philoponella* have been documented practicing this behavior, one wonders whether *P. oweni* might not join their ranks at some point in the future.

Aerial web construction may have evolved in response to insects taking to the air during the Carboniferous Period, 360 to 290 million years ago. The explosive adaptive radiation of flying insects opened up new niches for predators. Unfortunately, the delicate webs of spiders leave little in the way of fossil evidence to see exactly when they did appear, so much is still conjecture. Millions of years of evolution have tested and perfected this snare trap for flying insects and given us both a practical design and an object of beauty: the orb web.

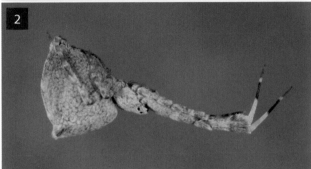

1 and 2. *Uloborus diversus* has several color morphs, including dark morphs and light morphs. However, the shape of the body, the "feathered" legs, and the cribellate orb web make this species easy to recognize.
The center of its orb web frequently has a thick ribbon of cribellate silk in the middle.

3. *Philoponella oweni* female with egg sac. This is a facultatively communal orb weaver found at higher elevations in the mountains of southern Arizona.

CHAPTER 14 Irregular Web Builders:
Theridiidae, Nesticidae, Leptonetidae, Diguetidae, Dictynidae, and Pholcidae

As a warning to any predator, a western black widow, *Latrodectus hesperus*, displays the distinctive red hourglass on its belly.

Theridiidae

Imagine an ant as she ventures off the beaten path scouting for food. In her exploration, she does not see several strands of silk anchored to the ground and stretching upward as they connect to an irregular spider web. She chances to brush up against one of the silken threads and is instantly stuck to the glue coating the silk. She struggles to free herself—and in doing so breaks the silk's connection with the ground. She is swiftly lifted off the ground as though she were on a bungee cord as the silk springs upward. Dangling helplessly and unable to gain any traction to pull herself free, she continues to struggle. But these efforts to escape have only succeeded in alerting the web's maker of her presence. A half-grown black widow, still brown and white in coloration, investigates the disturbance. She has captured many other ants from this colony over the past few weeks and is familiar with the pattern of vibration. She nimbly darts down the silk, kills the ant, and transports it up to her hiding place at the top of the web, where she can feed on it in privacy.

Black widows belong to the family Theridiidae, the comb-footed spiders. The hind legs have tiny, bristlelike setae that roughly resemble a comb. This structure helps the spider to manipulate the threads of silk as she wraps her prey. Black widow webs are fairly representative of most theridiid webs. The webs appear untidy, but they are actually well constructed. In fact, it is now believed that theridiid webs are derived from orb webs, not the other way around. The strong, sticky silk is attached under tension to the substrate. As soon as the connection with the substrate is broken, the elastic silk springs upward. The irregular webs are specialized for capturing arthropods that travel on the ground. Grasshoppers, crickets, beetles, other spiders, and even scorpions are captured. Ants, which are rarely taken by most other predators, are among the most common prey of theridiids. Black widows have perfected the capture of scorpions, including the bark scorpion *Centruroides sculpturatus*. As soon as a scorpion encounters the sticky silk of the black widow, the spider rushes down and selectively immobilizes the claws and the metasoma (tail) of the scorpion, wrapping silk around these weapons before delivering the killing bite to the trussed and helpless scorpion. Interestingly, in a contest between these two highly venomous creatures (*Centruroides* and *Latrodectus*),

the black widow always wins if she is on her home territory, the web.

The venom of the black widow is dangerous to humans. This venom affects the nervous system by causing the massive release and subsequent depletion of neurotransmitters at the nerve and muscle junction. This may produce symptoms of extreme muscle cramping, nausea, headache, vomiting, and possible respiratory failure. Oddly enough, among the more severely affected people may be very muscular, physically fit individuals, because of the severity of the muscle contractions. Hospitalization and administration of antivenin may be necessary. Fortunately, the black widow is not normally aggressive toward humans unless she is guarding her egg sacs.

The black widow has a slightly undeserved reputation as the ultimate femme fatale. Popular myth claims that she always kills and eats her spouse immediately after mating. Although this does occasionally occur, it seems to be the exception rather than the rule for western black widows. Even though the female black widow may not fully deserve her notoriety, the male *Latrodectus* is understandably cautious during courtship. He leaves his own web after his final molt, at which time he is fully mature. (In an interesting twist, male western black widows retain the juvenile coloration, in contrast to many other animals in which the adult females retain juvenile coloration.) He wanders in search of a female's web, which he recognizes by chemical pheromones in her silk that tell him not only that a black widow is present, but also the gender and maturity of the occupant. As he cautiously explores the web, he taps and plucks the silk with his front feet, which have rows of chemosensory hairs on them. At the same time, he sends a message to the resident female identifying himself as a suitor by vibrating his abdomen at a high frequency. Vibration of the abdomen is commonly used in spider communication. Theridiids may also be stridulating during this vibration, with the file and scraper located at the rear of the cephalothorax and the front of the abdomen. The message is transmitted to the female via the silk; she in turn may send a reply by vibrating her abdomen. Once the male has reached the vicinity of the female, he carefully snips most of the silk surrounding her, cutting off potential escape routes. With these preliminaries over, he starts to stroke first her legs, then her abdomen. He may wrap a bit of silk around her body, forming a sort of "bridal veil" before mating with her. This bondage silk could easily be broken by the large female, but she is passive. Perhaps there are male pheromones in his silk that help to inhibit her predatory response. Actual mating may take from 10 minutes to 2 hours.

After mating, the female produces several egg sacs a week or two apart near the retreat section of her web. Each egg sac contains from 125 to almost 400 eggs. As new egg sacs are added, she moves them all in close proximity to each other, leaving none behind. This may indicate that the female black widow may have a rough concept of numbers, at least as far as it concerns her egg sacs. A vigilant parent, she rushes to protect the eggs if there is any disturbance, touching the sacs with her front legs as though reassuring herself. Despite her best efforts, sometimes a tiny parasitoid destroys her eggs. Black widow egg sacs are sought out by a fly, *Pseudogaurex signatus*. Although the fly deposits her eggs on the exterior of the egg sac, the newly hatched maggots push their way through the silk and feed on the spider eggs. These flies even pupate within the egg sac before pushing their way out again as mature flies. Another egg parasitoid is a tiny wasp, *Baeus lactrodecti*. Less than a millimeter in length, the wingless female wasp must find a black widow egg case and chew an opening in the silk, through which she can gain entrance to the interior. She then proceeds to lay an egg on each spider egg. A single tiny wasp can destroy the entire contents of an egg sac. The female spider eventually discards the "dud" egg sacs, cutting them loose and allowing them to fall to the ground.

Occasionally the female black widow may kill and eat her mate. But frequently, the male black widow stays on the female's web, basically living the life of a kleptoparasite. Perhaps the tolerance which the female black widow shows the male serves as a preadaptation for other spiders to live in close association to her web, such as cellar spiders or the tiny thief, *Argyrodes*.

The name *Argyrodes* is derived from the Latin word for silver, *argentum*. Most of these tiny spiders are only a few millimeters long and have some silvery color on their abdomens, thus giving them the charming name "dewdrop spider." At rest, they tuck their legs close to their bodies, resembling a tiny, triangular bit of detritus in the web. As a group, *Argyrodes* are famous for inhabiting other, far larger spiders' webs, especially those of orb weavers and black widows. There, they

A half-grown western black widow, *Latrodectus hesperus*, has captured a snap-jawed ant, *Odontomachus clarus*. Bent down across its thorax, the head of the ant is larger than the whole body of the spider. Despite this size difference, the ant was no match for the widow.

live as kleptoparasites, pilfering food caught in their host's web. Some species of *Argyrodes* are gregarious, with several individuals inhabiting a single host web, especially if the host is a more aggressive species of spider. The presence of multiple spiders may confuse the host, as vibratory signals may come from several sections of the web. In one species of gregarious *Argyrodes*, one spider may distract the host while another steals the food. Some tropical species of *Argyrodes* have fine-tuned their theft technique. First, they move slowly within the orb web of their host, avoiding sudden vibrations that can draw the attention of their host. Then, while cutting prey from the web prior to feeding on it, they hold the radial lines of silk with their legs, preventing the sudden snap from the released tension. Before releasing the cut radial line, the *Argyrodes* stretches out its legs so as to very gently and gradually release the tension of the silk. The theft is thus imperceptible to the host spider. These little spiders have perfected the art of the sneak-thief.

In southern Arizona, *Argyrodes pluto* favors the webs of female black widows. Several may be found in a single web. Even when they are tiny spiderlings, they must find a host web to live in, since they never live independently.

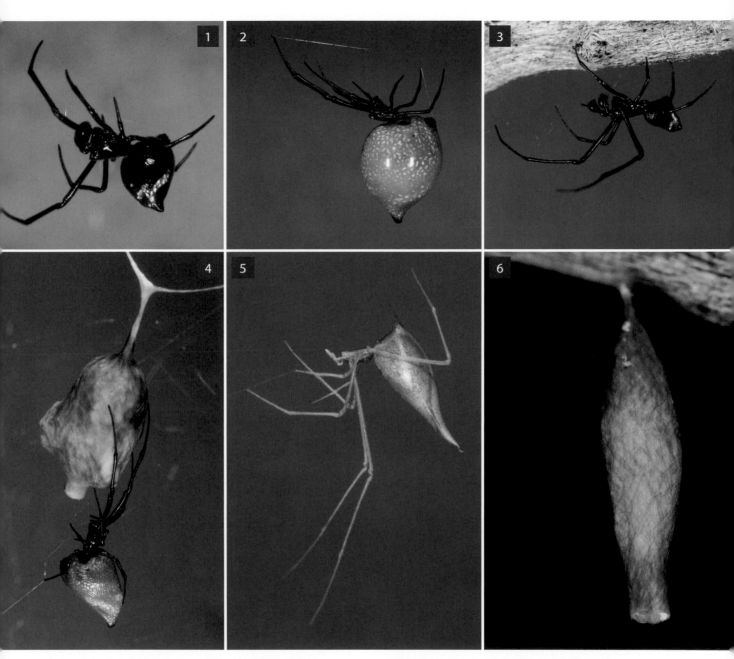

1. The tiny thief *Argyrodes pluto* displays a silvery band on its abdomen. Almost all spiders in the genus *Argyrodes* have some silvery color, thus giving them a name derived from the Latin word for silver, *argentum*. They are also known as dewdrop spiders.

2. A gravid *Argyrodes pluto* is so swollen with eggs that she resembles a tiny, sparkling Christmas tree ornament.

3. The mature male *Argyrodes pluto*.

4. Exhausted from her efforts, a female *Argyrodes pluto* rests after laying her eggs. The silken egg sac has a shape that is characteristic for each species of *Argyrodes*. When the spiderlings emerge, they must locate an actively occupied black widow web in order to survive. *Argyrodes* are obligate parasites in other spiders' webs.

5. *Rhomphaea fictilium* is closely related to the parasitic *Argyrodes* but is specialized in capturing and killing web-based spiders.

6. *Rhomphaea fictilium* egg sac.

Thus, they are obligate parasites in the webs of their host.

Kleptoparasitism may have evolved from a neonatal extension of juvenile behavior. Juvenile spiders may show greater tolerance of peers and may share a web. Tolerance of the kleptoparasite by the host spider may have its roots in maternal care, since mother spiders may tolerate their own offspring living in their web. There may have been selection for *Argyrodes* to behave in a juvenile manner and stay small (like a juvenile of the host spider).

One type of spider that used to be classified as *Argyrodes* has been moved into its own genus: *Rhomphea*. This spider superficially resembles *Argyrodes*, but it has a strange little pointed "tail" (of unknown function) at the tip of its abdomen. It has taken the ability to sneak into another spider's web to another level. No longer content to steal the occupant's food, it kills and eats the actual host spider itself. *Rhomphea* slips into the web of its victim, and once it is close enough, throws some silk over its target in order to capture it. Both *Rhomphea* and *Argyrodes* spin egg sacs shaped like inverted vases; the "mouth" of the vase forms the opening from which the spiderlings emerge.

Black widows are adept at capturing ants, but the real specialist in ant predation belongs to another member of the theridiid family: spiders in the genus *Euryopis*. Relatively few creatures prey on ants. Ants are formidable opponents, armed with strong jaws and the advantage of vastly superior numbers. Some may also be able to sting or may have spiny armor as added protection. Several *Euryopis* species in North America have developed special tactics for capturing particular types of ants. In Wyoming and Idaho, one species of *Euryopis* takes cover behind pebbles on harvester ant mounds, darting out to ambush and kill these very large and imposing ants, which are then dragged off attached to the silk spinnerets of the spider. Sometimes 10 or more spiders may be simultaneously hunting on a single ant mound, each picking off its own victim. Other species of *Euryopis* may hunt along a pheromone trail of small ants, stashing the captures until a sufficiently large pile has been accumulated for a feast. In the Arizona desert, *Euryopis* may be observed hunting small ants on barrel cacti in this way. *Euryopis scriptipes* captures large harvester ants, *Pogonomyrmex rugosus*, near the ant nest, but usually ascends some tall grass before feeding on its prey. Some *Euryopis* are nocturnal; some are diurnal. Some hunt on the ground,

and some set snare traps with sticky silk much like their distant cousin the black widow.

Many theridiids have some degree of sociality. In some species of theridiids, maternal care may include sending signals through the silk, warning the babies to stay away if dangerous prey is in the web, and subsequent "invitational" signals once the prey has been immobilized and is safe for the babies to approach. In some species, the young may even solicit food from their mother by begging, touching her mouth, palps, and legs until she regurgitates a meal for them. Both *Theridion cochise* and *Hentziectypus schullei* demonstrate maternal care of their young by sharing food with their offspring, and their young do cooperatively band together to some degree in order to kill prey. But perhaps the most notable example of sociality in a North American spider is found in *Anelosimus arizona*. This little brown spider lives in southern Arizona at higher elevations near mountain streams where small insect prey are abundant. A single adult female starts the colony. She shares food with her babies, and as they grow in size, the young spiders cooperatively kill prey and feed together. The web is a three-dimensional network of silk located in a tree or a bush. It can be anywhere in size from a fist to a football, depending on how many spiders are living in the colony and how large they are. As the web ages, dead leaves and other detritus may become entangled in the silk on the perimeter of the colony, giving it a characteristic untidy appearance. As soon as an insect (especially a fly) lands on the silk, it elicits a response from the nearest spider, and soon more and more spiders come darting out from their hiding places in the web and rush over to tackle the prey. It would appear that the presence of prey is communicated to the spiders of the colony through vibrations carried by the silk of the web. Sometimes a fly may be completely obscured by the mass of spiders attacking it. Once the prey has been subdued or killed, the spider siblings share the meal, any number of them feeding simultaneously.

For the subsocial (also known as periodic-social) spider *A. arizona*, reoccupied nests tend to have a greater success rate than do new nests, suggesting that staying in a proven good location is advantageous over trying other, more chancy new locations. Perhaps if prey availability were greater, the subsocial spiders of Arizona would become social (permanent-social) colonies.

Small but formidable, the ant hunter *Euryopis scriptipes* has captured the dangerous harvester ant *Pogonomyrmex rugosus*, which is roughly twice the spider's size. *Euryopis* is a solitary hunter.

1. Working together, these sibling *Anelosimus arizona* spiders have succeeded in overcoming a large fly.

All the spiders in this colony are offspring of a single female. Once these offspring are close to maturity, they will disperse.

Mating after dispersal ensures that outcrossing occurs, thus maintaining genetic variability.

2. A tiny kleptoparasite lives in the webs of *Anelosimus arizona*: the mirid bug *Ranzovius clavicornis*. Several individuals of this species can be found in any one web of the subsocial spider. Both the nymphs (as shown here) and the adults feed on prey captured by the spiders. Sometimes the bugs do not even wait for the spiders to finish before they start to feed on the prey.

It is not known whether this species of bug is an obligate kleptoparasite. It would be interesting to see if web-based social spiders in the tropics have any comparable kleptoparasitic bugs living in their webs.

In an interesting side note, a small mirid bug (*Ranzovius clavicornis*) has demonstrated a kleptoparasitic existence in the webs of *Anelosimus arizona*. Both nymphs and adults of this species of bug were observed to gather near captured prey, and would sometimes commence feeding on the prey even before the spiders were finished feeding themselves. As many as half a dozen nymphs of this mirid bug were observed in a single web, and almost no colonies lacked at least a couple of these bugs.

An average of 25 spiders live in each colony. As the siblings approach maturity, each goes its separate way. Consequently, this species is categorized as being subsocial (or periodic-social), as opposed to the fully social (or permanent-social) species, such as *Anelosimus eximius* native to Central and South America. Permanent-social species live together in a colony for multiple generations. Prey acquisition, web maintenance, and raising of young are all done communally in *Anelosimus eximius*.

Comparing the permanent-social spider (*Anelosimus eximius*) with the periodic-social spider (*Anelosimus arizona*) sheds light on the cost/benefit of dispersal versus staying within the natal nest. The principal cost of dispersal is increased mortality, resulting in fewer females successfully producing offspring. The principal risk of remaining within the natal nest is inbreeding, with a potential decrease in genetic fitness; however, resource competition within the natal nest may necessitate dispersal if food is not abundant enough to support the growth of the increasingly large spiders. It would appear that in temperate climates, the lower abundance of food favors dispersal. In one study, 41 percent of *A. arizona* females survived to produce successful nests. Virtually all these spiders dispersed from their natal nest. (A few did reoccupy their birth nest, but only as a single spider.) In contrast, less than 10 percent of solitary *A. eximius* females produced successful nests, suggesting that a solitary lifestyle may be evolutionarily advantageous in temperate climates but less so in the tropics; however, species of permanent-social spiders exist in apparent good health, despite inbreeding through multiple generations. Perhaps for a species, the dangers of inbreeding can be survived as long as the most damaging recessive alleles are eliminated from the gene pool. But in the long run, the loss of genetic variation traps the fully social species. They do not have the genetic raw material to change and adapt, should conditions change.

The theridiids may not be the most spectacular spiders in their appearance, but they are among the most intriguing family of spiders in regard to their diversity of lifestyles. From the notorious black widow to the tiny kleptoparasitic thief *Argyrodes*, from the ant hunter *Euryopis* to the array of subsocial and social spiders, this group presents endless challenges to our preconceptions and stereotypes about spiders.

Nesticidae: The Cave Cobweb Spiders

Spiders in the family Nesticidae share a great many characteristics with spiders from their sister family Theridiidae. Their body morphology and eye arrangement are almost identical. Members of both families have a "comb" of specialized setae on each fourth leg tarsus that assists the spider in manipulating silk. Perhaps most striking, many spiders in both families utilize similar web construction and capture strategies. This capture web consists of an irregular, three-dimensional network of intersecting sticky silk lines, some of which descend vertically and are attached to the substrate. If a walking insect such as an ant or a cricket brushes against one of these lines attached to the ground, it becomes stuck to the silk. As it struggles to free itself, the connection with the ground is broken, and because the line was under tension, it springs upward while still holding the insect. These traps are referred to as gumfoot webs. Theridiid webs have a simple, single line of silk making up each individual snare, whereas nesticid webs have bifurcated (split) lines leading to the substrate.

Both families prefer to build their webs in quiet, dark locations. But nesticids are a bit more delicate than theridiids and therefore require cooler, moister environments. In the arid southwestern United States, this requirement restricts nesticids primarily to caves, where the temperature and humidity remain almost constant. *Eidmannella pallida* is a species found throughout much of the United States, but in the arid deserts of southern Arizona it is restricted to only a few caves. Although some species of nesticids are surface dwellers living under stones, in leaf litter, or in the shade and protection of rocky canyon walls, a large number of

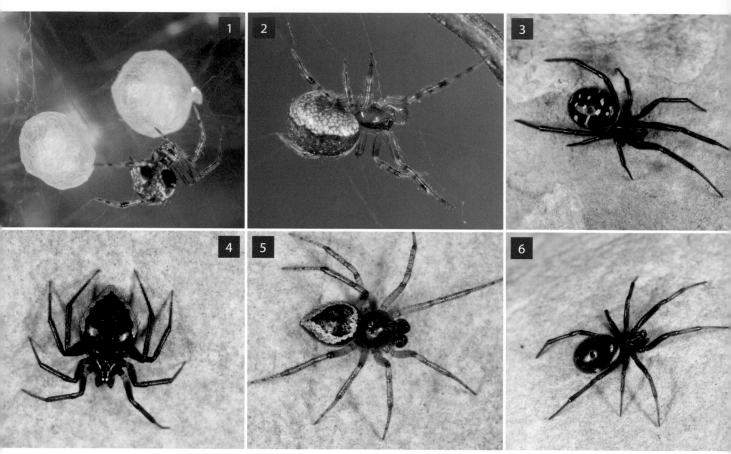

1. Easily overlooked due to its small size, an adult female *Hentziectypus schullei* may be only about a tenth of an inch (3 mm) in body length. She builds her tiny webs under rocks or other debris. This female was found in the twilight zone of a cave. Each egg sac contains only a few eggs. If the egg sac is built close enough to the substrate, the female spider will cover the outside of the egg sac with bits of debris from the ground. Most spiders in this genus are found in Central America.

2. *Anelosimus arizona* is a subsocial species found along riparian areas in the mountains of southern Arizona.

3. *Steatoda punctulata* is an extremely common species in the hot, lower-elevation deserts of the arid southwest. These small spiders build webs under rocks.

4. *Euryopis* near *funibris*. This attractive species has been found in southern Arizona in the lower-elevation deserts.

5. *Euryopis formosa* is found in cooler habitats along the California coast. It can also be found along riparian areas in the higher elevations of the mountains of southern Arizona.

6. *Asagena fulva* was previously known as *Steatoda fulva*. This small species, measuring 0.1–0.25 inches (2.5–6 mm) in body length, is found principally in the arid southwestern United States and in Mexico. The markings on the dorsal surface of the abdomen can be somewhat variable. Like *Steotoda*, this species builds webs under rocks or other sheltered locations. *Asagena fulva* has been documented as capturing harvester ants (*Pogonomyrmex*).

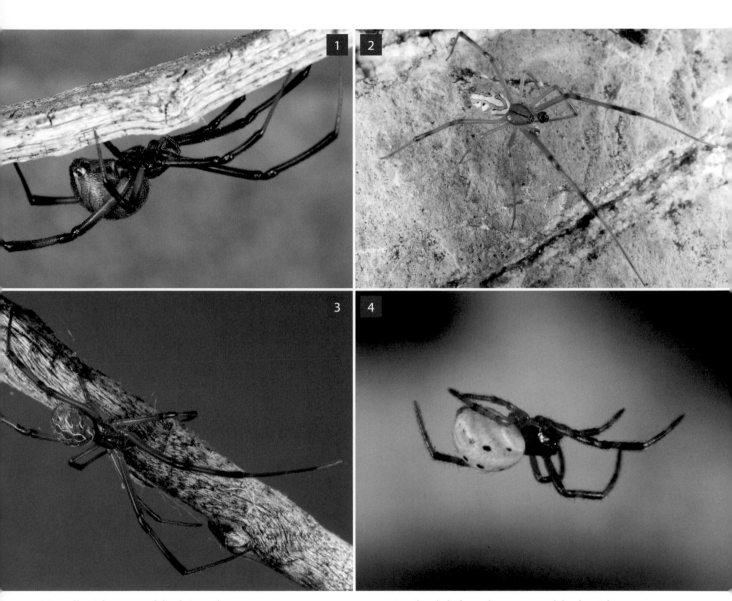

Not all widows are black in coloration.

1. A female brown widow, *Latrodectus geometricus*. This non-native species is associated with human habitation. Although both the brown widow and the black widow have potentially dangerous venom, reports of serious bites from the brown widow are extremely rare.

2. A mature male western black widow, *Latrodectus hesperus*. This species can be somewhat variable in coloration, but the mature males retain the coloration seen in juveniles.

3. A subadult female western black widow, *Latrodectus hesperus*. Within a couple of weeks, this individual molted into a mature, brown-black female, and the only red remaining was the hourglass shape on her abdomen.

4. A tiny spiderling of the western black widow, *Latrodectus hesperus*, has just emerged from the egg sac. The babies of this species immediately disperse, and each tiny spider is self-sufficient, armed with powerful venom.

The western black widow, *Latrodectus hesperus*, is able to capture even the highly venomous bark scorpion, *Centruroides sculpturatus*.

1. This female black widow rapidly wraps the metasoma ("tail") and the claws of the scorpion, thus immobilizing the most dangerous weapons of its prey. This process takes only a few seconds for the spider to accomplish.

2. Once the hapless scorpion is trussed and no longer a threat, the black widow administers the killing bite. A resident male black widow (seen as the smaller, lighter-colored spider) quickly rushes over and attempts to feed on the scorpion. In this instance, the female spider repeatedly chased the male away from her prize and, after carefully wrapping it in silk, laboriously transported the scorpion up to a sheltered area behind a porch light in which to feed in peace. Male western black widows can frequently be found in the webs of mature females, sometimes behaving as kleptoparasites.

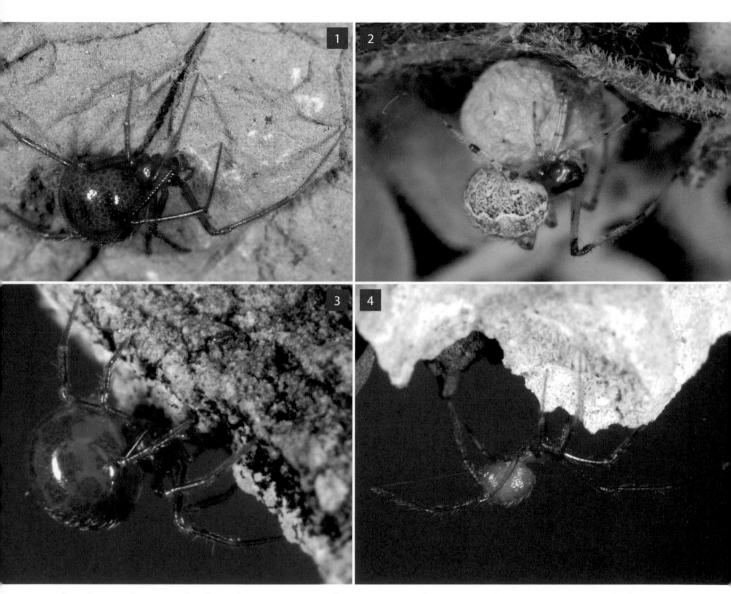

1. *Theridion cochise* may be found in caves in southern Arizona. This species demonstrates extended maternal care of young, and the young may cooperatively kill prey.

2. Only about a tenth of an inch (3 mm) in body length, this tiny *Theridion* protectively clutches its egg sac. The genus *Theridion* contains dozens of species that may be highly variable in coloration; therefore, they can be identified to species only with difficulty. This particular *Theridion* is commonly found in low-elevation desert trees such as mesquite.

3. Rarely photographed due to its diminutive size, *Thymoites maderae* builds its web under rocks where small species of ants have colonies. To the unaided eye, this spider appears to be a tiny bead of amber. It lives at higher elevations in the mountains of southern Arizona.

4. *Thymoites minero* was first collected from a mine shaft: hence its species name *minero*. However, this pale, delicate spider also inhabits cool, humid caves in the mountains of southern Arizona.

species are strictly cave dwellers. These include several endemic *Eidmannella* species limited to caves in Texas, and a diversity of *Nesticus* species found only in Appalachian caves. These troglobitic species frequently evolve troglomorphic physical characteristics, such as a reduction or complete loss of eyes as well as loss of pigment. Both extrinsic and intrinsic factors may limit movement between populations. The extrinsic factors include geological features that act as physical barriers, and intrinsic factors include narrow physiological limits of species, such as a requirement for a humid environment. In combination, these factors may isolate populations, thus preventing genetic exchange between populations. Consequently, new species evolve. The diversity of species may be greater than what is superficially apparent from only morphological data. Some populations that appear morphologically indistinguishable may actually contain a number of genetically different species. These are referred to as cryptic species.

Not much is known about the courtship and reproduction of nesticids, partly because of the difficulty in finding mature males (which probably have shorter lives and do not stay in a web). However, it is known that the mother spider carries her egg sac either with her chelicerae or attached to her spinnerets. (In contrast, theridiids attach their egg sacs to their webs, sometimes tucked away in a sheltered retreat.) Like that of so many other small, unobtrusive species, much of the natural history of nesticids remains to be discovered.

Leptonetidae: The Midget Cave Spiders

Building their small webs under rotting logs, in leaf litter, or in caves, leptonetids prefer dark, moist microhabitats. These tiny, pale spiders almost certainly require humidity in order to survive. With a body reaching only 0.04 to 0.12 inches (1–3 mm) in length, and possessing long, elegant iridescent legs, this delicate little spider would rapidly desiccate in a dry environment given its extremely high ratio of surface area to body mass. Consequently, in arid climates such as Arizona, these spiders are restricted to oak and sycamore leaf litter in riparian areas or to caves. In cooler, moister climates, they can be found in forests and foothills along the Pacific coast of Oregon and

northern California. The members of this family appear to have a disconnected distribution, being widespread in California but found only in isolated localities in Arizona, Texas, southern Mexico, and Panama, as well as in the southern Appalachian Mountains of the southeastern United States. They are also found in caves in the Mediterranean area and in eastern Asia.

The two sister families Leptonetidae and Telemidae share many characteristics. Both contain tiny, long-legged species that live in dark moist habitats. In fact, telemids are predominant in caves along the Pacific coast. Spiders in both families either have 6 eyes or may have no eyes in troglobitic, cave-restricted species. Finally, spiders from these two families have a series of unusual cuticular plates on the tibia and/or the patella of their legs. These plates contain the openings to glands that secrete a chemical with a repugnant smell. If the spider feels threatened, it drops to the ground, lying belly-up and with its legs folded up over its body. In this position, the glands (called Emerit's glands) may be able to protect the otherwise completely defenseless spider with a chemical repellant.

Like other long-legged web spiders, leptonetids hang suspended beneath their small webs. The male initiates courtship by plucking at the female's web, after which he slowly approaches and then mates with her. The male remains in the female's web for several days after mating. This behavior is seen in many species of web spiders and is probably a form of mate guarding, whereby a male tries to prevent rival males from mating with the female. Some species in this family produce a small hanging egg sac that has dirt or debris attached to the outer layer of silk, and other species produce a disk-shaped egg sac that is attached to the underside of a rock.

An assemblage of distinct, isolated species is found in caves. Species of leptonetids as well as spiders in the genus *Cicurina* (Dictynidae) and a variety of harvestmen, including *Sitalcina* and *Texella* (Phalangodidae), may contribute to this assemblage. Caves may be grouped into karst faunal regions comprising geologically distinct regions containing these endemic, troglobitic species. In some cases, the divergence of species can be correlated with geological characteristics such as faulting in the underlying rock. Some of these cave species are highly restricted, to such a degree that some species are known from only a single cave. The preservation of these vulnerable species can succeed only with the preservation of their cave habitats.

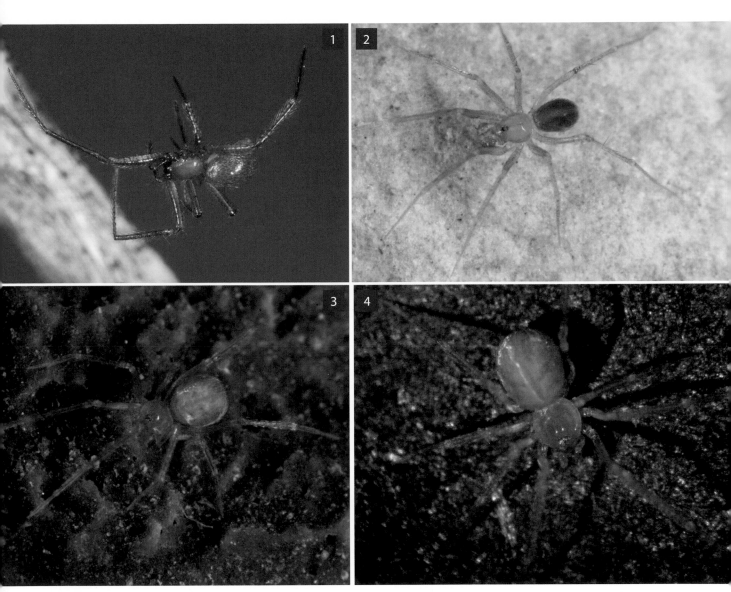

1. *Eidmannella pallida* is almost identical in appearance to *Thymoites minero*. However, this species belongs to the family Nesticidae instead of to the family Theridiidae. This species has a wider distribution, being found in humid environments in much of the United States as well as into Mexico. In the arid deserts of southern Arizona, *Eidmannella pallida* is restricted to caves such as Kartchner Caverns, where it is protected from desiccation.

2. Tiny and delicate, this *Ozarkia* species is found under rocks in forests and along riparian areas where it is shady and moist. This particular species occurs in the oak zone of the Santa Rita Mountains of southern Arizona.

3 and 4. Shimmering like a tiny opal, this undescribed species of *Darkoneta* has been found in only one small cave in southern Arizona. This minuscule spider is only about one 25th of an inch (1 mm) in body length and almost completely lacking in pigment. *Ozarkia* and *Darkoneta* each possess only 6 eyes; both belong to the family Leptonetidae.

Spiders in the genus *Dictyna* specialize in capturing flies. Their tiny webs are built where flies perch or where flies bask in the sun, such as on the tips of vegetation or on rock surfaces. The web is constructed of cribellate silk, which is mechanically "sticky." Unlike garden orb weaver webs, which owe their stickiness to moist chemical glue, cribellate webs can continue to capture prey for a long time even in very dry conditions. Consequently, these diminutive spiders are abundant even in the arid deserts of the southwestern United States. The same web is shown top and bottom.

Dictynidae: The Mesh Weavers

Although easily overlooked, dictynid spiders are truly ubiquitous in many environments. Their tiny cribellate webs are commonplace on the tips of grass stalks, on the ends of twigs, and on irregular rock surfaces. One species, *Dictyna calcarata*, may even form an entire colony along a sunny windowsill inside a house. These diminutive spiders can be seen only by carefully examining their webs. Tucked behind a small leaf, a flower, or a bit of debris in the web, the dictynid remains hidden from all but the most persistent searcher. This may be necessary for its survival. Hummingbirds are experts at capturing small spiders such as dictynids, which provide an important source of protein for both the adult birds and their chicks.

Other dictynids are even less conspicuous. The ecribellate members of this family live under rocks, in leaf litter, or under debris on the ground. Blending in with the colors of dead leaves and the soil, these small spiders do not even construct a cribellate web that would betray their presence. These delicate little ground dwellers require some moisture; therefore, in the lower desert elevations they are frequently restricted to riparian areas or to caves. In fact, dictynids in the genus *Cicurina* are not only commonly found in caves, but some have also become troglomorphic, losing both their pigment as well as their eyes. Some of the cave-adapted *Cicurina* also mature more slowly than their surface relatives, taking two years instead of one to reach maturity. A slower growth rate and a lower metabolism would confer a survival advantage in an environment such as a cave, which has so little available prey. Some caves may even possess more than one population of *Cicurina*. Spiders living in the "twilight zone" near the entrance may still have both pigment and eyes, while other *Cicurina* living deeper in the cave are true troglomorphs, pale and completely lacking eyes.

Among the cribellate dictynids are spiders in the genus *Dictyna*, *Mallos*, and *Mexitlia*. The webs of these spiders are irregular in structure, made of cribellate silk. Each cribellate silk strand is composed of hundreds of microscopic threads "roughed up" by the comblike calamistrum on the dictynid's fourth leg. This woolly silk mechanically catches the legs of small flies, rather than requiring chemical glues to make the silk sticky. Because they do not require moist sticky glue on their silk, these webs can effectively capture prey over a period

1. A typical *Dictyna* web at the tip of a stalk of grass.

2. A closer view of the same web reveals a pair of adult *Dictyna* sharing the web. A mated pair of *Dictyna* may cooperatively kill and share prey.

3. Extremely tiny, *Tivyna moaba* is a desert species.

4. *Mallos dugesi* female. This species frequently builds its web across a single leaf and then waits on the surface of the leaf just below the silk.

5. *Dictyna calcarata* is an extremely common, tiny cribellate dictynid. The males of this species have distinctive pointed pedipalps.

6 and 7. *Yorima* species or *Blabomma* species. These two genera are difficult to tell apart; they both include species of six-eyed, ecribellate spiders.

1. Blind cave *Cicurina*

2. Cave *Cicurina* with 8 eyes

These two *Cicurina* were living in the same cave. The blind cave *Cicurina* has no eyes and almost no pigment, while the eight-eyed *Cicurina* still has some pigment. The blind *Cicurina* may be better adapted for the depths of the cave, while the sighted *Cicurina* may inhabit the twilight zone of the cave close to the entrance. However, both spiders are restricted to living in the cave, since they would both die in the harsh, arid southern Arizona desert that surrounds their cave. Both species are currently undescribed.

Because populations of troglobitic *Cicurina* are geographically and genetically isolated, they are almost certainly distinct species. At least one other mountain range in southern Arizona also contains blind cave *Cicurina*, which suggests the possibility of karst faunal regions distributed throughout the sky islands of southern Arizona. These populations are extremely vulnerable to any disturbances or changes to their ecosystem, since they cannot disperse to other habitats.

Pack rats and bats contribute to the ecosystem of this particular cave by depositing droppings within the cave. These droppings support detritivores as well as fungus. The fungal growth is fed upon by booklice (psocids), which in turn are fed upon by a variety of small predators such as pseudoscorpions, young scorpions, *Sitalcina* harvestmen, and tiny spiders. A surprisingly diverse assemblage of small predators may be found in a single cave.

of days if not weeks, even if built in exposed locations. In drier habitats, this is an advantage compared with the chemically sticky silk of most orb weavers and theridiids, and might explain why dictynids have been so successful at exploiting their little niches.

Cribellate dictynids have become specialized in capturing dipterans (flies). These spiders construct their tiny webs in areas attractive to flies either as a perch or as a basking spot, such as the tips of twigs or a sunny rock surface. Some species include lines of silk extending beyond the web. These extension lines are also attractive to flies looking for a perch and effectively increase the capture area. The silk of the dictynid's web selectively transmits vibrations from flies better than other vibrations. The vibration frequencies produced by a fly's wingbeat elicits an immediate predatory response from the dictynid.

Some dictynids will tackle flies many times their own size. Considering the spider's tiny size of just a few millimeters, this may seem self-evident. If a dictynid could capture flies only its own size or smaller, it would have a very narrow range of prey available to it. Being able to kill flies larger than itself would give the spider a much greater range of potential prey. Some cribellate dictynids cooperatively kill and share prey, demonstrating a degree of sociality. This is commonly observed in male/female pairs, where a male cribellate dictynid will frequently cohabit in the web of a female. Working together, they can overcome prey that would be too large for either spider to capture alone.

Cribellate dictynids exhibit several degrees of sociality. *Dictyna calcarata* is a territorial communal species. Each female of this species defends her web against other females but may share it with a male. The webs within a colony may be immediately adjacent to each other, usually on the surface of a rock. In the mountains of southern Arizona, *Mixitlia trivittatus* may form colonies consisting of dozens to hundreds of individuals. One colony consisted of upward of 10,000 spiders. This species is also communal. Individual spiders defend their own webs and there is no mass attack on prey; however, in southern Mexico, one species of dictynid is fully social. *Mallos gregalis* may form large permanent colonies. Up to 30 spiders cooperatively kill and share prey. The webs of *Mallos gregalis* appear to selectively convey the vibrations of dipteran prey while attenuating the vibrations generated by the movements of other spiders. These spiders are especially attuned to

the vibrations caused by *Musca domestica*, the house fly. Like many other fully social species of spiders, *Mallos gregalis* is found in the tropics, where abundant prey can support a large aggregation of predators in a small area.

Diguetidae: The Desert Shrub Spiders

Summer in the desert can be a challenge. Temperatures can reach well over 110 °F (43 °C), and relative humidity can be as low as 4 percent. Most creatures living in this environment utilize sheltered microhabitats such as pack rat nests, become nocturnal, or require high energy and moisture resources (such as nectar from flowers or extra floral nectaries). A conspicuous exception is *Diguetia*, the desert shrub spider. The web of this spider is a common component of the deserts of the southwestern United States. This distinctive web is frequently built in the middle of a prickly pear cactus or in a thorny shrub. An irregular network of strong silk lines surrounds a central retreat. Tubular in shape with an opening at the bottom, this vertical retreat is decorated with bits of dead leaves and the remains of *Diguetia*'s captured prey. Sometimes the spider builds a sheet web that radiates out from the entrance of the retreat. The spider hangs suspended near the retreat, waiting for flying insects to blunder into the scaffold of capture lines. But if *Diguetia* sees a threat, the alert spider ducks into the retreat almost instantly.

Although most of its hunting occurs in the early morning or late afternoon, this spider may be seen outside its retreat even during the extreme heat of a summer day. The secret to its success lies in a special protein. This protein, called heat shock protein (Hsp 70), has undergone a higher rate of evolutionary change in *Diguetia* than in other spider lineages. Hsp 70s are used in basic, crucial functions within the cell, such as assisting in the proper folding, translation, and translocation of other proteins. This protein shows remarkably little variation between species. For example, species as divergent as fruit flies and humans may have 98 to 100 percent identical Hsp 70s. Being highly conserved is probably due to Hsp 70's importance to all living cells. However, these proteins can also be induced by heat stress and help to confer heat tolerance. Hsp 70s from *Diguetia* were compared

Having captured a golden-eyed lacewing in its web, *Diguetia albolineata* prepares to feed. Specialized heat shock proteins allow these spiders to thrive in low, hot desert environments.

with Hsp 70s from other families of spiders. Positive selection of the Hsp 70s was shown to have occurred only within the *Diguetia* lineage, thus conferring a greater degree of heat tolerance in this genus. Episodes of adaptive evolution probably occurred long ago that selected for altered Hsp 70 sequences, allowing these spiders to specialize in a microhabitat with little competition from other spiders.

Female *Diguetia* mature in early summer. The first egg sacs are produced in August, and the female continues to produce a succession of egg sacs through September. Each discus-shaped egg sac is neatly stacked within the tubular retreat. With up to 10 egg sacs produced per female, and each egg sac holding from 150 to 250 eggs, the number of eggs produced per mature female can be well over a thousand. This may seem to be a ridiculously large number of eggs, but there is a reason why producing high numbers of eggs has been selected for in this species. Ultimately, less than 1 percent of *Diguetia* eggs may produce offspring that will survive to reach maturity.

As long as the mother spider is still alive, she vigilantly guards her egg sacs. But once she dies,

another spider may move in. This interloper is one of two species of jumping spider: *Metaphidippus manni* or *Habronattus tranquillus*. These little salticids use the *Diguetia* retreat as a site for their own egg sacs. The salticid eggs are laid after the *Diguetia* eggs, but they are larger and develop more rapidly. Consequently, they hatch before the *Diguetia* eggs do. Both the adult jumping spiders as well as their offspring live in the retreat and feed on the *Diguetia* eggs as well as on the significantly smaller *Diguetia* spiderlings. Fortunately for the diguetid, as long as the mother spider is still alive, she can defend her eggs against the salticid brood parasites. Also, because the production of the *Diguetia* egg sacs occurred over a period of weeks, the hatching of the spiderlings is staggered. This strategy helps to "hedge the bet," so that the spiderlings do not emerge all at once, thus increasing the odds that at least one batch will hatch when conditions are maximally favorable for their survival.

Salticids are not the only danger to the eggs. The larva of a clerid beetle, *Phyllobaenus discoideus*, has also been documented as a probable egg predator in *Diguetia* retreats.

Diguetia is a generalist predator, taking prey from 10 insect orders. However, because the web is up off the ground, the predominant prey are insects that fly or are associated with living in bushes, including leaf hoppers (Homoptera), bees, wasps and ants (Hymenoptera), and beetles (Coleoptera). Diguetia initiates the attack on prey by first biting it. Then the spider waits for the venom to immobilize the prey before proceeding to wrap a bit of silk around the captured insect and settling in to feed.

Diguetia fills a niche for a heat-adapted spider in the desert and in turn provides a resource for other species. These spiders exemplify the adage that nature abhors a vacuum. Selection for advantageous changes in a heat shock protein allow this spider to survive and even to thrive in one of the toughest environments on Earth.

Pholcidae: The Cellar Spiders

Commonly found in human habitations, cellar spiders in the family Pholcidae are a frequent sight hanging from the ceiling of a basement, a porch, or a garage. The morphology of these spiders exemplifies traits for a life spent upside down. With a somewhat globose body and long, spindly legs, the cellar spider moves with grace and alacrity while it hangs from its irregular cobweb. A living pendulum, the pholcid converts gravitational potential energy into kinetic energy as the weight of its body swings forward at the end of the long legs. The longer the trajectory, the greater is the energy efficiency in this style of locomotion. Consequently, disproportionately long legs have been selected for in these spiders that hang suspended from their silken webs. The primary limitation on impossibly long legs is the necessity for the spider to occasionally walk upright as it searches for a new home or for a mate.

The typical pholcid web is an irregular network of silk forming a "cobweb" that somewhat resembles the irregular webs of theridiids. This web may be built either close to the ground with silk extending down to the substrate, or it may be higher up on a ceiling or on a natural rock overhang. Smaller species build their webs under rocks. These spiders are not especially well adapted to extreme heat; therefore, in general, they do not build their webs out in exposed open areas but instead choose areas that provide some shelter and protection from heat and sunlight. Consequently, they are commonly found in caves and in buildings, which for the spider are the man-made equivalent to caves.

1. The genus *Diguetia* has 6 eyes in a unique arrangement, making this genus easy to identify.

2. *Diguetia* constructs its irregular three-dimensional webs in small trees, shrubs, and cacti. A silk retreat incorporating leaves and debris is at the center of the web. If the spider feels threatened, it rapidly disappears into the hollow retreat.

Pholcids are generalist predators that can take a wide variety of prey, including the highly venomous bark scorpions in the genus *Centruroides*. Some species of pholcids are araneophages, attacking and feeding on other spiders. One species, *Pholcus phalangioides* (longbodied cellar spider), has been documented using aggressive mimicry much the way that some pirate spiders do. It first plucks the web of the intended victim in a fair simulation of struggling prey. When the resident spider approaches, *Pholcus phalangioides* captures it with silk before biting and killing it. Then the invader opportunistically eats the resident spider as well as any eggs the victim spider had been carrying or insect prey still in the web. This invasive European species as well as another pholcid introduced from Eurasia, the marbled cellar spider (*Holocnemus pluchei*), might be displacing native pholcid species partly because of their talent for preying on other spiders.

Among the native pholcid species, *Physocyclus* may associate loosely with other spiders. One or more may be observed near the edge of a black widow web, and in some cases, several pholcids may live in such close association with each other that they appear to be on the verge of living communally. Many times, a mature male pholcid may be seen cohabiting in a female's web. The male of some species has been documented as exhibiting "chivalrous" behavior, deferring to the female if prey is available. In fact, in some species the male not only cedes captured prey to a nearby female, but he may also go a step further. In this case, the male plucks the female's web before depositing wrapped prey for her, or he may even carry the prey over to the female. Sometimes the male may be literally starving, but he still defers to his mate. Not all species have this behavior, but the selective advantage for it appears obvious. A well-fed female is likely to be a fecund female. She is also likely to stay in a favorable location. Meanwhile, the male might be able to monopolize mating privileges with that female, thereby passing his "chivalrous" genes on to the next generation.

The male *Physocyclus* courts the female by stridulating. The male's chelicerae have a series of pronounced ridges that can be used as a file for stridulation. Pholcid spiders are haplogyne in regard to their reproduction. This means that in the female, the copulatory duct does double duty as the fertilization duct. Therefore, the last male to mate with a haplogyne female is most likely to fertilize the eggs. (This is in contrast to entelegyne spiders, in which the first male to mate

The longbodied cellar spider, *Pholcus phalangioides*, is a synanthropic species distributed outside its native Europe to human habitations throughout the world. It frequently outcompetes native spiders due to its talent for killing and eating other pholcids.

has the advantage.) Some male pholcids may actually remove a predecessor's sperm before transferring their own sperm to the female.

The female pholcid produces a cluster of eggs that are only lightly wrapped in silk. She carries this egg sac in her chelicerae, never feeding during the incubation of the eggs. Once the spiderlings emerge, they disperse to fend for themselves.

Pholcids are exceptionally resourceful spiders that have been adapting to many different environments for an extraordinarily long time. Early pholcids were already established before the supercontinent Pangaea began to break apart. About 200 million years ago, they had started to diversify, and by the mid Jurassic around 170 million years ago, the subfamilies as we know them today had already appeared. Because these spiders do not balloon, the drifting apart of continents fragmented populations, with the result that different lineages are found on different continents. However, lack of ballooning has not stopped some species from being able to extensively disperse. A number of synanthropic species have become common throughout the world due to their association with humans. Once contemporary with *T. rex* and now with *Homo sapiens*, these delicately graceful arachnids are surprisingly tough survivors.

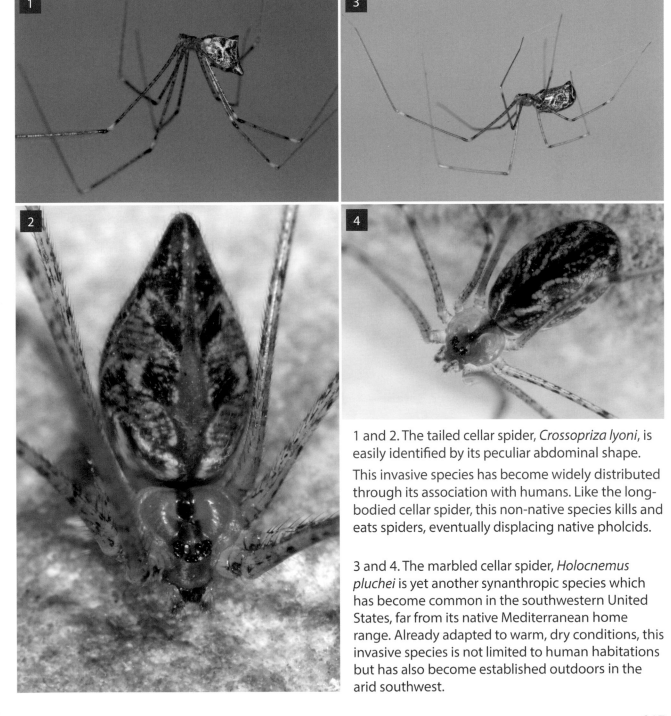

1 and 2. The tailed cellar spider, *Crossopriza lyoni*, is easily identified by its peculiar abdominal shape.

This invasive species has become widely distributed through its association with humans. Like the long-bodied cellar spider, this non-native species kills and eats spiders, eventually displacing native pholcids.

3 and 4. The marbled cellar spider, *Holocnemus pluchei* is yet another synanthropic species which has become common in the southwestern United States, far from its native Mediterranean home range. Already adapted to warm, dry conditions, this invasive species is not limited to human habitations but has also become established outdoors in the arid southwest.

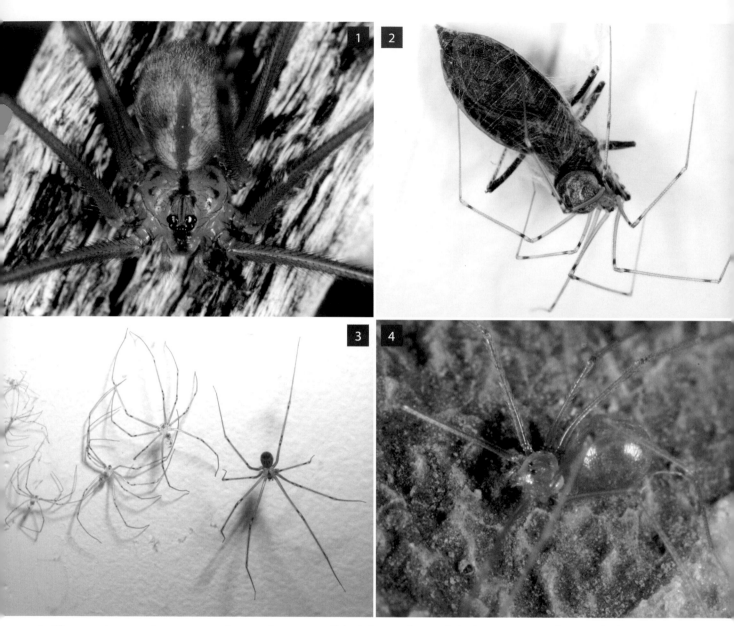

1. *Physocyclus enaulus* are conspicuous because of their large size and their presence in human habitations. In many areas, these native pholcids have been displaced by invasive species of spider-hunting pholcids.

2. Having captured a kissing bug (*Triatoma rubida*), this *Physocyclus* settles down to a feast. *Physocyclus enaulus* is a welcome addition to any home. Although capable of capturing and killing bark scorpions and kissing bugs, these large spiders are completely harmless to humans. Kissing bugs are large, blood-feeding nocturnal bugs. About 41% of the kissing bugs from a southern Arizona area were found to carry the trypanosome that causes Chagas disease, *Trypanosoma cruzi*.

3. *Physocyclus* demonstrates site fidelity when it molts. It holds onto the previous molt each time it is ready to molt again.

4. Small and delicate, this cave *Psilochorus* has almost no pigment and eyes that are greatly reduced in size. This undescribed troglobitic species was found in a cave in southern Arizona.

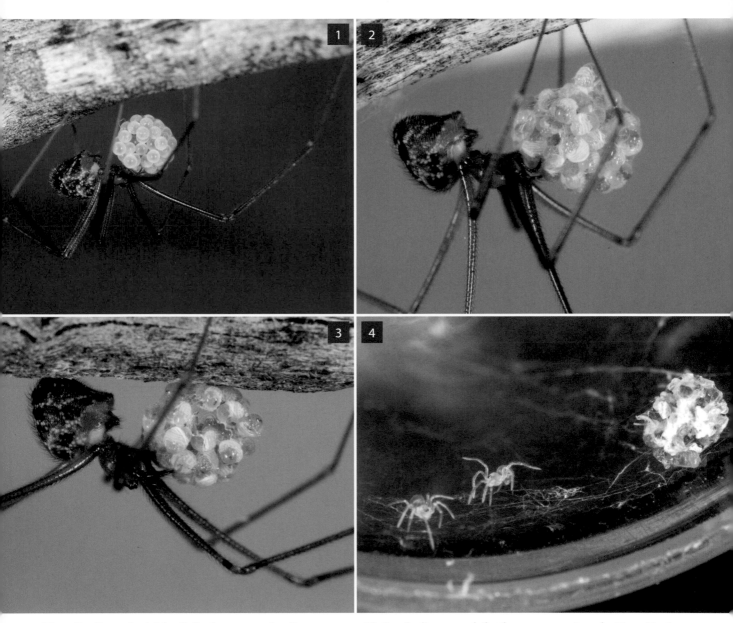

Like all other pholcids, *Psilochorus* carries its egg sac with its chelicerae while the eggs are incubating. During this time, the mother spider must fast. These small pholcids are extremely common under rocks even in the low, arid desert environments of the southwestern United States.

1. On the first day of incubation, the loosely wrapped eggs appear as simple white spheres.

2. On the sixth day of incubation, they show developing legs.

3. On the eighth day of incubation, the embryos have legs and eyes that appear as a cluster of small red spots.

4. On the tenth day of incubation, the spiderlings emerge from the egg sac, ready to disperse and begin their lives as independent, solitary hunters.

CHAPTER 15 Crevice Weavers, Ground Weavers, and Sheet Web Builders: Sicariidae, Filistatidae, Plectreuridae, Zoropsidae, Titanoecidae, Amaurobiidae, Oecobiidae, Linyphiidae, and Agelenidae

Closely related to the notorious brown recluse, the Arizona brown spider (*Loxosceles arizonica*) has the same eye arrangement (3 pairs) and the same necrotizing venom component. Not all individuals have the violin-shaped dark area on the cephalothorax.

220

Sicariidae

The notoriety of the recluse spider rivals that of the infamous black widow. Gruesome tales involving spider bites, gangrene, and amputations generate both horror and fascination. But the true story of these spiders and their venom is in fact as intriguing as a mystery novel. In the case of the recluse spiders, the clues are still being teased out of a story that began millions of years ago.

The family Sicariidae contains two genera, *Loxosceles* (the recluse spiders) and *Sicarius*, the six-eyed sand spiders. The family name derives from the Latin word *sicarius* meaning "dagger-man" or assassin, while the related word *sicae* means a concealed dagger. Spiders in the genus *Sicarius* live up to their name by burying themselves in sand as they wait to ambush prey. Their camouflage is all the more convincing because like *Homalonychus* in the southwestern United States, *Sicarius* has specialized setae (hairs) to which sand particles adhere. As an insect wanders near, *Sicarius* explodes out of the ground, seemingly out of nowhere, and rapidly kills its prey. These spiders are found in South America and Africa.

The genus *Loxosceles* includes the famous brown recluse spider, *Loxosceles reclusa*, which lives in the southeastern United States. A handful of related species live in the arid southwestern United States, all of which resemble *Loxosceles reclusa*. There are also additional species of *Loxosceles* native to Europe, Asia, Africa, Central America, and South America. All the *Loxosceles* spiders have essentially the same kind of venom. These spiders are frequently found in homes and garages, sometimes in surprisingly large numbers; however, as indicated by its name, the recluse spider is not aggressive and instead actively seeks a secluded and dark hiding place in which to live. These hiding places may be behind a bookcase, under a cardboard box, or any other quiet little spot where this shy spider can construct its ground-hugging, irregular web and wait for prey. Unfortunately for both the spider and humans, occasionally the refuge chosen by the spider may be under the sheets of an unmade bed or in some clothing left lying on the floor. When the spider feels threatened by the sudden appearance of a human appendage, it may bite in self-defense. This is where the venom comes into the story.

Both *Loxosceles* and *Sicarius* have an unusual component of their venom that is found, as far as we know, only in these spiders and in a few species of bacteria. That component is called sphingomyelinase D, and its toxicity lies in its ability to break down tissue, even including fat tissue. In addition, it causes cell death in skin epidermal cells (keratinocytes) and the breakdown (hemolysis) of red blood cells. Sphingomyelinase D very likely came about from a mutation in a gene for a very common housekeeping enzyme, GDPD (glycerophosphodiester phosphodiesterase), which is involved in the metabolism of lipids. The lipids this housekeeping enzyme "cleans up" are found in the membranes of all cells, and presumably the normal enzyme is needed to break down and recycle damaged cell membrane lipid layers. But sphingomyelinase D cleaves the phospholipid molecule in an unusual spot: at the D site between the choline and the phosphate, leaving the lipid molecule headless. This linear strand of lipid then bends around and forms a cyclic molecule. For reasons that are not understood, the immune response of the victim is then activated, initiating the complement-dependent hemolysis of red blood cells and the further breakdown of tissue. As the destruction progresses, the body attempts to wall off the affected area by cutting off the blood supply, resulting in death of the tissue. Secondary bacterial infections can then invade the necrotic site, causing additional damage. Fortunately, in most human cases of envenomation, the damage is limited to a small area that eventually heals by itself. Many bites seem completely innocuous, perhaps reflecting less venom injected at the time of the bite (possibly analogous to the "dry" bites occasionally associated with the defensive bites of rattlesnakes). But in rare cases, the bite can cause a large area of necrosis that may require skin grafts to repair. And in very rare cases, a systemic reaction may occur, resulting in death.

Like a good mystery novel, the story of the recluse spider has its share of "red herrings," clues that lead the investigator down the wrong path. In this case, many conditions unrelated to spider bites are blamed on spiders. *Staphylococcus aureus*, especially methicillin-resistant *Staphylococcus aureus* (commonly referred to as MRSA), causes lesions that are frequently misdiagnosed as spider bites. (In one case, a jail was repeatedly sprayed with pesticides in an attempt to stop an outbreak of "spider bites" that was later discovered to be an outbreak of MRSA.) Adding to the confusion, many other species of spiders are misidentified as *Loxosceles*. If there is an unexplained lesion that might be a spider bite, and any of the "usual suspects" (meaning any brown-colored spiders) are seen in the general vicinity, immediately there is a strong and unyielding suspicion that the perpetrator is a brown recluse. Consequently, much of the evidence regarding recluse spider bites is seriously suspect. But despite all the confusing and misleading data, the fact remains that the spiders in the family Sicariidae have a serious bite with a potent and highly unusual venom containing sphingomyelinase D.

It is suggestive that only one other group of organisms are known to possess sphingomyelinase D in an identical structure to that found in the venom of the spider family Sicariidae. These are bacteria in a group loosely referred to as diphtheroids, and one species in particular, *Arcanobacterium haemolyticum*, is a pathogen that causes acute and severe pharyngitis in humans. Perhaps, millions of years ago, there was a lateral transfer of the genetic material to code for the production of sphingomyelinase D between the bacterium and the common ancestor of *Sicarius* and *Loxosceles*. But no one knows which had the sphingomyelinase D first, the spider or the bacterium.

Where did these spiders come from, and where and when did they acquire sphingomyelinase D? The mystery of these spiders takes us back 100 million years to the supercontinent Gondwana. But to get there, we have to travel even further back in time, to the dawn of the age of the dinosaurs, 250 million years ago.

Two hundred and fifty million years ago, at the start of the Triassic Period, a huge supercontinent called Pangaea existed. The north, called Laurasia, was composed principally of what is now North America and Eurasia. The south, called Gondwana, was made up mostly of what is now South America, Africa, Antarctica, Madagascar, Australia, and the Indian subcontinent. During the Jurassic, Laurasia and Gondwana separated as a result of plate tectonics. Also starting in the Jurassic, about 180 to 167 million years ago, Gondwana started to break up. First, the western half (Africa and South America) separated from the eastern half (Madagascar, India, Australia, and Antarctica). Beginning about 140 million years ago, Africa broke away from South America. Other parts eventually followed, with Australia being one of the last to separate about 50 million years ago. Though traveling only at the speed at which fingernails grow, continents have drifted thousands of miles, carried on the huge plates making up the Earth's outermost crust.

As the continents moved apart, populations of plants and animals also became separated by the growing expanse of ocean. This is an example of vicariance, a process in which a single taxon or species becomes split into discontinuous fragments by a geographical barrier that prevents gene exchange. A good example of this was the formation of the Isthmus of Panama about 3 million years ago. Fish on the Atlantic side became genetically separated from fish on the Pacific side, and now there are pairs of species, one for each side of the isthmus, where once there were single species. After a population is isolated, it continues to evolve, but its evolutionary path will diverge from its nearest relatives to some degree. Small mutations continuously occur in any population of organisms, and by measuring genetic differences between two discontinuous populations, a "biological clock" can estimate how long ago a vicariance may have occurred to separate them. And so the concept of "species" is complicated, involving morphology, geography, and genetic data.

The most recent common ancestor of both *Sicarius* and *Loxosceles* has been placed on western Gondwana more than 95 million years ago. This common ancestor must have already acquired the rare enzyme sphingomyelinase D, for both *Loxosceles* and *Sicarius* inherited the capability to produce it. *Sicarius* occurred on western Gondwana, and so its descendants are found in South America and Africa. *Loxosceles* is now widespread, found in Eurasia, Africa, South America, North America, and Australia. The history of this genus is extremely complicated, because a multitude of dispersal events have occurred over a wide range of time. Neither *Loxosceles* nor *Sicarius* balloon as spiderlings, so dispersal must depend on means other than the wind. Part of the puzzle is to determine what those other means might be.

In the New World, *Loxosceles* was limited to South America until fairly recently, geologically speaking. The radiation of the *reclusa* group, which includes all the species native to the United States, is estimated to be at least 33 million years old. *Loxosceles* fossils have been found in 20-million-year-old Dominican amber from the Greater Antilles in the Caribbean. Therefore, the incursion of *Loxosceles* to North America considerably predates the formation of the Isthmus of Panama, ruling out that land bridge as their route of dispersal. A more likely candidate as a dispersal route would be the proposed bridge called GAARlandia, the Greater Antilles and Aves Ridge. This was a bridge in existence 35 to 33 million years ago that may have provided a discontinuous

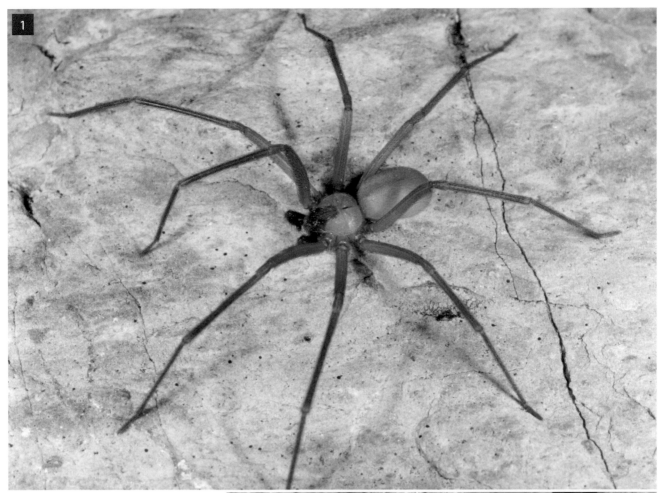

1. *Loxosceles arizonica* adult male. This species is common in southern Arizona, living under debris in low-elevation desert. It can also be found living in houses.

2. *Loxosceles sabina* is a rarely encountered species. Although one specimen was found in Sabino Canyon of southern Arizona, the only population known is in a small cave in the Rincon Mountain foothills. This individual has killed a cave scorpion.

"stepping stone" connection to North America by way of the Caribbean. Another potential path may have been even more recent, via the Panama Island Arc during the late Tertiary, 15 million years ago. But the presence of *Loxosceles* fossils in Dominican amber is fairly strong evidence supporting the GAARlandia hypothesis. The *reclusa* group, including the native U.S. species, and the *laeta* group from northwestern South America are each other's closest relatives, also known as sister taxa.

In historical times, the South American species *Loxosceles laeta* has become introduced into North America, Australia, and even Finland. Its dispersal to these new areas has been by hitchhiking among human possessions. This large and rather dangerous spider is so far limited to fairly small pockets in the United States, mostly in California and the east coast. Potentially lethal systemic reactions are more common following bites from *Loxosceles laeta* than from bites of spiders in the *reclusa* group. *Loxosceles rufescens*, native to Europe, has been introduced throughout the world. This cosmopolitan spider is well adapted to cohabiting with humans and consequently travels about the world wherever humans take it. It is now found throughout much of the southern United States and has also been found in New York.

The family Sicariidae is an ancient and resourceful group of spiders. They have survived from the Cretaceous, 100 million years ago, to the present. Despite the fact that they produce a potent and unusual venom, many members of this group coexist with humans (albeit unknown by many of their human housemates) throughout much of the world. Perhaps the secret of their success lies in the fact that they are shy and stay hidden, true recluses.

Filistatidae: The Crevice Weavers

Kukulcania, named after the Mayan feathered serpent god Kukulcan, does indeed seem ancient. These enigmatic spiders are found near the base of the araneomorph family tree and retain some characteristics that are considered primitive. The large, velvety black female could even be mistaken for a small tarantula. Her eight eyes are clustered together in a tight grouping, and her pedipalps are large and leglike, similar to those of a tarantula. Very young *Kukulcania* have posterior

book lung leaves. In addition, female *Kukulcania* can live for 8 to 10 years and can molt after reaching sexual maturity. These characteristics are consistent with the mygalomorphs, those spiders such as tarantulas and most of the trapdoor spiders. But her fangs are pincerlike, putting her in the araneomorph clade along with orb weavers, jumping spiders, and crab spiders. (The mygalomorphs have downward-pointing fangs like those of a saber-toothed cat.) Consequently, the family Filistatidae, to which *Kukulcania* belongs, is on a little branch of the araneomorph spider family tree that split off shortly after the point where the Araneomorphae split from the Mygalomorphae.

Filistatids are known as crevice weavers. *Kukulcania* build webs in crevices under rocks, fallen logs, dead agaves or yucca, and in the debris of pack rat nests. Human habitations and structures provide other good sites for *Kukulcania* to live; hence, their other name is "southern house spider." They construct a distinctive, lacy-looking web using woolly, cribellate silk. The catching area looks something like a lace doily with a tunnel retreat at its center. The silken retreat is constructed in a crevice or other protected spot where the spider can wait in safety. Trip lines radiating out from the center of the web signal the presence of potential prey to the patient spider. The velvety black or gray/tan female may only be glimpsed as she pops out from her refuge to capture prey, which she then carries back into her retreat.

The male *Kukulcania*, on the other hand, may be more conspicuous as he wanders about in search of females. The mature male looks so different from the female that they almost appear to be two different species. (This different appearance between sexes is known as sexual dimorphism. In fact, a number of spiders, especially jumping spiders, were initially classified as two species because of how different the males looked from females.) The male *Kukulcania* is tan in color, with a darker area on his cephalothorax. Unfortunately for the harmless male *Kukulcania*, this gives him a superficial resemblance to the dreaded brown recluse and frequently earns him the death penalty. However, even a fairly quick examination can distinguish him from other spiders. He has a single tightly clustered group of eyes, unlike the three separate pairs of eyes of the recluse spiders, and he also has enormously long, straight palps that are folded in half and held extended together in front of him as he walks, as though pointing the way forward.

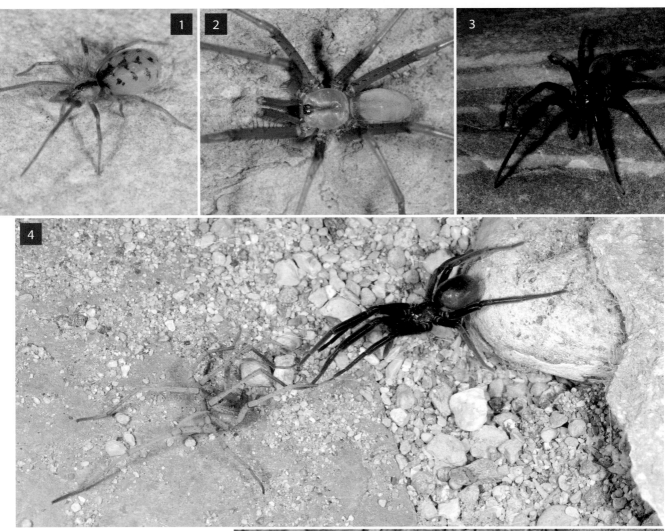

Family Filistatidae:

1. *Filistatoides insignis* (presumptive).

2. *Kukulcania arizonica* mature male

3. *Kukulcania arizonica* mature female

4. Before mating, the male *Kukulcania* hooks his tarsal claws with the female's tarsal claws and proceeds to pull her toward him.

5. The male uses all eight of his legs either to hold the female or to brace himself while he mates with the female.

As soon as a wandering male *Kukulcania* encounters the lacy web of a mature female, he starts to court her. In the first phase of his courtship, the male adds a large number of his own silk threads to the female's web, briefly pausing at times in order to vibrate his abdomen. These non-cribellate silk threads may contain a sex pheromone communicating his intentions, or may possibly contain a chemical that calms the female and reduces her aggression. It is conjecture at this point how this silk helps the male. Certainly, it is not uncommon for the female to initially attack the male. There is the chance that the female may even kill the male during this introductory phase of courtship. But if the male is bold enough to continue his pursuit and if the female does not kill him immediately, then the courtship proceeds. As the male adds more threads and periodically vibrates his abdomen, the female usually ceases her aggressive attack and returns to her tunnel retreat. When she moves again, it is to slowly approach the entrance of the retreat. In doing so, she is apparently communicating to the male that she is receptive to his advances.

In the next phase of courtship, the male starts to tap the web, and the female responds by tapping back in answer. The courting couple advances toward each other while maintaining their dialogue of tapping. Once the male's legs make contact with the female, she raises her first two pairs of legs up high over her cephalothorax. The male then intensifies his tapping and leg waving. Soon after, the male slides the tarsi of his first pair of legs (and sometimes also his second pair) along the corresponding legs of the female. As he slides the tip of his tarsi along her tarsi, he hooks the female's claws with his own tarsal claws. As the male walks backwards, he pulls the female toward him. Her first pair of legs stretches out in front of her since her tarsal claws are still hooked together with the male's claws. Sometimes the female resists the pull of the male, in which case he pulls harder. This prenuptial dance may last for only a few minutes but probably serves a useful purpose in that the female can evaluate the male's strength during this dance. Despite being a unique behavior among spiders, similar courtship dances have been observed in many other groups of arachnids, including scorpions, pseudoscorpions, uropygids, and amblypygids.

The male then rubs the female's body (and sometimes her legs) with his third pair of legs, while still maintaining his connection with her via their interlocked claws. At this point, the female signals that she is ready to mate by raising the front of her body. The male continues to face the female as he assists in lifting up her cephalothorax. Once in position, he quickly mates with her, inserting each pedipalp in turn for only a few seconds. After the male finishes mating, he quickly releases the female and makes a hurried escape. If he lingers for even a few seconds, he is attacked by the female.

After mating, the female *Kukulcania* constructs an egg sac, which is carried into her silken tunnel to incubate. After about 80 days of incubation, the spiderlings emerge. The mother spider assists her offspring by sharing food with them. In some species, this sharing goes considerably beyond merely tolerating the presence of the spiderlings feeding on her kills; on the contrary, she actively encourages her young to feed while she herself feeds only sparingly, leaving almost all the food for her young. The mother spider continues to care for her offspring for as long as they remain on her web, into the second- or even third-instar stage of their lives.

Social behavior is also demonstrated among the spiderlings themselves. The spiderlings cooperatively capture prey that is considerably larger than that which could be captured by any one individual. They then collectively feed on the prize.

Plectreuridae

Species from the family Plectreuridae are currently found primarily in the southwestern United States, Mexico, Costa Rica, and Cuba. In addition, a fossil specimen of *Plectreurys* has been described from 16-million-year-old Miocene amber from the Dominican Republic. Two other fossil spiders from this family have been described: one is from Eocene Baltic amber found in Europe, and the other is a fossil from China dating from the Jurassic Era. This 165-million-year-old specimen displays an uncanny resemblance to the modern members of this family, demonstrating evolutionary conservatism. By one measure, any creature that has survived essentially unchanged from the time of *Tyrannosaurus rex* may be considered an evolutionary success; however, it is also of interest that this family of spiders was once far more widespread than it currently is. Apparently, spiders from the family

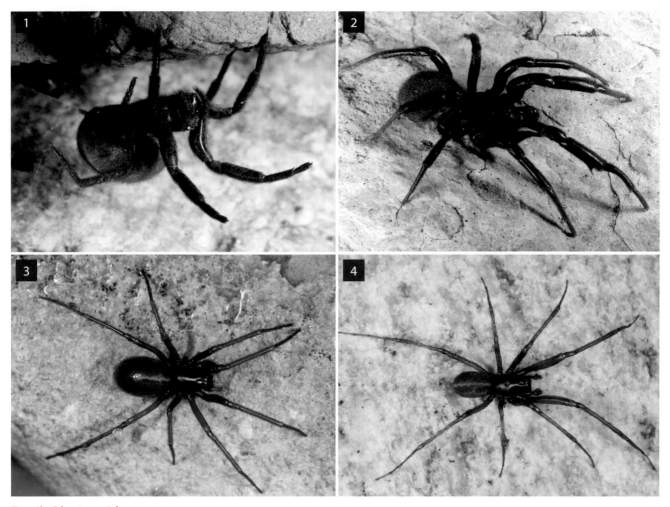

Family Plectreuridae:

1. *Plectreurys tristis* adult female. Although this spider superficially resembles *Kukulcania*, the eye arrangement is different.

2. *Plectreurys tristis* adult male. At maturity, the male has a spinelike clasping spur on each front leg tibia. Presumably, this is used to hold the female during mating.

3. *Kibramoa* species, subadult. The females and subadults are frequently reddish in color.

4. *Kibramoa* mature male. Like others in the family Plectreuridae, *Kibramoa* builds its web under rocks or other shelters. This genus is found in Arizona, California, and Nevada.

Plectreuridae have been extirpated from much of their former range, and the remaining species may represent the last relics of this group.

Plectreurids are somewhat primitive haplogyne spiders that live under rocks and dead vegetation and in the debris of pack rat nests. An irregular fairly flat catch-web close to the ground under the rocks or debris provides the hunting area. The spider waits in a silken tunnel located in a crevice or other protected niche adjacent to the catch-web. With the exception of mature males, these spiders are content to remain in their silken tunnel for long periods of time, rarely venturing out from their safe refuge. They can survive long periods of fasting.

These spiders may be mistaken for *Kukulcania* at first glance. They are fairly large black spiders from half an inch to two-thirds of an inch in body length (11.5–17.0 mm) and are found in similar niches as

Kukulcania; however, they do not spin cribellate silk, and therefore their web does not have the lace doily appearance of *Kukulcania*'s web. The 8 eyes of plectreurids are arranged in 2 rows across the front of their cephalothorax instead of clustering tightly together on the dorsal surface of the cephalothorax. Finally, the mature male *Plectreurys* has a conspicuous spine on each tibia, presumably used in clasping or holding up the female during mating. In fact, the name *Plectreurys* means "having a broad cock's spur" in reference to this tibial spur. The spur may imply a shared common spurred ancestor in the distant past, or it may be an example of convergent evolution in parallel with other spiders such as the mygalomorphs, which also have tibial spurs used to hold the female upright during mating.

Little is known about the natural history of this retiring spider.

Zorocrates (Zoropsidae)

These attractive silvery gray and brown spiders are found in the southwestern United States and throughout Central America. They are fairly common at higher elevations, such as in the oak zone, especially in riparian areas or just inside caves. They construct an irregular, somewhat tangled web of cribellate silk under rocks or other debris. The woolly silk may be a startling sky blue in color. In this family, the calamistrum, used in "roughing up" the cribellate silk, is a characteristic oval patch on the metatarsus of the fourth leg.

Little is known about the natural history of this family of spiders.

Titanoecidae

These small, dark spiders are found throughout the United States but are more common in arid (xeric) environments. They resemble a smaller version of *Amaurobius* in their general appearance. Even their cribellate webs are similar to the webs of their larger cousin, as they are constructed under rocks or other debris on the ground. In the arid regions of the southwestern United States, they are more frequently found in riparian areas.

Little is known about these small, velvety spiders.

1. *Zorocrates* species. This spider can produce an irregular web of unusual blue-colored cribellate silk under rocks or debris.

2. *Zorocrates* courtship is a very active affair. The male is on the right.

3. *Titanoeca* species. Only 0.16–0.31 inches (4–8 mm) in body length, this velvety black spider is commonly found in riparian areas.

Amaurobiidae: The Lace-Weaver Spiders

Broad, black chelicerae and a stocky build give members of the family Amaurobiidae a powerful appearance suggestive of bulldogs. The imposing physique of these spiders contrasts with the delicate beauty of their webs, built of woolly cribellate silk. These characteristic webs give this family its name "lace-weaver spiders" or "hacklemesh weavers." The cribellate silk of the web radiates out from a tunnel retreat, where the spider lies in wait. Insects become tangled in these hackled threads, giving time for the spider to rush out from its hiding place and catch the prey.

Amaurobius has a truly remarkable natural history in regard to its reproduction. In the case of *Amaurobius ferox* (an introduced species from Europe reportedly found in southern California), the mother spider lays her first clutch of eggs and stays nearby to guard them until the babies hatch. This maternal behavior is consistent with many other species of arachnids and is therefore hardly noteworthy. But then it gets interesting. After the spiderlings emerge from the egg sac, they interact with their mother, and she is induced by this interaction to lay a second clutch of eggs. This time, the eggs are laid before they are mature, and consequently this clutch of eggs serves as food for the spiderlings that hatched from the first clutch. This is referred to as trophic egg laying and is a strategy employed by several other species of animals, including some species of poison dart frogs. The spiderlings that receive this food are heavier and have a higher survival rate than spiderlings deprived of the trophic egg meal.

But the mother spider's sacrifice does not end there. The mother and her offspring interact further, and this time she actually appears to solicit her babies to feed on her. This the babies do, collectively killing and feeding on their mother. This matriphagy appears to be regulated by the life stage of the spiderlings, the reproductive state of the mother, and the behavioral interactions of the mother and her young. The spiderlings derive considerable benefit from this behavior. They are heavier and larger at the time they disperse compared with spiderlings deprived of their mother as food, including groups of spiderlings given an abundance of other prey to eat.

Finally, the matriphagous spiderlings have a longer period of social behavior, compared with the

Callobius arizonicus. Callobius are found in cool, forested areas, such as the mountains of Arizona and the forests of California.

nonmatriphagous spiderlings. The subsocial spiderlings live together through several instar stages on the web of their mother, cooperatively killing prey and sharing it. The maternal web appears to provide a superior platform for the offspring to detect and cooperatively kill prey, as compared with webs that the spiderlings construct themselves. Prey that would be too large for one or two spiderlings to overcome is killed by groups of spiderlings, and that prey is shared even with those that did not take part in that particular kill. The spiderlings appear to use coordinated teamwork to subdue prey that is up to 10 times the size of any individual spiderling. Cooperative prey capture increases predation efficiency and survival of all the spiderlings.

Most of the members of the family Amaurobiidae live in cool, moist habitats, making their webs under debris, in caves, or in the nooks and crannies of trees. California boasts the greatest diversity of these spiders in the United States. In the arid southwestern states, *Callobius arizonicus* lives principally at higher elevations where it is cooler and moister. Found under rocks and dead wood, several individual *Callobius* spiders may share a single shelter. Only centimeters may separate their webs, indicating some degree of tolerance between individuals of the same species. Also like *Amaurobius*, *Callobius* guards her egg sac, which is produced in the shelter of her refuge. But in the case of *Callobius*, it is unknown whether the mother spider feeds her young.

1. *Oecobius* species. With a body length of only 0.04–0.12 inches (1–3 mm), these unobtrusive little spiders are easy to overlook. Masonry surfaces provide tiny nooks for their webs. Here, an *Oecobius* is hauling away a captured moth on a doorstep.

2. *Oecobius* can capture large ants. First, it races around the ant, hobbling it with silk. Then it throws more silk over the ant before killing it.

Oecobiidae: The Wall Spiders

Tucked in the corner of a room or a hallway, *Oecobius* spiders make themselves quite at home in human habitations. In fact, their name derives from the Greek meaning "living in a house." Their silken retreats are distinctive, being composed of two small silk sheets superimposed over each other. The upper sheet appears to be stretched taut by the signal lines radiating out from the edges. The *Oecobius* spider rests sandwiched between the upper and lower sheets, able to dart out in a flash from between the silk layers should the trip lines signal the presence of potential prey. Prey is immobilized by the spider rapidly racing around it, first hobbling the prey with silk, and then enclosing it with sheetlike layers of silk. This precaution is necessary because a captured ant can still inflict damage by biting or stinging unless it is completely immobilized. Finally, a killing bite can be safely delivered. Some prey may be dragged behind the spider into the silken retreat for consumption, while other prey may be wrapped and stashed near the retreat. Although ants are commonly captured by *Oecobius*, these little spiders will take a variety of prey, including moths and other spiders, even the stray pirate spider, which is itself a spider hunter.

In nature, surfaces with irregular cavities provide suitable sites for oecobiid spiders to build webs. A large number of individual webs may be found in good habitat, sometimes forming dense aggregations. In some communal species, such as *Oecobius civitas*, individual spiders display an unusual strategy in competition over web sites. If one *Oecobius* invades the site of another one, the resident spider simply leaves without putting up any fight. The displaced spider may seek out an empty web or crevice, or it may in turn displace yet another spider without a fight. This nondefense of a territory has been referred to as a "paradoxical strategy," but it actually may have several advantages. First, the complete absence of an agonistic response avoids any possibility of a physical fight which could cause injury or death to either party. Second, the displacement of individuals might facilitate a greater emigration and immigration of spiders. This might ensure sufficient genetic outcrossing to maintain healthy populations both within the colony and within a species. Finally, it may encourage dispersal of the species, promoting colonization of new locations.

When a male *Oecobius* courts a female, he first signals to her by depositing some silk while seductively waggling his abdomen. If she is not receptive and he

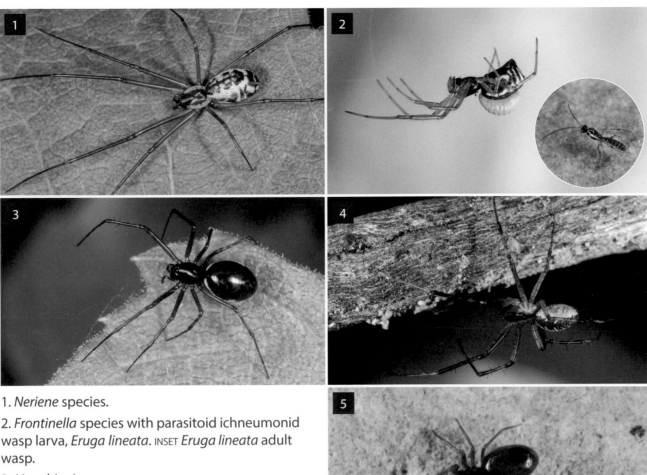

1. *Neriene* species.

2. *Frontinella* species with parasitoid ichneumonid wasp larva, *Eruga lineata*. INSET *Eruga lineata* adult wasp.

3. *Linyphia rita*.

4. *Pityohyphantes* species. This spider builds webs in conifer trees. All these species are found in cooler, moister environments, such as up in the mountains.

5. *Tennesseellum formica*. This is one of many small brown erigonine linyphiids.

is persistent, she may throw silk over him. A receptive female will allow the male to construct a cylindrical silk "honeymoon suite" on top of or adjacent to her own web. Once she enters this special web, they mate. The significantly smaller male is at risk during this time, and he might be killed at any time during or after mating. A female that eats her mate receives nutrition that is then directed toward the production of eggs, thus promoting that male's genes (albeit in a somewhat less than romantic context). Some days after mating, the female constructs the first of a series of egg sacs. Each is quite small, holding only 3 to 10 eggs. She stashes these egg sacs in her retreat, but otherwise exhibits no further maternal care. The tiny

spiderlings are on their own as soon as they emerge from the egg sac.

Linyphiidae: Sheet Web Weavers and Dwarf Spiders

In North America, the family Linyphiidae contains more species than any other family of spiders, even surpassing the incredibly species-rich Salticidae. Close to a thousand kinds of linyphiids from North America have been described, and there are almost certainly more awaiting description. Despite the challenge of possessing such an overwhelming diversity of species, linyphiids can be divided into two broad categories: the larger members that construct conspicuous sheet webs (belonging to the subfamily Linyphiinae) and the dwarf spiders that live on the ground (belonging to the subfamily Erigoninae).

Spiders from the subfamily Linyphiinae include both the bowl and doily spider (in the genus *Frontinella*) and the filmy dome spider (in the genus *Neriene*). Although the larger linyphiids resemble theridiids, their webs look quite different from the irregular gumfoot trap of the classic theridiid web. *Frontinella* constructs a bowl-shaped sheet web with a scaffold of silk threads both above and to a lesser degree below the bowl. The spider hangs on the underside of the bowl, waiting for insects to fall onto the sheet after encountering the network of knock-down lines above the sheet. When an insect falls onto the sheet of silk, the spider bites it from below the sheet and then pulls it down through the web. The damage to the sheet is later repaired. *Neriene* builds a dome-shaped sheet web with a tangle of knock-down lines above it. Like *Frontinella*, *Neriene* hangs below the dome and attacks her prey from below.

Males and females from the subfamily Linyphiinae are frequently found living together in the female's web. The male may take up residence even before he mates with the female, and he may stay for several days after mating. This is probably a form of mate guarding, whereby the resident male attempts to maintain exclusive breeding rights to the female. A certain degree of tolerance must exist between the pair; otherwise there would be casualties due to predation. Some days after mating, *Frontinella* produces an egg sac. She places her thinly wrapped egg sac on the ground, protected by moist soil or litter. She usually dies before the spiderlings emerge.

The spiders from the subfamily Erigoninae are obscure in every sense of the word. Not only are they minuscule (some are less than 0.04 inches or 1 mm in body length), but they also live on the ground in leaf litter or under rocks or other debris. Not all build webs, but if they do, the webs are quite small and close to the ground. In general, these little spiders require some moisture in order to survive. Therefore, in the arid southwestern United States they are most commonly found either at higher elevations or in more mesic riparian areas. In some species, the males have peculiar pits or protuberances on their heads. The female

may grasp the male's head with her fangs during the courtship and mating of these spiders, indicating that the structures on the male's head have some important function (possibly involving pheromones). In England, these tiny spiders are supposed to bring good luck; hence they are called "money spiders."

Agelenidae: The Funnel Web Spiders

A sheet web makes an excellent trap for catching insects. The funnel web spiders in the family Agelenidae (such as *Agelenopsis aperta*) have mastered the construction and use of this system so well that their webs rival orb weaver webs in abundance. In the arid southwestern United States, these spiders are especially numerous in riparian areas, building their sheet webs along the grassy banks of a stream or in the partial shelter of a fallen log or an exposed rock. There they can be seen, patiently waiting in the silken tunnel that opens out onto the catching surface of the sheet web. If any insect chances to fly directly over the sheet web, it may encounter the irregular scaffold of silk lines suspended over the trap. These lines interrupt the flight of the insect, causing it to fall onto the sheet web trap. Because this trap is not sticky, the agelenid must dart out onto the sheet at lightning speed in order to capture the momentarily delayed insect. The spider must instantly evaluate the type of insect and its suitability as prey. The spider can estimate the size of the insect based on the degree that the sheet web bends under its weight. Chemosensory hairs on the spider's legs and palps help it to identify whether the fallen insect is desirable prey, such as a grasshopper, or potentially dangerous, such as a wasp. After overpowering the insect, the agelenid quickly retreats to the safety of the funnel to feed.

The sheet web and its tunnel are situated in a choice location for the spider. In a good location, not only does the sheet web provide a means of obtaining food for the agelenid, but it also provides protection against excessive heat and desiccation. The resident spider inspects the sheet web each day as dusk falls, repairing any damage and adding silk to the structure. The long, flexible spinnerets of the agelenid allow it to apply silk in a swath both to the sheet web and to the tunnel retreat. In a case of convergent evolution, this is much the same way that the unrelated diplurid spiders

Agelenopsis aperta. Funnel web spiders must be fast if they are to catch their prey. The sheet web is not sticky but serves as a "catching platform," transmitting vibrations to the waiting spider. A tunnel retreat provides a back door for the spider's escape should a visitor prove dangerous.

construct their sheet webs and tunnel retreats. When it needs to defecate, the agelenid spider fastidiously holds its abdomen over the edge of the web so that it avoids soiling the silk.

In some habitats, there may be more spiders than there are good locations for webs. Therefore, wandering, homeless agelenids may attempt to usurp ownership of a web. The rule that possession is nine-tenths of the law appears to apply to spiders, as long as the trespasser is not significantly larger than the resident spider. Each of the contestants can evaluate the size of its opponent by the change in the tension of the sheet web as it bends under the weight of the spider. Bouts in the contest consist of opponents orienting toward each other, signaling with threat displays, and possible physical contact. One of the contestants may be forced off the edge of the sheet web. Fortunately, confrontations rarely progress to physical combat, which can be costly if one or both parties are injured during the conflict. An old Chinese proverb

states that when two tigers fight, one is killed and the other is mortally injured. Agelenids, on the other hand, avoid direct physical combat if at all possible. In fact, fewer than 1 percent of agelenids involved in disputes die as a result of injuries sustained in a fight. Each spider adjusts the level of aggression relative to the value of the contested real estate as well as the size and aggressiveness of its opponent, thus optimizing benefits versus costs. If the resident spider loses, it beats a retreat through its back door. The tunnel is open at the back, providing a handy escape route. This spider consequently becomes a homeless spider, searching for a web which it in turn can usurp. As many as 35 percent of the agelenids may be homeless at any one time. In areas with many agelenids competing for prime web locations, turnover rates may approach 10 percent a day. This implies even more confrontations, since some of the contests do not result in a change of ownership. It is evident then that there is a clear selective advantage to a nonviolent resolution

Agelenopsis aperta mature male. *Calilena* and *Agelenopsis* appear almost identical, but *Calilena* lacks the "hoop embolus" of the male *Agelenopsis* pedipalp.

of these extremely common confrontations. Sun Tzu, in *The Art of War*, states that "To subdue the enemy without fighting is the acme of skill." Unlike some other species, agelenids seem to have mastered this concept.

The competitive advantage to obtaining a good location becomes evident in the spider's reproductive success. Site microhabitat is critical. Spiders in excellent locations capture more prey and ultimately may gain up to 13 times the reproductive potential compared with spiders in poor locations. This reproductive success then selects for those spiders that can identify and defend a good web site.

In contrast to the solitary, territorial agelenids of North America, two species found in Africa are fully social. These spiders, *Agelena consociata* and *Agelena republicana*, are both found in the tropical rain forests of Gabon. A colony of *Agelena consociata* may consist of a giant web complex containing more than a thousand adult spiders. These spiders cooperate in web maintenance, killing of prey, and rearing of young. Multiple generations live in the permanent colony, and dispersal occurs only if disaster befalls the colony with destruction of the web by a storm or other event. Although in theory the ability to cooperatively overcome large prey benefits these social spiders, one study found that as a colony grew in size, less food became available for any one spider; however, a selective advantage still remained because of the decreased costs of web maintenance for any individual spider. The shared cost of maintaining the web within a colony is much less per spider than the cost of each spider having to build its own individual web, especially in a climate where rain routinely damages webs. In addition, dispersal is extremely hazardous where these spiders lived. One experiment documented a 10 percent mortality rate within the first hour of dispersal. This mortality during dispersal selects for survival of individuals who stay with the colony, rather than those who wander. Finally, the ability of the colony to exist year-round because of a mild tropical climate allows a continuity of multiple generations within the colony. This continuity is virtually impossible in temperate climates, including the southwestern United States, where seasonal differences might not permit a colony to exist year-round.

It may appear paradoxical that spiders within this single family can demonstrate such dramatically opposite social behaviors even while still utilizing almost identical sheet web capture strategies. Comparing the fiercely territorial and solitary *Agelenopsis aperta* of the southwestern United States with the fully social *Agelena consociata* of the African rain forest illustrates the success of natural selection. Territoriality may be the best solution in one situation, and sociality may be the best solution for a different set of circumstances. The beauty of variation and natural selection is that one family, Agelenidae, can display the diversity to meet such different challenges.

1. *Agelenopsis aperta*. Guarding the egg sac provides some protection against egg predators such as grasshoppers. (Note the long spinnerets.)

2. *Hololena* species. *Agelenopsis* and *Hololena* are sometimes found in the same habitat.

CHAPTER 16 Jumping Spiders:
Salticidae

The large anterior median eyes provide a detailed image for this inquisitive *Phidippus*. The smaller flanking anterior lateral eyes provide depth perception, a necessity for judging distance before initiating a leap.

Jumping spiders have so many pleasing qualities that it would be difficult to decide what is most admirable about these delightful little creatures. Shimmering iridescence and rich, velvety colors equal the beauty of birds and butterflies. Their fearless capture of prey and their acrobatic leaps surprise and astonish us. Their complex courtship song and dance pique our curiosity. But perhaps the most endearing aspect of jumping spiders is their enormous, forward-facing eyes, gazing at us with every appearance of intelligence and inquisitiveness.

These large, forward-facing eyes, called anterior median eyes, are indeed the key characteristic of this diurnal hunter. Jumping spiders locate their prey visually, stalking and pouncing on it like tiny cats. Unlike vertebrate systems in which one pair of eyes handles depth perception, motion detection, and detail resolution, the spider's 4 pairs of eyes divide these tasks. Collectively, their 8 eyes create a visual system that rivals any other arthropod's vision.

In hunting prey, the first order of business is being able to detect motion in the field of view. This is accomplished largely by the small posterior lateral eyes, set about halfway back along the top of the cephalothorax. These eyes each have a 120-degree field of view encompassing almost all the area to the side and even behind the spider. Only about 30 degrees directly behind the spider remains a blind spot. As the hunter detects motion, it orients its body so as to face the object of interest. At this point, both the large, forward-facing anterior median eyes and the smaller flanking anterior lateral eyes are utilized. The anterior lateral eyes each have a 55-degree field of view, and because the fields overlap each other they give the spider binocular vision, necessary for depth perception. Simultaneously the large anterior median eyes, the trademark of the jumping spider, are put into play. Spiders, like other arachnids, have an exoskeleton, and the eyes are covered with a rigid cuticle. Consequently the lens of the eye is locked into place. But the eyes of the jumping spider are still movable. They are shaped like conical tubes, and the tapered base of the eye has a harness of muscles that can move it. In addition, muscles can move the retina. As the spider focuses on an object of interest, it "scans" the object, moving both the base of the eye and the retina back and forth in alignment with each other. This achieves the highest definition, allowing the spider to see detail. Finally, the conical shape of the eye has the

optical properties of a telephoto lens, magnifying the image. It is only after the spider has scanned an insect that it proceeds to stalk it.

The leap of a salticid is powered by a hydraulic system. A sudden increase in hemolymph (blood) pressure straightens the last pair of legs, launching the spider into the air. The spider lands with a sure-footed grip because its feet are equipped to cling to the surface via tiny scopula hairs on its tarsi. Each scopula hair in turn has hundreds of microscopic cuticular extensions called "end feet" that increase the area for gripping the surface. (Under an electron microscope, the hair has the appearance of a hand broom of the type used with a dust pan; scopa is the Italian word for broom.)

The force of the grip is due to physical adhesion, not to suction cups or electrostatic forces. If two glass slides are overlapped with a thin film of water between them, they are difficult to pull apart because of the capillary force of the water. The scopula hairs of the spider utilize these extremely strong capillary forces. Apparently the water available in the atmosphere and on surfaces provides the necessary thin film for the end feet to grip the surface. This explains why spiders with scopula hairs can walk sure-footedly on vertical surfaces and upside down, even on glass surfaces. In jumping spiders, the tips of the tarsi (feet) have such dense claw tufts of scopula hairs that they appear to have fuzzy "toes."

The appearance of intelligence in salticids has some foundation in reality. Many jumping spiders must use some form of working memory during the normal course of their hunting. A key component of a jumping spider's life is that its style of hunting requires the active pursuit of its prey. Many other species of spiders are essentially "sit-and-wait" predators, which take up a hunting position and then wait for prey to come close enough to capture. In contrast to passively waiting for an opportunity, jumping spiders must identify prey from a distance, stalk it, and then attack it. During the approach and pursuit, the jumping spider must contend with a three-dimensional obstacle course of vegetation that may obscure the prey from the spider's direct line of sight. Therefore, the successful hunter must be able to store information in its working memory regarding the position of the target relative to the surrounding landscape, and must devise an approach that brings it within striking distance of the prey. The jumping spider may also factor in the advantages of positioning itself above or at an equal level with the prey, thereby

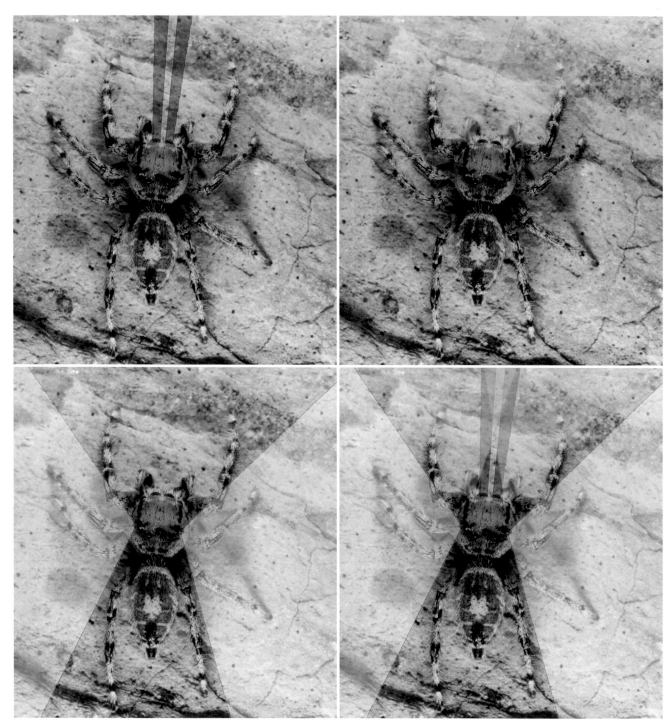

The 4 pairs of eyes provide different aspects of vision for the jumping spider. The large forward-facing anterior median eyes see only a narrow field of view but provide magnified high-resolution images. Flanking the anterior median eyes are the anterior lateral eyes, each of which has a 55° field of view. These fields of view overlap, providing the spider with binocular vision and hence depth perception. The posterior lateral eyes each have about an 120° field of view, mostly to the side and behind the spider. These provide motion detection over a very large area, important for alerting the spider to both prey and predators. The role of the tiny posterior median eyes is so far unknown. The spider's only major blind spot is directly behind it.

The leap of a jumping spider is powered by a hydraulic system. A sudden increase in "blood" pressure (actually hemolymph pressure) straightens the last pair of legs, launching this *Phidippus apacheanus* into the air. If it falls short of its target a safety rope, in the form of a silk dragline, allows it to climb back up and try again. Photo by Bruce D. Taubert.

avoiding the difficulty of working against gravity as it makes its assault. These diminutive spiders demonstrate a surprising degree of intelligence for their small size.

A male jumping spider may also have to use similar skills and strategies while pursuing a fleeing female as she takes evasive action to elude her suitor.

Jumping spiders make up the most diverse family of arachnids in the world, with approximately 6,000 species described so far. As one might expect with such diversity, some jumping spiders have evolved behaviors that fill extremely specialized niches. One jumping spider in Africa, *Evarcha culicivora*, prefers to feed on blood-filled female mosquitoes, especially mosquitoes of the genus *Anopheles*, famous for spreading malaria. Another salticid, *Phyaces comosus* from the bamboo areas of Sri Lanka, specializes in predating the eggs and hatchlings of other jumping spiders. It is so tiny and so closely resembles a bit of dirt or debris that it can sneak into the nest of another jumping spider undetected. Yet another species, *Bagheera kiplingi* from Central America, has a primarily vegetarian diet—unique in the spider world. It lives in acacia trees that produce little nubbins of

protein and fat from their leaf tips, as well as nectar from the base of the leaves. These provide food for the ants that in return guard the tree from caterpillars and other herbivores. The jumping spider steals the nubbins and nectar despite the ant patrols, living almost entirely on this vegetable source of protein.

Even North American salticids demonstrate tremendous diversity. An obvious factor in this diversity involves specialization in particular habitat niches. Within any given area, there may be grasses, shrubs, and trees, as well as rocks or other debris on the ground. Different species of jumping spiders have evolved specializations for each of these environmental microhabitats. For instance, *Metacyrba* species live in tight spaces, such as under the bark of trees, and so have evolved morphological adaptations that permit them to move easily in those areas. Like a pseudoscorpion, *Metacyrba* can move backward as easily as it moves forward, clearly a great advantage for maneuvering in confined spaces. *Marpissa* specializes in living and hunting on grasses. Its body is elongated so that when aligned with a blade of grass, it becomes

A tale of the "toes":

1. *Hentzia palmarum* has no difficulty walking on a vertical surface.

2. Jumping spider tarsus ("foot"). Jumping spiders, tarantulas, and some other spiders can walk on vertical surfaces or upside down thanks to their "fuzzy toes." Their feet, or tarsi, have many scopula hairs, each of which has hundreds of microscopic cuticular extensions. These give the spider a sure grip on almost any surface.

3 and 4. Highly magnified scopula hairs.

almost invisible. The genus *Habronattus* has more than 100 species, many residing in the southwestern United States. Even accounting for microhabitat specialization, how can there be such a diversity of species in this one genus?

Part of the answer may lie in their complex courtship ritual. Male jumping spiders not only perform a ritualized visual display or dance, but some species also produce a seismic "song" while courting females. The dance component of the ritual involves waving their legs in a stereotypical pattern, almost akin to using semaphore flags for signaling. The male's legs, especially the front ones, frequently have tufts or plumes of longer hair and flashes of color, including stripes and iridescence. The face of the spider may also have distinctive colors and patterns. During the visual display the male draws near the female in a slightly zigzag approach, directly in front of her large forward-facing anterior median eyes. This ensures that she is able to see his colors and signals with the greatest

degree of detail, increasing the likelihood of a positive identification of a suitor versus that of a prey item.

As the male signals the female, he sends her seismic signals as well. These consist of thumps, scrapes, and buzzes. Drumming the ground with the front legs produces the percussive thump, stridulation involving the back of the cephalothorax and the front of the abdomen produces the scrapes, and oscillations of the abdomen produce the buzz.

Habronattus pugillus lives at higher elevations in the "sky island" mountain ranges of Arizona. Because these populations are isolated from each other, regional differences in both their songs and visual displays have evolved. Experiments testing males and females from different mountain ranges placed together revealed that females actually preferred the "foreign" male, suggesting a female bias toward complex and novel courtship displays. The females selected for ever more elaborate secondary sexual characteristics, including decorations and behaviors as expressed in the courtship "song and dance." Perhaps these populations are driven to diversification by female sexual selection.

After mating, a female jumping spider lays its eggs in a silk nest where it then stands guard, fasting the whole time. The babies hatch after about three weeks, but remain in the nest until they molt two or three weeks afterward. If the female is able to obtain enough food after the babies' exodus, she produces another clutch of eggs, albeit a smaller one. The cycle may be repeated again if she can once again find enough food. A medium-sized jumping spider such as *Phidippus californicus* may produce more than 40 babies in the first clutch and more than 30 in a second clutch after a single mating. The babies hatch close to the summer monsoon season at a time of abundant prey availability. Staggering the production of egg clutches increases the probability that the emergence of at least one clutch of babies coincides with optimal prey availability.

The males tend to mature faster, in 5 to 7 molts, and are smaller than the females, which require 6 to 8 molts to mature. This difference in rate of maturing reduces the chance of inbreeding by assuring that males and females from the same clutch of eggs will not reach breeding age at the same time. The male is forced to wander in search of mature females.

As jumping spiders hunt during the day, they are in turn potential prey for diurnal predators such as birds and lizards. Mimicry may provide some protection against these other visually oriented predators. A number of jumping spiders, especially in the genus *Phidippus*, have rich red or yellow velvety abdomens. Since research has shown that jumping spiders probably cannot see the color red, it is unlikely that such markings are of any benefit in courtship displays; however, these colors match the markings of some species of velvet ants found in the same geographical areas as the spiders. Velvet ants are actually wasps, and the fuzzy, wingless females pack a very respectable sting. Vertebrate predators learn to avoid velvet ants, and by extension to also avoid their look-alikes. Resemblance to velvet ants is an example of Batesian mimicry, in which a relatively harmless animal benefits from its resemblance to a more dangerous one.

Other salticids resemble ants. Some have shiny abdomens that are somewhat elongated and may include a slight constriction, thereby creating the illusion of a distinct thorax, petiole, and abdomen; light areas at the constrictions contribute to the illusion. Instead of

Phidippus species, adult male. Bold markings and tufts of setae may be important during the courtship displays of male jumping spiders.

a typical spider gait, they adopt the walking gait of an ant. As a final touch, the spider holds either the first or second pair of legs forward and waves them around like the antennae of an ant. Since ants have relatively few predators, this gives some protection to the spider. Ironically, some species of ant mimics may actually prey on ants themselves, while most ant mimics take a variety of prey. In some tropical species of jumping spiders, a number of different species of ants serve as the models for a single species of jumping spider. With each successive molt, the jumping spider mimics a completely different species of ant that correlates in size to the growing spider.

One species of African jumping spider, *Myrmarachne melanotarsa*, has taken ant mimicry to yet another level. Not only do the individual spiders resemble ants, but these spiders also form aggregations, thereby exhibiting a form of collective mimicry. Since ants are rarely found away from other ants, the presence of many ant-mimic spiders in close proximity to each other is a more convincing display and more likely to deter potential predators.

In conclusion, jumping spiders rival any other group of creatures for their beauty, diversity, and complex behaviors. A combination of natural selection with sexual selection has produced an array of stunningly beautiful and surprisingly intelligent predators. The world is a richer place thanks to these diminutive gems.

The trials of courtship:

1. In some cases, the courting male jumping spider must pursue a reluctant female. This *Phidippus tux* was able to catch up with the female quickly. However, some male jumping spiders are led on a long chase where the female leaps from leaf to leaf, actively evading her ardent suitor. Leading the male on a merry chase may assist the female in selecting a "fit" suitor. Only an observant and agile male would be able to follow and mate with such a female. Producing eggs is a tremendous investment for the female; therefore, she must select only the most capable males to be her mates.

2, 3, and 4. Once the female accepts the male, he inserts the embolus from first one pedipalp into her epigynum, then from the second, in order to transfer the sperm. The pedipalps were loaded with sperm prior to courtship. The female obligingly tips up her abdomen to assist the male.

1. Waving his legs in a ritualized courtship dance, a male *Phidippus asotus* (right) approaches the female. He stands directly in front of her within the field of view of her high-resolution anterior median eyes.

2. Boldly advancing, the male continues to court the female. He must closely approach the female while courting her. This requires some bravery, since this also puts him within striking distance of a formidable predator.

3. Baring her fangs, this female has rejected her suitor. The male did not require more persuasion to abandon his suit; he beat a hasty retreat before she could kill him.

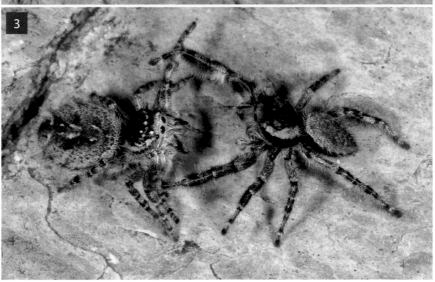

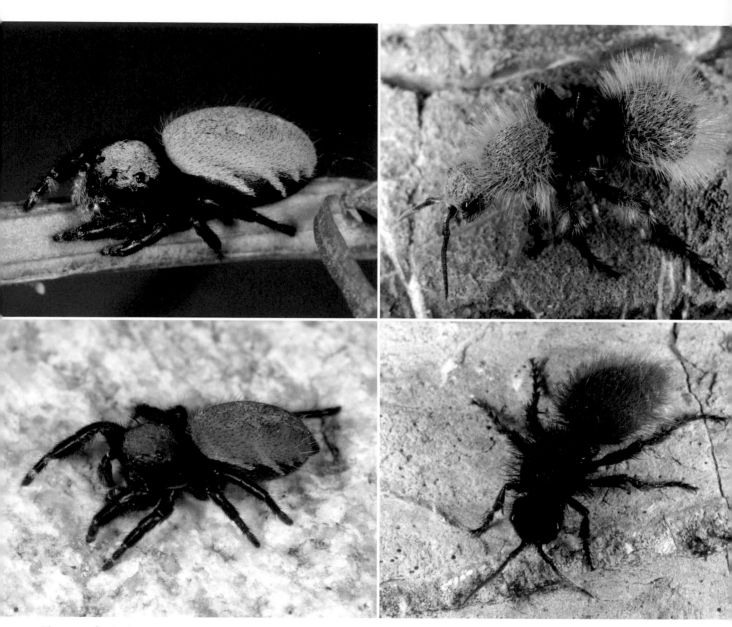

The art of mimicry:

Jumping spiders are diurnal hunters and are therefore vulnerable to sharp-eyed predators such as birds and lizards. One form of protection is Batesian mimicry, in which a harmless animal resembles a dangerous one. Many stinging insects are conspicuously colored in red, orange, yellow, and black. These warning colors are called aposematic coloration. *Phidippus apacheanus* has not only adopted aposematic coloration, it actually also has two coexisting color morphs that resemble two different female wasp models present in their environment. Although small and fuzzy, the female velvet ant wasp can deliver an extremely painful sting, so predators quickly learn to avoid anything resembling a velvet ant. Humans also use red and orange as warning colors.

1. Another form of mimicry is crypsis, or camouflage. This *Paramarpissa* species lives in trees and has evolved coloration that closely resembles the coloration of tree bark, even including dark areas simulating cracks in the bark. The front legs are fringed with hairs that obscure their outline, further enhancing the effect of invisibility.

2. Small, black, and shiny, this *Peckhamia* jumping spider is an almost perfect ant mimic. The illusion of three body segments is produced by a lighter coloration where the abdomen is slightly constricted. In addition, the second pair of legs serve as a facsimile of antennae, held aloft as the spider runs with an antlike gait. Few creatures eat ants, and therefore, this little spider can run about in plain view as it hunts on foliage.

1 and 2. *Habronattus dossenus* adult male.
Found in southern Arizona near the Mexican border, this species hunts on the ground.

3. *Habronattus clypeatus* penultimate male. The balloonlike pedipalps of this spider signify that he is almost mature. Paradoxically, the penultimate male of this species is more colorful than the mature male.

4 and 5. *Habronattus clypeatus* mature male.
The mature male has furry-looking pedipalps as well as distinctive facial markings and decorative setae on the front legs.

All these markings help the female to identify him during courtship.

6 and 7. *Habronattus conjunctus* mature male.
Although both these species of *Habronattus* are frequently found in the same habitat, *Habronattus clypeatus* tends to remain on the ground whereas *Habronattus conjunctus* tends to be found in low bushes, such as desert broom and seep willow. Therefore, they are rarely in direct competition with each other.

8. *Habronattus conjunctus* female with prey.

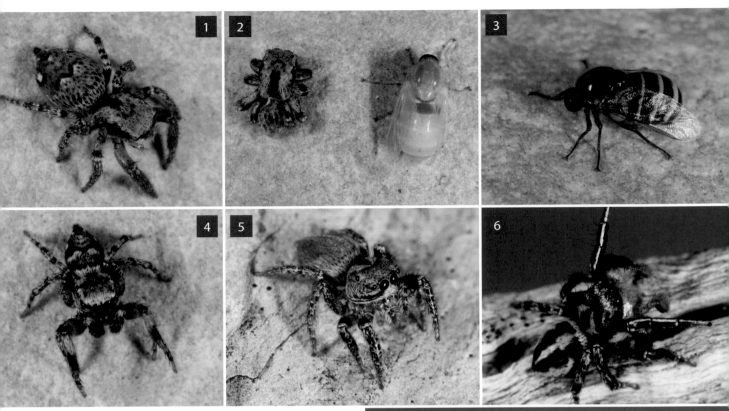

1. *Habronattus geronimoi* female.

2. Remains of the same *H. geronimoi* female and freshly matured small-headed fly, *Ogcodes* species. The maggot that emerged from the abdomen of the jumping spider was approximately the same size as the freshly matured fly. The maggot pupated for about a week as it metamorphosed into the mature fly.

3. Small-headed fly, *Ogcodes* species. This parasitoid lays its eggs on fallen logs or other areas where jumping spiders frequently hunt. Immediately after hatching, the maggot penetrates the cuticle of the spider at a leg joint and will grow within the spider's abdomen.

4. *Habronattus geronimoi* male. This species is found at higher elevations, usually in the oak zone of the mountains of southern Arizona. It can be found among the dead leaves of the forest floor or on dead logs.

5. *Habronattus hallani* female.

6 and 7. *Habronattus hallani* male. This species of *Habronattus* is found throughout a large part of the western United States and into Mexico. It is also primarily a ground hunter.

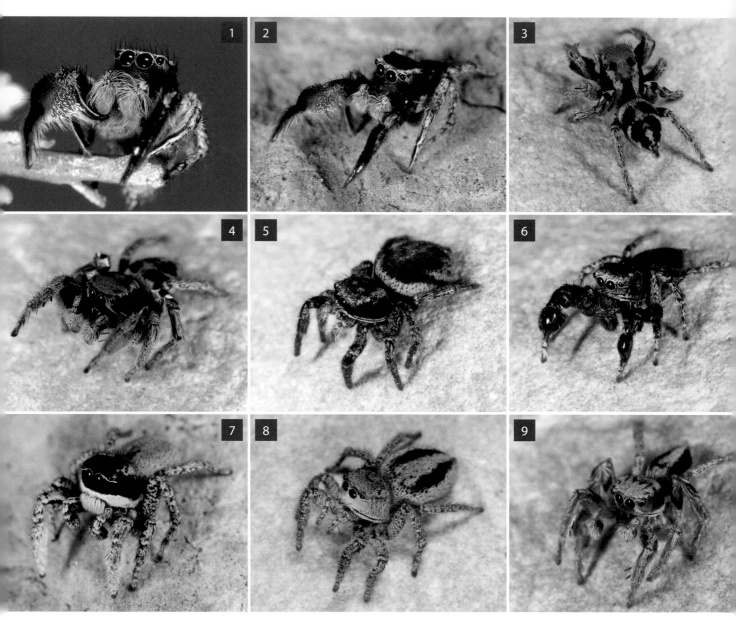

1 and 2. *Habronattus hirsutus* male. This species' name is descriptive of the long, feathery hairs on the front legs of the mature male. The yellow "racing stripe" on the inner surface of the leg also helps the female recognize a courting male. Not all populations have the red face of this male; some have black faces.

3 and 4. *Habronattus* near *festus* male. This may be one of several undescribed species within this extremely diverse genus.

5. *Habronattus oregonensis* female.

6. *Habronattus oregonensis* male. This is a mountain species, found in the pine zone.

7. *Habronattus pugilis* male.

8. *Habronattus pugilis* female. This montane species shows considerable regional variation. Each population is isolated from other populations, facilitating further diversification and possibly speciation within the group.

9. *Habronattus virgulatus* male. This species is found almost exclusively in areas where ocotillo grows.

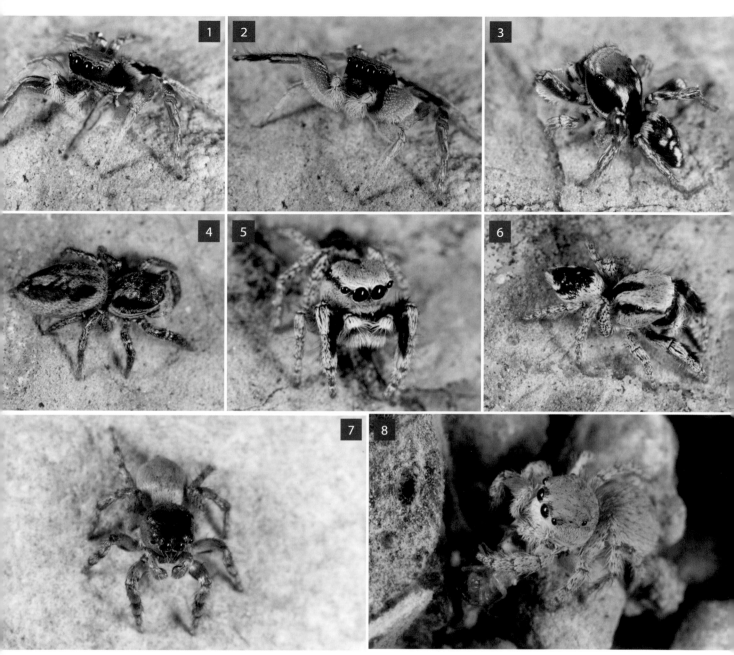

1, 2, and 3. *Habronattus pyrrithrix* male.

4. *Habronattus pyrrithrix* female. As the male raises his legs, the distinctive coloration and ornamentation of the underside of the front legs is displayed. This species is found primarily near water in grass or other very low vegetation.

5 and 6. *Habronattus tranquillus* male. This species is sometimes a brood parasite, eating the eggs and spiderlings of the desert shrub spider *Diguetia*.

7. *Habronattus ustulatus* male. This is the brown color morph of this species. A blue-gray color morph also exists in the males.

8. *Habronattus ustulatus* female. This tiny species hunts on the surface of the soil, where its drab coloration blends in perfectly. This female has captured a termite.

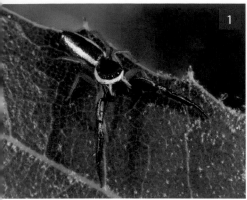

1. *Hentzia palmarum* male.

2. *Hentzia palmarum* female.

3. *Hentzia palmarum* mating. Found primarily east of the Rocky Mountains, *Hentzia palmarum* has an extremely limited distribution in the arid southwestern United States. This species lives in oak trees in Arizona close to the border with Mexico.

4. *Marpissa pikei* male. This species of jumping spider hunts exclusively in grasses. Its elongated body is compatible with the narrow blades of grass on which it sits.

5. *Marpissa pikei* female.

6. *Zygoballus rufipes* female. Like *Hentzia palmarum*, this species is found primarily in the eastern United States; however, it is also found in the same area of southern Arizona as *H. palmarum* but occupies lower vegetation instead of oak trees.

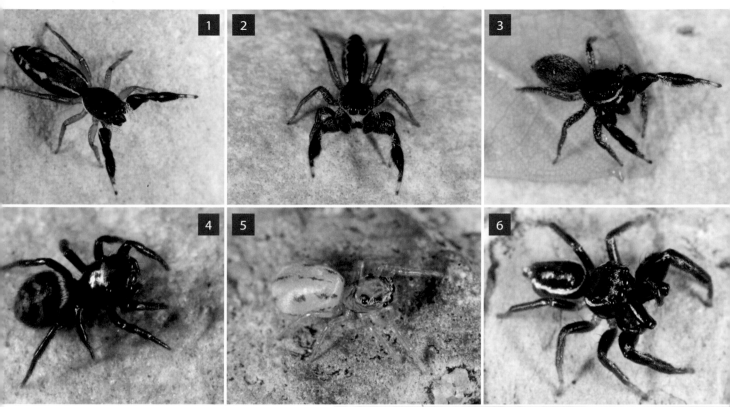

1 and 2. *Metacyrba floridana* female and male. This species has been only rarely documented in the western United States, possibly in part because of its habitat preference, which is primarily under the bark of trees.

3. *Metacyrba taeniola similis*. Like *M. floridana*, this species has evolved a morphology and lifestyle for tight spaces. These are also found under the bark of trees as well as under rocks. Both species share a resemblance to pseudoscorpions and, like pseudoscorpions, can move backward as rapidly as they move forward.

4. *Chalcoscirtus diminutus*. This tiny jumping spider is only 0.08–0.12 inches (2–3 mm) in body length.

5. *Cyllodania* species. This small jumping spider runs rather than jumps. It has been found in grasses at about 4,000 ft. (1,220 m) elevation.

6. *Messua limbata* male.

7. *Messua limbata* female. Living at higher elevations, this species is frequently found in oak trees.

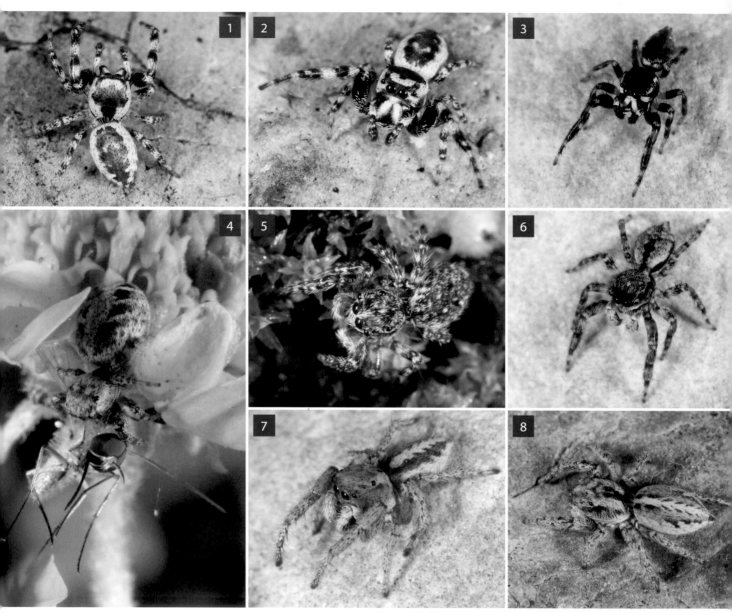

1 and 2. *Metaphidippus chera* males. *Metaphidippus chera* are commonly found in low-elevation desert trees and shrubs, such as mesquite and desert broom.

3. *Metaphidippus manni* male. As adults, these are often found in trees. However, there is some evidence that this species is a brood parasite on the desert shrub spider, *Diguetia*, living in the silk refuge of *Diguetia* and feeding on the eggs and spiderlings in the nest.

4. *Metaphidippus* female. These can frequently be found in the spring, hunting small insects on wildflowers such as brittlebush.

5. *Mexigonas* species female.

6. *Mexigonas* species male. This species is found in the cooler, higher-elevation oak and pine zones of the mountains. This particular female was hunting on a mossy rock. The size of the moss gives scale to this small spider.

7. *Pellenes limatus* male.

8. *Pellenes limatus* female. This genus is usually found in grassland, hunting on the ground or in low grasses.

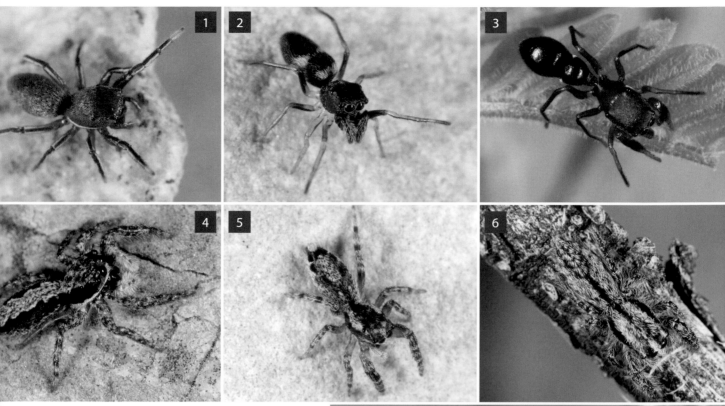

Convergent evolution: More than one species may share a similar strategy for survival.

Three genera of ant mimics:

1. *Tutelina* species.

2. *Sarinda hentzi.*

3. *Peckhamia* species.

Tutilina, *Sarinda*, and *Peckhamia* are all ant mimics, waving their legs in imitation of antennae and running with an antlike gait. This is an example of convergent evolution. Few predators feed on ants, so there is a selective advantage in resembling an ant for a diurnal hunter such as a jumping spider.

Three genera of bark mimics:

4. *Platycryptus californicus.*

5. *Phanias* species.

6. *Paramarpissa* species juvenile.

7. *Paramarpissa* species male. All three of these elongated, cryptically colored species are found primarily on the bark of trees such as oak and mesquite.

1. *Pelegrina furcata* female.

2. *Pelegrina furcata* male.

3. *Pelegrina furcata* with captured fly. These small jumping spiders can be found at higher elevations ranging from the oak zone up into the pine zone. They are frequently seen on wildflowers.

4. *Pelegrina* female, possibly *P. orestes* or *P. bunites*. It is often difficult to identify these spiders to species unless microscopic structures are examined.

5. *Terralonus* species. This small ground-dwelling spider is sometimes found on lichens, where it is well camouflaged. Some species of *Terralonus* are found primarily on beaches.

6. *Salticus palpalis* female.

7. *Salticus palpalis* male. These small jumping spiders are almost always found in trees, especially near a source of water where gnats and small flies are abundant. Their coloration may mimic the coloration of sweat bees.

1. *Colonus hesperus* male.

2. *Colonus hesperus* female.

3. *Colonus hesperus* female with prey. This species of jumping spider is commonly found near human habitations. This particular hunter captured her fly on the wall of a house. A carpenter bee mimic, the *Copestylum mexicanum* fly is far larger than the *Colonus*.

4. *Sassacus vitis* female. This species is frequently found on bushes and trees near a source of water.

5. *Sassacus papenhoei* mature male.

6. *Sassacus papenhoei* female. This species is abundant in low desert environments, inhabiting desert broom, mesquite, and other woody shrubs and trees. Small and with a metallic iridescence, these spiders may benefit from their resemblance to toxic leaf beetles.

1. *Phidippus asotus* juvenile.

2. *Phidippus asotus* male.

3. *Phidippus asotus* female.

This species is frequently found in juniper trees at elevations of approximately 4,000 ft. (1,220 m). This juvenile has captured a caterpillar that was feeding on juniper.

4. *Phidippus ardens* female.

5. *Phidippus audax* subadult. This widespread species is somewhat variable in coloration.

6. *Phidippus apacheanus* early instar. The juveniles of this species develop more orange coloration with each successive molt.

7. *Phidippus apacheanus* adult male.

The warning colors of the adult mimic the warning colors of female velvet ants.

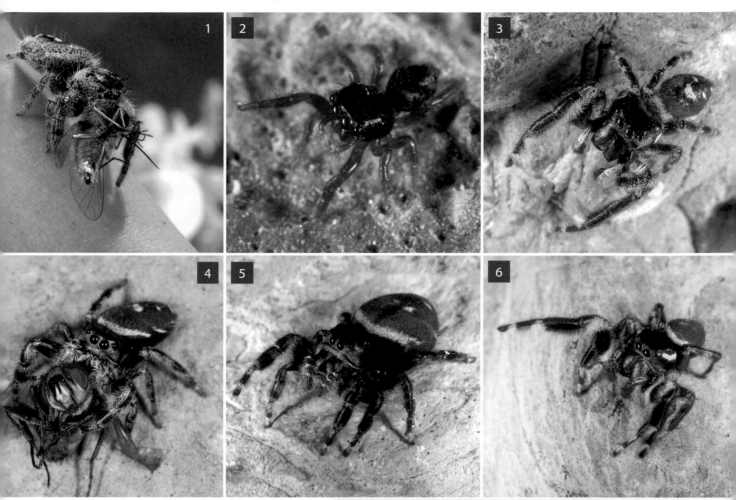

1. *Phidippus californicus* subadult hunting flies on a palo verde tree.

2. *Phidippus californicus* spiderling. Newly emerged from the egg sac, this tiny baby shows no hint of the beautiful colors it will develop later in life.

3. *Phidippus californicus* mature male.

4. *Phidippus carneus* subadult. Hunting for flies on a favorite sunning rock has paid off handsomely for this fierce little predator.

5. *Phidippus carneus* mature female. At 0.7 inches (18 mm) in length, this large jumping spider hunts mostly on the ground.

6. *Phidippus carneus* mature male.

7. *Phidippus carneus* juvenile. Many species of *Phidippus* have aposematic coloration, combining red or orange with black.

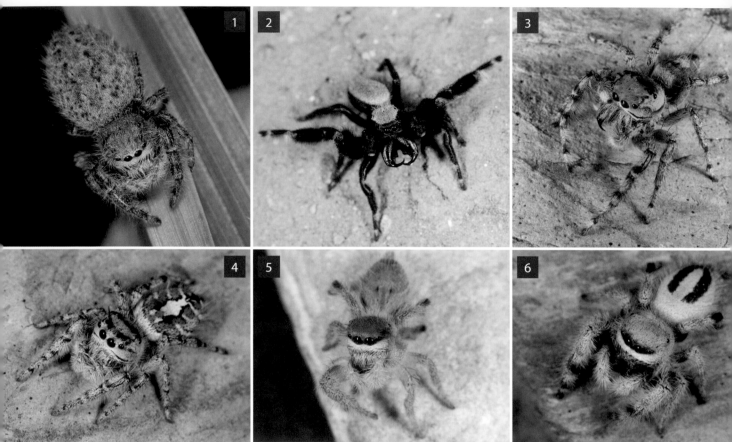

1. *Phidippus octopunctatus* female. The female is very large, about 0.7 inches (18 mm) in body length. These spiders live in grass and low vegetation, making a silk nest that can be 2 inches (51 mm) across.

2. *Phidippus octopunctatus* male. This male is doing a threat display, extending his fangs with a bitter drop of venom hanging from the tip of each fang.

3. *Phidippus phoenix* male.

4. *Phidippus phoenix* female. These are generally found on the ground in low desert.

5. *Phidippus tux* female.

6. *Phidippus tux* male. This species is found primarily in grasses and low vegetation in the oak zone in southern Arizona. The type locality is in southern Mexico.

7. *Phidippus tigris* male. These spiders are found at higher elevations primarily in the pine zone of the Huachuca Mountains of southern Arizona.

Habitat partitioning:

1. *Paraphidippus basalis*. This species is found in the cooler, higher elevation of the oak zone along with *Paraphidippus aurantius*. However, *Paraphidippus basalis* is found almost exclusively on plants such as agave and desert spoon, both of which form large rosettes of stiff, spiny leaves.

2. *Paraphidippus aurantius* female subadult.

3 and 4. *Paraphidippus aurantius* adult females.

5. *Paraphidippus aurantius* mature male. This species is found on leafy, herbaceous plants in the same area in which *Paraphidippus basalis* lives.

Consequently, these two species coexist by occupying different microhabitats. The mature females of *P. aurantius* show some variation in color, with either white or orange on their abdomen. The metallic green or bronze is surprisingly effective camouflage in leafy vegetation.

CHAPTER 17 Lynx Spiders:
Oxyopidae

More tropical than temperate, the green lynx spider (*Peucetia viridans*) belongs to a genus with worldwide tropical and subtropical distribution. It can be found throughout the south in the United States.

Gila monsters, baby quail, and turkey vultures herald the return of spring in the Arizona desert. Less conspicuous is yet another sign of spring in the arid southwest: green lynx spiders lurking in prickly pear flowers as they ambush unwary pollinators. Ambush hunting depends upon being unnoticed in one's environment, and it is striking how well *Peucetia viridans* utilize the shapes and colors of the flowers and leaves on which they wait. Lynx spiders are visual hunters, having 6 fairly large eyes distributed in a roughly hexagonal arrangement on the cephalothorax, along with 2 smaller eyes facing forward. When an insect comes close enough, it is captured with a quick lunge and a grab with the spider's long, spiny forelegs.

Tiny, recently hatched lynx spiderlings start their hunting life after ballooning away from the maternal nest in late summer or fall. They will then make their homes among vegetation, ambushing small flies. This arrangement continues during the winter and into the spring as long as temperatures remain mild enough. As spring progresses to summer, the spiders grow in size and their choice of prey shifts. Warmer weather brings a bounty of insects, and the spiders' larger size enables them to take full advantage of the wider range of prey. These generalist spiders opportunistically eat whatever they can catch. Larger prey such as bees replace the small flies taken in early spring. The slow-moving non-native honeybees are seemingly preferred over the alert and nimble native bees. Still later in summer, grasshoppers, bugs, and even other spiders (including members of their own species) are captured and eaten.

During the summer, the female constructs an egg sac a foot or two (30–60 cm) above the ground, fastened to vegetation with silk. Sometimes the female builds a shelter of leaves and silk over the egg sac, which may hold from 130 to 600 eggs. During construction of the sac, or even after it has been completed, a tiny intruder may appear. This is the 0.04-inch (1 mm) first-instar larva of a mantidfly, also known as a *Mantispa*. These insects are in the same order as lacewings and ant lions—the Neuroptera. Resembling an unlikely insect chimera, the adult mantidfly has both the delicate gossamer wings of a lacewing and the heavy raptorial forelegs of a praying mantis. Neuropteran larvae are best characterized by their large, sharply pointed mandibles with which they grasp their prey. The minute *Mantispa* larva enters the egg sac of the lynx spider (or any of several

other species of spiders) and proceeds to devour the eggs and spiderlings. It then spins a pupation cocoon while still in the egg sac and eventually emerges as an adult mantidfly. In some cases, all the eggs and baby spiders are consumed, while in other instances, some of the spiders may survive despite the presence of their rapacious roommate. Meanwhile, the mother spider watches over the egg sac, diligently guarding it.

The female spider is the model of dedicated motherhood. She carefully shifts the position of the egg sac to ensure that it is heated or cooled as necessary, maintaining the best incubation temperatures possible. She must guard the eggs and newly emerged spiderlings against invertebrates such as ants and other spiders and must protect herself against being eaten by birds or lizards. If she were to die before the babies hatch, their chance of successfully emerging from the sac drops from about 70 percent to about 10 percent. Behavioral plasticity allows the green lynx spider to deploy a variety of completely different defense strategies against a diversity of threats.

Beetles and grasshoppers may eat the eggs, and grasshoppers are capable of polishing off an entire egg sac and its contents in just half an hour. Ants are a major threat to eggs. They can chew a hole through the egg sac and carry off the eggs one by one until none are left. If the mother spider discovers that ants are attacking the egg sac, she will first attempt to drive off the ants by biting them even as they counterattack and bite her. Sometimes after such a battle, a spider may have the bodiless head of an ant still clamped onto a leg or chelicera, and in extreme cases the spider may voluntarily amputate (autotomize) her own leg when an ant does not release it. If she is not able to drive the ants away from her egg sac, she has a couple other strategies to which she may resort.

She can suspend the sac out of reach of the ants, or she may relocate the sac to a safer location.

Relocating is a laborious process, involving either moving the sac to a new site on the same plant or even transferring it to an altogether different plant. While many spiders carry their egg sacs in their palps or chelicerae, the lynx spider does not. Instead, she will alternatively cut old silk support lines that attach the sac to the original location and fasten new lines leading to the new location. Thus the sac is supported on silk lines while the spider guides it from one location to the other. The move is made all the more complicated by

1. A female green lynx spider (*Peucetia viridans*) guards her egg sac against both vertebrate and invertebrate predators.

2. A common brood parasite for spider eggs is *Mantispa cinticornis*, the mantidfly. The larva of this lacewing cousin infiltrates the egg sac and devours eggs and spiderlings. It then pupates in the egg sac.

3. The mother green lynx spider remains close to the egg sac and babies until her offspring have dispersed. She captures food during this time but does not share it with her offspring.

4. Newly emerged green lynx spiderlings.

Oxyopes tridens. The female trident lynx spider constructs her egg sac in low vegetation such as grasses and then guards it.

the fact that the female spider still has to fight off the attacking ants while she is doing this.

Suspending the sac out of reach of the ants is accomplished by cutting most of the threads of silk that attach the sac to the nest plant, leaving the sac dangling by only one to three silk lines. The mother spider then climbs on top of the sac and guards it. The ants are thus restricted to approaching from a very narrow front, which the spider can easily defend.

Different species of green lynx spiders are found throughout the tropics, suggesting that *Peucetia viridans* of the southern United States probably has tropical ancestry. Ants are a major component of tropical ecosystems, and the defensive strategies described above are very likely a legacy of the green lynx spider's coevolution with tropical ants.

The sac spider, *Cheiracanthium inclusum*, is a nocturnal predator on the eggs of the green lynx spider. The sac spider builds a little silk retreat immediately adjacent to the lynx spider's egg sac. It then chews a hole in the egg sac and feasts on the eggs and spiderlings, after which it may use the empty egg sac as a temporary retreat. Not surprisingly, this predation seems to occur more commonly with unguarded egg sacs.

The green lynx spider has a unique defense against vertebrate predators: she sprays venom from her fangs. The venom tastes bitter, and if it lands in the predator's eyes it can irritate them for a couple days. The spider can spray her venom for an impressive distance of 8 inches (20 cm).

After the babies hatch, the mother opens the sac for them. Presumably the babies cue their mother when they are ready to emerge because empty sacs or sacs that contain only a mantidfly are never opened by the mother spider, even though she may still be guarding the sac. The babies stay close to the egg sac until they molt. During this time the mother stays near the babies and catches insect prey, but she does not share the food with the babies. A few days after molting the babies are ready to disperse by ballooning. Climbing to the top of a leaf or twig, a baby spider stands on tiptoe, raises its abdomen to the sky, and releases a strand of silk. This strand of silk catches the breeze, and in the blink of an eye the baby spider is gone—off to start a new life wherever the wind may take it.

Two other members of the lynx spider family are *Hamataliwa grisea* and *Oxyopes tridens*. Little is known about these two species, not because they are particularly rare, but because they are difficult to observe. Both have the typical eye arrangement of their family: a hexagonal pattern of 6 large eyes and the 2 small forward-facing eyes. Both share the tendency of other family members to rapidly jump up and down when threatened. *Hamataliwa* frequents acacia and

Peucetia viridans can frequently be found ambushing pollinators as they visit flowers.

mesquite trees and ocotillos, for which it is perfectly camouflaged. *Oxyopes tridens* lives in the grasses—a small-scale grassland predator able to run down and pounce on its prey like a miniature lion. A young *Oxyopes tridens* may tackle prey much larger than itself using a technique similar to that of oecobiid spiders. To bring down a large ant, for example, the *Oxyopes* races around its prey, encircling the ant's legs with silk, completely immobilizing it before moving in for the kill.

All three of these genera may inhabit the same locality; however, they appear to partition the habitat by selecting different vegetation types and heights in which to live. The *Hamataliwa* species lives high up in the branches of trees and ocotillos. *Peucetia viridans*,

the green lynx spider, lives in midlevel shrubs such as desert broom or in prickly pear cacti. And the *Oxyopes tridens* is found close to the ground in grasses, weeds, or other low-growing vegetation. Thus a single locality may support a surprising diversity of lynx spiders.

Other species of *Oxyopes* are also found in this region, including the striped lynx spider. This spider is usually encountered where temperatures remain more moderate, such as the mountainous "sky islands" and riparian canyons of the arid southwestern United States.

Oxyopes spiders construct a lumpy egg sac on leaves or other vegetation and guard it until the babies disperse. Like their cousins the green lynx spiders, the babies throw their lives and luck to the wind, ballooning away to a new place and a new life as little hunters.

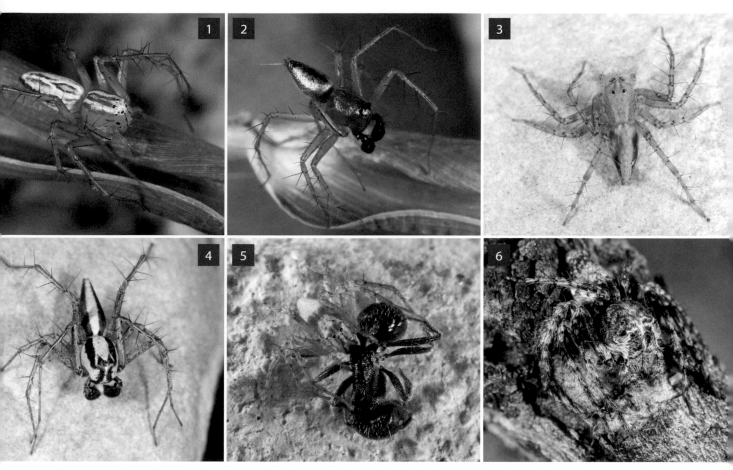

1. *Oxyopes salticus* female. The striped lynx spider can be common in tall grass.

2. *Oxyopes salticus* male. The male has completely different coloration from the female. It is not known why this species demonstrates such extreme sexual dimorphism.

3. *Oxyopes scalaris* subadult. The western lynx spider.

4. *Oxyopes tridens* mature male. The trident lynx spider

5. *Oxyopes* juvenile. This tiny lynx spider captured a large ant by racing around the ant in circles, hobbling it with silk before moving in for the kill.

6. *Hamataliwa grisea* is perfectly camouflaged when sitting on tree bark. It is found almost exclusively in trees and ocotillos.

Crab Spiders:
Thomisidae, Sparassidae, Philodromidae, Selenopidae, and Zoropsidae

A perfect match in color, this white-banded crab spider (*Misumenoides formosipes*) is camouflaged as it waits in ambush for unwary insects on a yellow cowpen daisy.

Crab Spiders: Thomisidae

Bribery, deception, and murder are commonplace in a patch of wildflowers. Plants bribe pollinators to visit by providing nectar and pollen rewards, advertising with conspicuous floral displays. Lurking within the flowers are the opportunistic ambush hunters, the flower spiders. Belonging to the family Thomisidae, flower spiders have taken the art of camouflage to a whole new level; they can actually change the color of their bodies to match the color of the flower in which they sit.

Juvenile flower spiders have almost no pigment, and a translucent cuticle covers their body. Their body color changes according to the color pigments found in their prey; eating a green insect may produce a green spider. With other food items, they may turn pink, orange, brown, yellow, or white. This may help them blend with their environment, but perhaps more importantly it makes the spider's appearance unpredictable, making it more difficult for a predator to develop a reliable search image of the spider.

As the spider matures it adds the ability to produce its own pigment in response to the color of the flower it occupies. The color change is not instantaneous but takes place over the period of a few days. An adult flower spider hunting on white flowers will appear white from the light reflected off the guanine crystals beneath the cuticle. But if that spider relocates to a patch of yellow flowers, the spider will slowly release a yellow pigment into its hypoderm, turning the body and legs yellow. The longer it remains among the yellow flowers, the brighter yellow it becomes. It is not yet known how the spider knows to change colors, nor how it chooses flowers with which it blends, but one study showed that 85 percent of the white spiders found were in white flowers, and 85 percent of the yellow ones were in yellow flowers. Matching the flower color protects the spiders from potential predators, allowing crab spiders the chance to exploit the rich abundance of diurnal pollinators.

Flower spiders are attracted to flowers, but how they find and recognize flowers is not yet known. There is some preliminary evidence that they may be attracted to some flower compounds and to ultraviolet light reflected by flowers. Pitfall traps baited with floral compounds and set for catching other arachnids have drawn flower spiders. Several families of diurnal hunting spiders, including the crab spiders and jumping spiders, are known to be able to see ultraviolet light. Adult flower spiders have turned up at ultraviolet-lit sheets set out as insect traps, even when the sheets are away from vegetation in which flower spiders may be living. Many flowers reflect ultraviolet light in patterns that guide pollinators to the nectar and pollen. It is possible that the spiders can utilize these nectar guides in positioning themselves for ambushing pollinators drawn to the flowers.

Among the favorite prey choices of flower spiders are flower flies (Syrphidae) and bee flies (Bombyliidae) as well as honey bees and butterflies. The spider venom is extremely effective, allowing the flower spiders to tackle large and potentially dangerous prey such as honey bees, which are quickly subdued with a bite between the head and the thorax.

Male flower spiders are significantly smaller than females and in some varieties, such as *Misumenoides,* may be darker or colored differently. There is a popular myth that adult males feed exclusively on nectar, but this is not true. Although they do drink some nectar from the flowers on which they hunt, they have been observed to readily kill and eat small flies as well. Both males and females possess the characteristic crablike body shape, including large raptorial front legs used for grasping their prey. They further resemble their namesake in that they move sideways like a crab. Sitting motionless and unnoticeable on a flower with front legs spread wide to welcome an unwary pollinator, they are the embodiment of the ambush hunter.

Several varieties of flower spider may share the same habitat. Arizona boasts species of both *Mecaphesa* and *Misumenoides.* Different maturation times for these two genera may allow them to effectively partition their habitat, separating them in time rather than space. *Mecaphesa* may hatch in early summer, overwinter as late-stage juveniles, and mature in early spring. In contrast, the white-banded crab spider *Misumenoides formosipes* matures in late summer or autumn, and the young overwinter in the egg cocoon. This sequence allows two species occupying similar niches to avoid direct competition.

Flower spiders construct a silk nest in which to lay their eggs. The white-banded crab spider sits in her nest guarding her eggs but dies before the eggs hatch. Perhaps one of the most touching accounts of maternal devotion in spiders was written by J. Henri Fabre in the

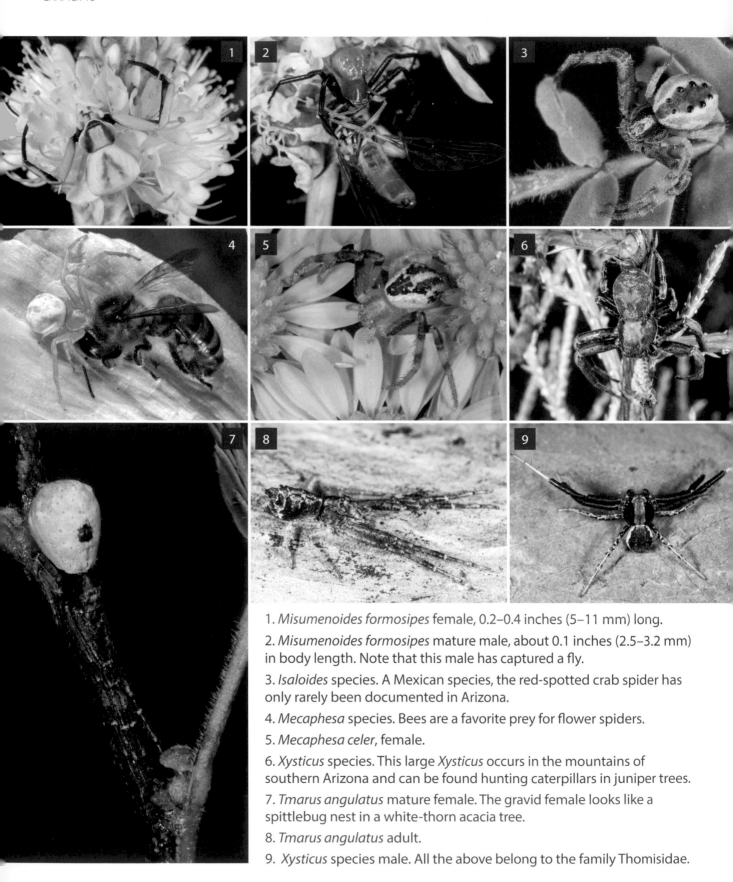

1. *Misumenoides formosipes* female, 0.2–0.4 inches (5–11 mm) long.

2. *Misumenoides formosipes* mature male, about 0.1 inches (2.5–3.2 mm) in body length. Note that this male has captured a fly.

3. *Isaloides* species. A Mexican species, the red-spotted crab spider has only rarely been documented in Arizona.

4. *Mecaphesa* species. Bees are a favorite prey for flower spiders.

5. *Mecaphesa celer*, female.

6. *Xysticus* species. This large *Xysticus* occurs in the mountains of southern Arizona and can be found hunting caterpillars in juniper trees.

7. *Tmarus angulatus* mature female. The gravid female looks like a spittlebug nest in a white-thorn acacia tree.

8. *Tmarus angulatus* adult.

9. *Xysticus* species male. All the above belong to the family Thomisidae.

early 1900s. Describing a flower spider common to the French countryside in *The Life of the Spider*, he writes:

> *The work of laying is finished by the end of May, after which, lying flat on the ceiling of her nest, the mother never leaves her guard-room, either by night or day. Seeing her look so thin and wrinkled, I imagine that I can please her by bringing a provision of Bees, as I was wont to do. I have misjudged her needs. The Bee, hitherto her favorite dish, tempts her no longer. In vain does the prey buzz close by, an easy capture within the cage: the watcher does not shift from her post, takes no notice of the windfall. She lives exclusively upon maternal devotion, a commendable but unsubstantial fare. And so I see her pining away from day to day, becoming more and more wrinkled. What is the withered thing waiting for, before expiring? She is waiting for her children to emerge; the dying creature is still of use to them....*
>
> *The fabric is too thick and tough to have yielded to the twitches of the feeble little prisoners. It was the mother, therefore, who, feeling her offspring shuffle impatiently under the silken ceiling, herself made the hole in the bag. She persists in living for five or six weeks, despite her shattered health, so as to give a last helping hand and open the door for her family. After performing this duty, she gently lets herself die, hugging her nest and turning into a shriveled relic.*

The courtship of most flower spiders is fairly straightforward and uneventful. The small male simply walks over to the female and, with a minimum of introductory caresses, mates with her. However, the courtship of another thomisid, *Xysticus*, is more complex and unusual. As the male *Xysticus* woos the female, he spins a silken "bridal veil" around her, encircling her and tying down her legs with many strands of silk. Despite the romantic name, the "bridal veil" may have a pragmatic purpose. The silken bonds may give the male spider time to mate and escape before the potentially dangerous female frees herself. Several other species in other spider families also utilize this technique.

Another North American member of the family Thomisidae is *Tmarus*. This angular spider can be found with its long front legs stretched out along a twig of white thorn acacia or other tree. While the body of the male or immature spider appears to be a bit of protruding bark, a mature female looks for all the world like a spittlebug nest, nestled among the same twigs in which the frothy white insect nests are found. Like other members of this family, this predator is accomplished at hiding in plain sight.

It is worth mentioning that the family Thomisidae includes three species of social spiders, all living in Australia and belonging to the genus *Australomisidia*. In the case of *Australomisidia ergandros*, each colony is founded by an adult, mated female. She constructs a home base by fastening eucalyptus leaves together to form a nest in which she incubates her egg sac. Her offspring do not immediately disperse the way most other thomisids do; instead, they remain in the maternal nest, and as the spiderlings grow larger, they communally add more leaves to enlarge their shelter. The mother spider captures prey and shares it with her offspring. As the mother spider grows older and weaker, her offspring proceed to feed on her. After mating, the female offspring leave the natal nest and start their own colonies.

Selenopid Crab Spiders: Selenopidae

Selenopid crab spiders are extremely flat, mostly tropical spiders. Adapted to hiding in crevices and under bark by day, they cruise around on vertical surfaces at night hunting for insects. They are commonly found in caves in the twilight zone, where they may hunt camel crickets. In turn the spiders are hunted by a wasp. Several kinds of wasps are adapted to hunting in the twilight zone of caves, including cricket hunters (in the family Sphecidae) and the spider wasp *Ageniella evansi* (in the family Pompilidae). This wasp chews off the legs of the spider before carrying it off to lay an egg on it as food for its larvae.

Houses and garages in many ways resemble man-made caves, providing a relatively constant temperature and many vertical surfaces on which to hunt. Selenopid crab spiders are quick to adapt to this artificial habitat and make themselves at home. Since they are not aggressive or dangerous to humans, these add an interesting and beneficial component to the ecosystem of a home, hunting crickets and other insects.

Lauricius hooki (**Zoropsidae**)

Spiders in the family Zoropsidae are not easily distinguished as a group. In fact, many of the spiders in this group have been placed in other families at one time or another. Microscopically, many share the characteristics of both a vestigial third claw along with claw tufts or scopula hairs on their tarsi. Although this may discourage field identification, one species in this family is quite distinctive and may be easily recognized. *Lauricius hooki* is a handsome spider found at higher elevations in the mountains of Arizona and New Mexico. This spider bears a slight resemblance to the selenopid crab spiders, possessing laterigrade legs and a mottled tan and brown coloration. But they are generally not quite as extremely flat to the surface as are the selenopid crab spiders, and their eyes are arranged in a different pattern. During the day *Lauricius hooki* may be found under debris on the ground from the juniper zone up to the pine zone, from 4,000 to 6,000 ft. (1,220–1,829 m) in elevation or even higher. At night, they wander as they hunt. These spiders are also frequently found in caves; hence they are troglophilic, or "cave-loving." Consequently, they can also be found in "man-made caves," such as the concrete buildings serving as rest rooms in hiking and camping sites.

Running Crab Spiders: Philodromidae

Philodromid crab spiders are most conspicuous hunting for moths at night on the outside of windows. Also called running crab spiders, they have long legs that allow them to move rapidly on vertical surfaces such as windows, walls, and plants, where they hunt their prey. During the day they conceal themselves in vegetation or under debris. One genus, *Tibellus*, has an elongated body. During the day it aligns its body along a plant stem with two pairs of legs lying stretched out in front and one pair lying flat behind it (while one pair securely embraces the vegetation). Coupled with

its cryptic coloration, it is well hidden from birds and other predators.

Many philodromids occupy the same areas as the flower spiders, but each group partitions the habitat based on their circadian rhythms. Flower spiders hunt principally by day, whereas philodromids are active primarily at night.

An identifying characteristic of Philodromidae is that the second leg is longer than the first. Unlike thomisids, most philodromid spiders also possess claw tufts and scopula hairs on their tarsi.

Giant Crab Spiders: Sparassidae

Giant crab spiders, *Olios giganteus*, may be responsible for more startle reactions than most of the other spiders put together. Like their cousin, the famous banana spider *Heteropoda venatoria*, their size alone elicits excitement and consternation, with a leg span that may exceed 2 inches (5.1 cm). Coupled with shiny black chelicerae and beady black eyes, they appear imposing, especially while nonchalantly strolling along the ceiling over someone's bed. However, they are not aggressive toward humans, and once an individual is seen repeatedly and becomes a familiar presence in the house, an appreciation for these handsome spiders may develop. Unlike many spiders that live outside, an indoor spider may be observed over a long period of time, giving special insights into its life history. A particularly interesting aspect of spider biology may be witnessed in this way: the regeneration of missing limbs.

As long as a spider has not reached maturity, it can regenerate legs to replace those that are lost. Legs are frequently damaged in the course of the spider's life of hunting. If a leg becomes badly damaged, or worse yet, envenomated, it may be best to self-amputate (autotomize) the limb. Orb weavers do this if bitten by venomous ambush bugs; they may die if they do not. The life of a hunter is full of dangers, and if a predator or rival seizes a spider's leg, losing the leg may be a small price to pay to escape with one's life. Consequently many spiders are missing one, two, three, or even four limbs, and are still able to hunt. The leg is voluntarily broken off between the coxa and the trochanter; however, there would be a strong selective advantage in being able to regenerate legs. Many arachnids can

Olios giganteus, the giant crab spider. The lower left leg has been regenerated; therefore, it appears a bit thinner than the other legs.

in fact regenerate limbs as long as they have not yet undergone their final molt. The new limb is contained inside the old coxa, highly folded and completely hidden until the molt, at which time it is freed from its confinement. Somewhat thinner and smaller than the original, it is still functional. Of course, the molt also gives a fresh cuticle cover to the eyes, the fangs, and all the other body surfaces. The shiny appearance of spiders' eyes is due not to the presence of moisture, as is the case with birds and mammals, but to the cuticle that protects the spider from desiccation. As a consequence, spiders cannot blink.

Another noteworthy aspect of giant crab spiders is that they chew their food. Many spiders simply suck out the partly digested inside of an insect, leaving the empty exoskeleton more or less intact. But some, like the giant crab spiders, have cheliceral teeth with which they first masticate their prey before sucking down the predigested, liquefied material. Any solid particles larger than 1 millimeter are filtered out in the pharynx and regurgitated as a little pellet, somewhat analogous to an owl regurgitating the hair and bones of its prey.

Other spiders that share this technique are *Zorocrates* and the wolf spiders.

Some of the most extraordinary spiders of the world belong to the family Sparassidae. Although from a morphological viewpoint these spiders may not appear very different from each other, their behaviors are diverse and highly distinctive.

A spider from southern Africa, *Leucorchestris arenicola*, shows an amazing ability to navigate in its environment and to "remember" the location of its home burrow. It is a mystery how the mature male spider can make a nighttime round trip of as much as 2,625 feet (800 m), wandering here and there as he searches for females, and then proceeding not only to find his home burrow but also to return to it in a straight line.

Another sparassid from the southern part of Africa has evolved a specialized style of locomotion. Living in the deserts of Namibia, *Carparachne aureoflava* can actually cartwheel down sand dunes in order to escape from pompilid wasps. The spider first tips its body to the side and bends its legs before rolling down the dune

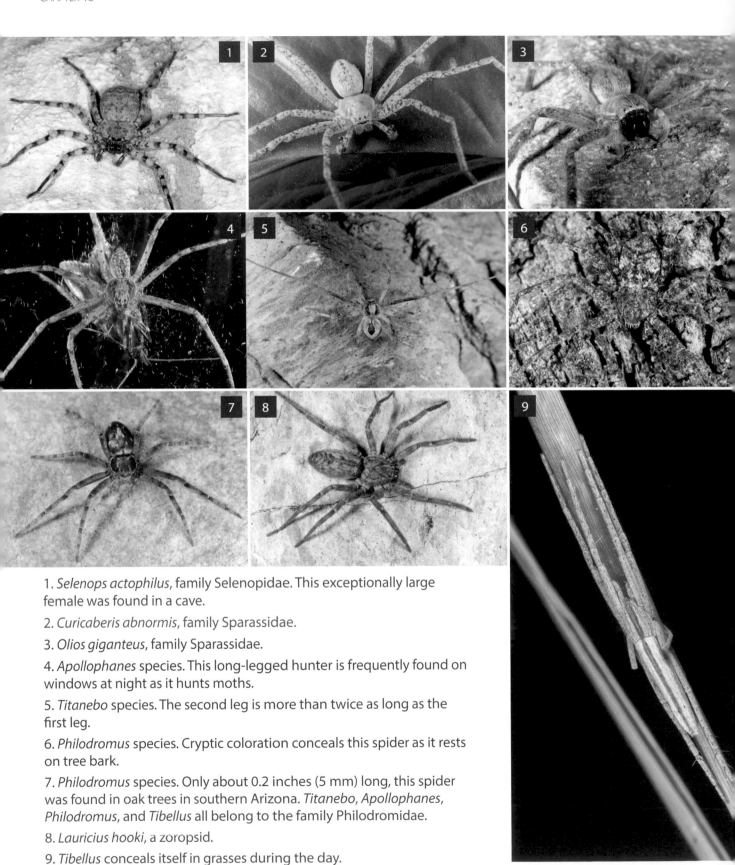

1. *Selenops actophilus*, family Selenopidae. This exceptionally large female was found in a cave.

2. *Curicaberis abnormis*, family Sparassidae.

3. *Olios giganteus*, family Sparassidae.

4. *Apollophanes* species. This long-legged hunter is frequently found on windows at night as it hunts moths.

5. *Titanebo* species. The second leg is more than twice as long as the first leg.

6. *Philodromus* species. Cryptic coloration conceals this spider as it rests on tree bark.

7. *Philodromus* species. Only about 0.2 inches (5 mm) long, this spider was found in oak trees in southern Arizona. *Titanebo*, *Apollophanes*, *Philodromus*, and *Tibellus* all belong to the family Philodromidae.

8. *Lauricius hooki*, a zoropsid.

9. *Tibellus* conceals itself in grasses during the day.

like the wheel of a car. The spider may achieve speeds of more than 3 feet (1 m) per second, a mere blur as it spins down the dune.

Perhaps the most remarkable sparassid is *Delena cancerides*, from Australia. Contrary to all expectations, this is a subsocial spider. Most social spiders are web-based, but these huntsman spiders are wandering hunters. Each colony may include as many as 300 individual spiders. The colony lives under the exfoliating bark of trees. Prey is shared among members of the colony, benefiting smaller, younger spiders. Although cannibalism is almost nonexistent between members within a colony, older spiders from other colonies are not tolerated. However, unrelated young immigrant spiders are accepted into a colony, thereby permitting outcrossing in this species.

A grassy meadow is rich habitat for *Tibellus*. At night, this spider can be seen actively hunting on vegetation, but by day, it remains motionless as it lies along a blade of grass.

Sand Spiders and Wolf Spiders:
Homalonychidae and Lycosidae

Shiny black eyes gaze out from the sand-covered visage of this *Homalonychus selenopoides* spider. The cuticle covering the eyes and the leg joints lacks the specialized setae that cover the rest of the body; consequently, the eyes and the leg joints remain free of sand after a dust bath.

Sand Spiders (Homalonychidae)

This long-legged wandering hunter lives only in the deserts of the southwestern United States, northwestern Mexico, and Baja California. Like many other spiders, *Homalonychus* is camouflaged to blend in with its environment. But unlike most other spiders, it achieves this effect by taking a dust bath.

Upon encountering soil of a suitable consistency, a freshly molted *Homalonychus* spider will start to dig furiously. The spider bears an almost uncanny resemblance to a terrier burying a bone as it rapidly scoops dirt in a pile behind itself. Suddenly, the spider stops digging, rears up on its hind legs, and tips over backward, falling upside down onto the dirt pile. There, it wriggles and shivers its body and legs in the dirt before righting itself. It then proceeds to do a belly flop onto the dirt pile. Lying flat with its legs radiating straight out, it again shudders and squirms in the dust for several seconds. This behavior is repeated in its entirety, including the digging, as many as 30 times. Each repetition coats the spider more thoroughly in dust particles. At the conclusion of this process, the spider is completely coated; only the eyes and the thin cuticle at the leg joints are spared.

Homalonychus spiders have a specific adaptation allowing them to acquire this camouflage. Specialized hairs, or setae, attract and hold the dirt particles. Electron microscopy reveals several long, thin hairlettes protruding from each of these setae. As dirt particles contact these microscopic hairlettes, they "stick" to them, thus conferring crypsis to the spider.

In an elegant example of convergent evolution, a completely unrelated group of spiders from South America and Africa in the genus *Sicarius* has independently evolved an almost identical strategy for becoming camouflaged. These spiders actually bury themselves in sand. (In the reptile world, a number of vipers from Africa and sidewinders from the American deserts are known to use this technique, called cratering.) Like *Homalonychus*, the setae of *Sicarius* become covered in sand. Under the electron microscope, the setae of *Homalonychus* and *Sicarius* are virtually identical. These hairs are unlike any other spiders' setae. Although camouflaging the body with dirt is seen in some other arthropods (such as assassin bug nymphs), this strategy is extremely rare

Homalonychus theologus (above) is found in southern California and into Mexico. During the day it can be found in chaparral under debris such as dead wood and rocks. At night it comes out to hunt.

among spiders. It is probably no coincidence that both *Homalonychus* and *Sicarius* are found in arid climates, where fine dry particles of dirt are readily available.

However, although females and immature males conceal themselves in this way, mature males do not. This is where another form of camouflage may come into play: mimicry.

As the sand spider rests during the day, it finds refuge under dead wood, decomposing cacti, or rocks. While at rest, the spider holds its legs together in pairs pointing forward and backward so that the four pairs of legs collectively form a single X shape. In this resting position, the spider resembles a cluster of cactus spines, a common enough remnant left behind after cacti disintegrate, and hardly an edible morsel by most predators' standards.

Homalonychus hunts at night using a technique similar to that of the wolf spider. It runs down an insect and holds it in a "basket" formed by its legs. The proportions of the legs balance the need for strength with the need to keep dangerous prey "at arm's length."

Homalonychus spiders perform an unusual courtship. First of all, after his final molt, the mature male does not take a dust bath, perhaps assisting the female in recognizing the male as a suitor. As the male approaches the female, he drums on the ground with his legs and his pedipalps. The receptive female abruptly lowers herself to the ground and lies motionless and quiet. The male becomes very animated as he circles the passive female, repeatedly touching her with his front feet. Soon, he

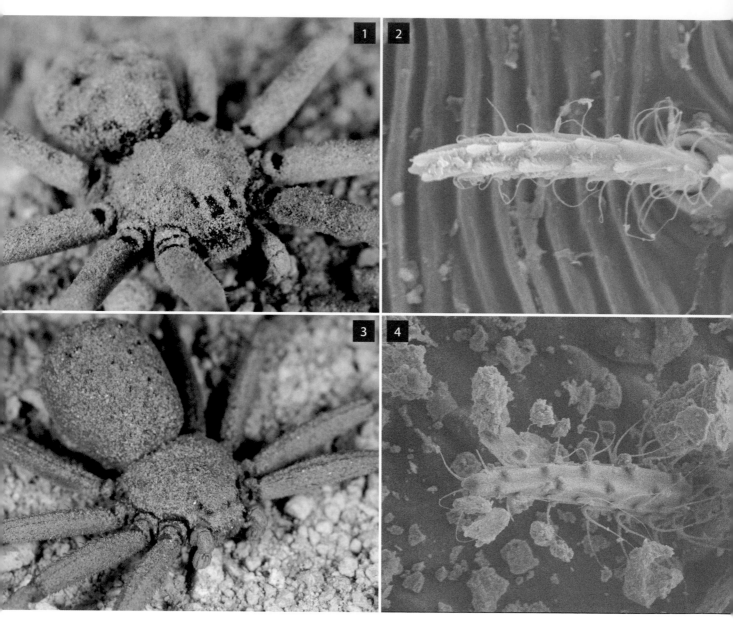

An elegant case of convergence:

1. *Homalonychus* species occur in the arid southwestern United States, northwestern Mexico, and Baja California.

2. The specialized setae of *Homalonychus* possess tiny hairlettes. These setae function as "dirt holders."

3. *Sicarius* spiders occur in the deserts of Africa and South America. These spiders bury themselves in sand, suddenly popping up as they ambush unwary prey.

4. The setae of *Sicarius* bear an uncanny resemblance to the setae of *Homalonychus*. The tiny hairlettes attract and hold fine particles of sand, as seen in this electron micrograph.

Homalonychus and *Sicarius* belong to two families of spiders: the eight-eyed Homalonychidae and the six-eyed Sicariidae; however, these unrelated families of spiders have independently evolved the same structures given the same challenge, namely the ability to use a dry, dusty substrate as camouflage. These specialized hairlettes are found only on these desert spiders.

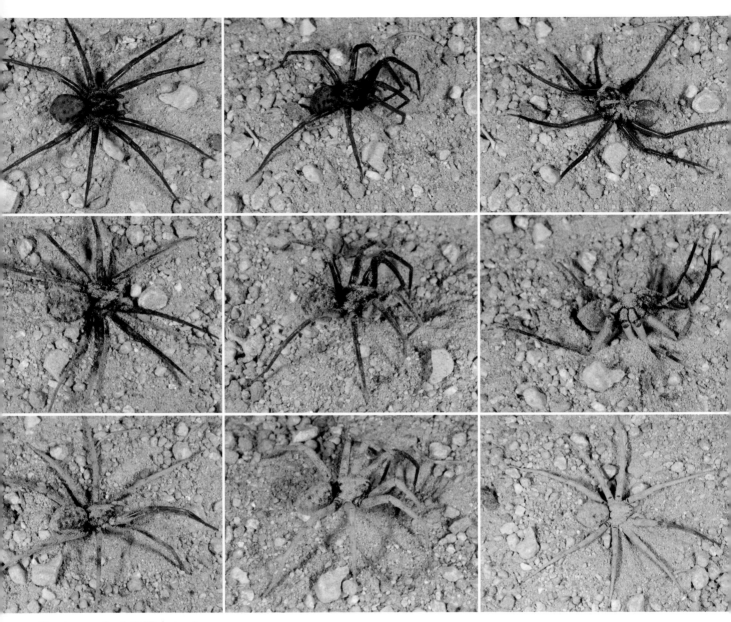

Dust as an invisibility cloak:

A freshly molted sand spider (*Homalonychus selenopoides*) takes a series of dust baths until its cuticle is thoroughly coated with dirt. First, the spider vigorously digs at the substrate, forming a little pile just under its body. Second, the spider steps forward over the pile and then falls over backward onto the dirt. Lying on its back with its legs stretched out flat, it shivers and wiggles all parts of its body and legs in the dust. Third, the spider rights itself and then proceeds to do a belly flop on the dirt pile. It shivers and wiggles again, while lying flat on the dust with its legs stretched out straight. After a few seconds of rest, the spider stands up and repeats the process. The entire sequence may be repeated from six to as many as thirty times until the sand spider is satisfied.

Sand spiders take dust baths after each molt throughout their lives. However, after his final molt, the male sand spider does not cover himself with dust. Perhaps this helps the female recognize a potential mate.

1. Wrapped in a "bridal veil," the female *H. selenopoides* lies still as the male prepares to mate with her. A receptive female lies down almost immediately once she is aware of the courting male. He then bustles about, repeatedly applying silk to her cephalothorax before mating with her. After the male mates, he wanders off, and the female quickly pulls free of her silken bonds. Although she can rapidly escape from the silk, the slight delay might buy an alert male time to escape should the female suddenly become aggressive.

2. The egg sac is hidden behind a "beaded curtain" of dusty silk strands attached to the underside of a rock. The tiny emerging spiderlings have already taken dust baths.

starts to cover the front end of the female in silk and may spend several minutes adding layer upon layer to this "bridal veil." She continues to lie still while he bustles around her. Eventually, her cephalothorax is covered in a thick sheet of silk, while her abdomen is uncovered. At this point, the male reaches under her with each pedipalp in turn, and mates with her. Shortly thereafter, he wanders off. The female continues to passively lie under the silk for a short while. After a few seconds to a minute, she suddenly and effortlessly extracts herself from under the silk.

This behavior is all the more interesting in light of the fact that *Homalonychus* does not use silk in capturing prey and is one of the few spiders known that does not use silk even as a dragline. The most obvious benefit to the male in this elaborate courtship might be protection against a potentially aggressive mate. There has been speculation that the male's silk puts the female into a sort of trance, or may tie her so well that he is safe from attack; however, there is documentation of the female suddenly pulling out from under the silk and killing the male, casting doubt on both these hypotheses. Instead, it is more likely that although the female is fully capable of quickly freeing herself, perhaps the second or two that it takes to do so gives an alert male a chance to escape should she reconsider her decision to mate. Another hypothesis suggests that the silk acts as a sort of aphrodisiac for the female.

After mating, the female constructs an egg sac in a sheltered spot such as the underside of a rock. She then hangs a large number of silk threads from the same rock. Each silk thread is covered with grains of sand, so the collective effect of all the hanging threads is of a beaded curtain sheltering the egg sac. The spiderlings stay near the egg sac for a short time before dispersing. By the time of dispersal, each tiny spider has taken a dust bath and is beautifully camouflaged, blending in with the surrounding soil.

Wolf Spiders: Lycosidae

The family name Lycosidae is well chosen, *lycos* being the Greek word for wolf. These formidable hunters do indeed remind one of a wolf. A pelt of short, furlike setae in shades of gray and tan covers their bodies. Their fairly large eyes reflect light at night with a green shine, a characteristic they share with other nocturnal predators such as cats and wolves. Their relatively long, powerful legs allow them to rush out from an ambush site and swiftly chase down their prey, much like their namesake the wolf. These spiders rely on speed, strength, and agility for hunting; stealth and snares are not for them. Wolf spiders are the most numerous of the guild of wandering hunters. In some ideal habitats, hundreds if not thousands may inhabit a few square meters of ground, their eyes shining at night like the stars in the sky.

Given the speed and prowess of these predators, it is not surprising that lycosids have evolved an elaborate method of acoustic communication. This allows them to recognize a potential mate in situations where they may not have the luxury of visual recognition at a safe distance. Three types of sound production are used by spiders: percussion, vibration, and stridulation. Stridulation is produced by the action of a file on a scraper. The loud, chirping song of a cricket is a familiar example of arthropod communication using stridulation. The

"songs" of stridulating spiders may be undetectable by most human ears but may be no less common than other, more conspicuous arthropod songs.

Each species of wolf spider produces its own signature "song" or pattern of acoustic sound recognizable to others of its species, which helps to differentiate it from similar species of wolf spiders that may share the same habitat. In some cases "cryptic" species that are essentially morphologically identical and coexist in a shared location may be able to avoid mistaken identities and accidental hybrid offspring through these species-specific acoustic communications.

The male wolf spider first detects the presence of a female by the pheromone, or chemical signal, in the dragline silk that she releases behind her wherever she goes. As soon as a mature male spider encounters the female's silk, he starts to follow it. As he explores along the trail, he periodically raises his front legs and waves them in the air. This may serve two functions: the female may see the leg-waving and recognize his amorous intentions, and he may be able to detect more chemical pheromone molecules in the air via the tarsal organs (little pitlike structures) on his legs by waving them in the air. Once he perceives the female, he may drum on the substrate, using the hard, cuticular plate on the underside of his abdomen. This drumming produces a purring sound that may be audible several meters away. He stridulates using the file and scraper at the tarsal/tibial joint of his pedipalps and may also tap or drum on the ground using his pedipalps.

The female wolf spider hears the airborne sounds via "ears" on her legs. These consist of two specialized structures: trichobothria and slit sensilla. A trichobothrial hair consists of a long, thin, specialized hair that sits in a cup-shaped socket. It bends in response to the slightest air movement, including vibrations from sound. Several dendritic nerve endings at the base of the hair transmit particular information regarding the hair movement, such as the direction from which the air movement originated. Even a blind spider can find a buzzing fly based on the input from the trichobothria. Additionally, spiders have slit sensilla on the forward-facing part of their tarsus, or foot. Each slit sensilla is a deep slit in the cuticle covered by a thin, easily deformed membrane and contains two dendritic neurons. The tarsal slit sensilla are in fact sensitive to airborne sounds. So, as the amorous male wolf spider stridulates with his pedipalps and drums with his pedipalps and abdomen, the song is carried by both ground vibrations and air vibrations to the object of his desire, the female. Knowing his identity in advance lessens the chance that she may mistake him for a cricket or other prey as he approaches her during courtship.

After they mate, she produces a silk egg sac that she attaches to her spinnerets. This silk does not allow water to penetrate to the inner layer, protecting the developing eggs even if the mother spider runs across a surface of water (as wolf spiders are adept at doing). The mother spider seems to know exactly how to best regulate the incubation temperature of her eggs. On cool days, she climbs onto exposed rocks and basks with her egg sac in the sun, thereby increasing the incubation temperature. The mother spider is devoted to her egg sac. If the sac is forcibly removed from her, she will spend hours searching for it.

After a couple of weeks of incubation, the mother spider appears to receive some signal from the babies within the egg sac that they are ready to emerge. If a younger egg sac is substituted for her original egg sac, she holds it for a longer time and opens it only when those babies are ready. The mother spider cuts open the egg sac with her fangs in order to release her offspring. The spiderlings scramble up onto the abdomen of their mother, where they cling to specialized knobbed hairs. There they stay for about a week as they absorb stored yolk and undergo further development. They may occasionally descend to get a drink of water during this time, but they clamber back up onto her abdomen by way of their draglines as soon as their mother starts to move. After about a week, the spiderlings are ready to leave their mother and live independently.

Wolf spiders chase down their prey and hold it until it is subdued. Favorite prey animals include orthopterans, such as crickets and grasshoppers, which have a powerful kick and may also bite. As a consequence, wolf spiders have developed features of their anatomy and hunting techniques to cope with these dangerous adversaries. They can run and leap with great speed and agility. The power for extending their appendages derives from a hydraulic system of increased blood pressure to their legs. The power for holding prey once it is caught comes from strong flexor muscles in their legs. However, the wolf spider does not pull its prey too close to its body, but holds it in a sort of leg basket "at arm's length," a technique that minimizes the risk of damage from a serious kick from a cricket or

a grasshopper. The spines on the wolf spider's legs help in grasping and holding its prey while simultaneously preventing the prey from being able to bite the spider.

The length of wolf spider legs may be a perfect compromise between the strength needed to hold prey (favoring shorter legs) and the ability to hold prey far enough from the body to avoid a damaging kick (favoring long legs). In addition, the speed and leaping ability of both the wolf spider and its orthopteran prey may be an example of coevolution analogous to that of the rabbit and the fox. Each developed speed because of the other.

Wolf spiders have large eyes with good night vision. Their two large forward-facing eyes reflect light from a layer of crystalline deposits in the eye called the *tapetum lucidum*, which reflects light back through the retina. Many other nocturnal hunters have similar "eye shine"; it is thought to increase the ability of these hunters to see in the dark. Strong chelicerae and large fangs with which to hold their prey complete this portrait of a highly effective predator.

Many species of wolf spiders live the life of a wandering hunter, but there are exceptions. Among these exceptions is the giant wolf spider, *Hogna carolinensis*. This impressive animal has a 2-inch (5 cm) leg span and takes three years to reach maturity. Perhaps because of its large size, it has adopted a lifestyle more like the sedentary, burrowing tarantula than like its cousins, the smaller wolf spiders. It digs a burrow about 10 inches (25 cm) in depth. Added to the top of the burrow is a little turret built of silk and debris, which may serve several useful purposes. The turret may deflect dirt, water, and leaves from falling into the burrow, and it may provide a nice lookout spot for the spider to sit above the hot ground surface just after sundown. Silk lines radiate out from the burrow onto the ground, giving the occupant early warning of anything walking on the lines. The giant wolf spider stays close to its burrow, rushing out in short bursts to capture prey. After reaching maturity, the males leave their burrow and wander in search of females. Female giant wolf spiders may live as long as 4 years, unusual for a non-mygalomorph spider.

Another exception from the typical wolf spider is *Sosippus californicus*, a species that constructs and lives in sheet webs with a funnel retreat. This web is very similar in construction and function to that made by the agelenid funnel web spiders. A flat, nonsticky sheet web is the spider's hunting platform. As soon as any unlucky

Carrying a load of babies, the mother wolf spider (*Pardosa* species) is still able to run about with surprising rapidity. The babies will ride on her for approximately a week before dispersing. Specialized knobbed setae on the mother spider's abdomen assist the young in holding onto their mother.

insect chances to fall upon the sheet web, *Sosippus* darts out from its funnel-shaped hiding place and captures it. *Sosippus californicus* reaches an impressive size, roughly equal to that of the giant wolf spider, and is able to capture large prey such as grasshoppers. Its spinnerets are long, an adaptation in common with other sheet web spiders such as agelenids and diplurids. In an interesting twist, wolf spiders in the genus *Sosippus* are subsocial. Spiderlings may stay in their mother's web for an extended period of time, not only tolerant of each other but even cooperatively killing large prey together. They may undergo several molts before leaving their mother's web. The genus *Sosippus* probably originated in Central America, and a few species are found in the southern part of the United States, where they have diverged sufficiently from their ancestral type to become distinguishable species in their own right; however, since most social and subsocial spiders live in tropical areas where prey is very abundant, perhaps the subsocial

behavior of *Sosippus californicus* is a legacy of a tropical ancestry.

The lycosid spider epitomizes an arachnid version of its namesake, the wolf. A combination of good eyesight, lightning-fast reflexes, speed, and agility make her a successful hunter. As Jean Henri Fabre wrote, "In the case of the *Lycosa*, the job is riskier. She has naught to serve her but her courage and her fangs and is obliged to leap upon the formidable prey, to master it by her dexterity, to annihilate it, in a measure, by her swift-slaying talent."

As dangerous predators with fast reflexes, wolf spiders evolved a system of acoustic as well as visual communication as a necessity for courtship and mating. At the same time, mother wolf spiders also show a touching devotion to their eggs and young, an attribute that we more closely associate with vertebrates. A complex mixture of fierce predator and devoted parent, the wolf spider is difficult to paint in words. Perhaps her name, "wolf spider," really does say it all.

1. *Allocosa* species. This genus has a distinctively shiny, dark cephalothorax. It can commonly be found in moist areas, such as along the banks of streams.

2. *Arctosa littoralis* (presumptive). Found along a desert stream, this wolf spider's coloration was an almost perfect match with the grains of sand in the streambed. This species is found throughout North and Central America.

3. *Camptocosa parallela*. Similar in appearance to the genus *Schizocosa*, mature males have a thick brush of black setae on their front legs, making them more conspicuous during courtship displays.

4. *Hogna carolinensis* mature male. The giant wolf spider can be found over much of North America. It tends to favor areas in which the soil is suitable for digging an underground burrow.

5. *Hogna carolinensis*. The burrow has a characteristic turret around the entrance: in this case, made of dead leaves and flowers.

6. *H. carolinensis*. A more typical example of the turret.

7. *H. carolinensis* has a *tapetum lucidum*, giving this nocturnal predator the "eye shine" similar to a wolf's eyes.

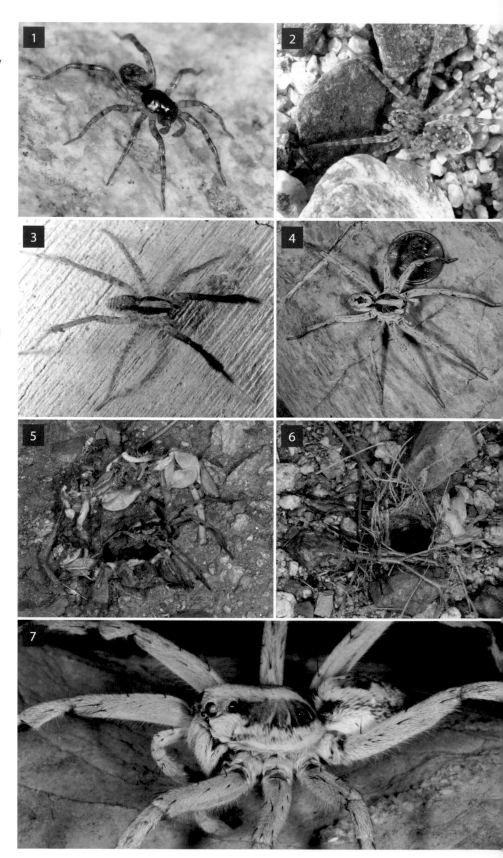

1: *Pardosa* species, adult male. There are approximately 80 described species of thin-legged wolf spiders (*Pardosa*) in North America. As with other spiders, wolf spiders can be identified to species by examining microscopic reproductive structures.

2: *Pardosa* species, adult male. This tiny male was only about 0.2 inches (5 mm) in body length. Typical for many wolf spiders, this individual was found along the moist margin edging a pool of water. The edges of pools and streambanks provide ideal habitat for thin-legged wolf spiders. Abundant prey and ground cover (in the form of grass or leaves) allow large numbers of wolf spiders to survive. Hundreds of individuals may be found in a few square feet of prime habitat.

3: *Pardosa* species, possibly *P. valens*. Most species of female wolf spiders carry their egg sacs attached to their spinnerets. They actively practice thermoregulation of the incubating eggs, basking in the sun to warm them, if necessary. The egg sac is waterproof, so the mother spider can run about even on the surface of water without harming the developing embryos. When the babies are ready to emerge from the egg sac, they cue their mother and she proceeds to open it to release her babies.

4: *Sosippus californicus*. This genus builds a funnel web similar to the web of agelenid spiders. *S. californicus* can grow almost as large as the giant wolf spider. The young may stay with their mother through several instar stages, cooperatively killing and sharing prey.

5: Pompilid wasp with prey. Wolf spiders themselves fall victim to another predator, the spider wasp. The paralyzed spider (*Arctosa littoralis*) is lifted and carried to a burrow, where the wasp will lay an egg on the spider. The wasp larva feeds on the still-living spider until it kills it just before pupating.

Ghosts, Goblins, Pirates, and Other Wandering Hunters:
Mimetidae, Corrinidae, Trachelidae, Gnaphosidae, Caponiidae, Dysderidae, Oonopidae, Anyphaenidae, Miturgidae, Eutichuridae, and Liocranidae

A *Metapeira* met a pirate. Incredibly slow and stealthy, the female pirate spider *Mimetus hesperus* slipped into the web of a small *Metapeira* orb weaver, taking more than an hour before getting close enough for the final lethal attack. The pirate spider then fed at her leisure.

The category of wandering hunters comprises spiders that do not use webs or ambush sites in order to capture prey. Included in this group is an assortment of families that have evolved different strategies and adaptations for a variety of niches. There are ground hunters and foliage hunters, as well as diurnal and nocturnal spiders. Some species are specialists, hunting only a particular type of prey, and others are generalists. Consequently, a surprising diversity of wandering hunters may be found in a single habitat.

Mimetidae: Pirate Spiders

With infinite patience, the pirate spider *Mimetus* slips into the web of an orb weaver and approaches its prey. Its progress is almost imperceptible as it pauses for long minutes between each stealthy step. But eventually the pirate spider is within striking distance of its quarry. With a sudden lunge, *Mimetus* attacks the target, using its long chelicerae to administer the lethal bite. Immediately, it releases the victim and waits a bit, making sure that the spider victim is dead. It does not have long to wait; the orb weaver dies almost instantly from the potent venom,

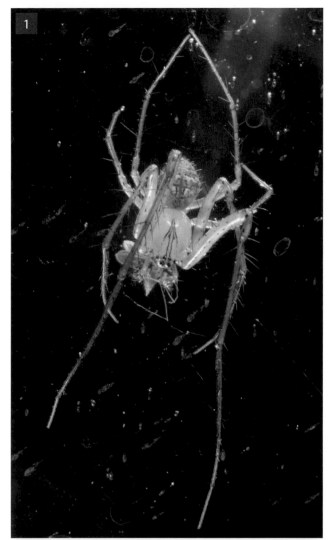

1 and 2. *Mimetus hesperus* male. Like most other male spiders, the male pirate spider must wander in search of a female. Mature male pirate spiders are not strict spider hunters; they readily capture and feed on gnats and other small prey. The male at right was one of four male pirate spiders that were hunting simultaneously on a sliding glass door at night, capturing small gnats.

The extremely long first pair of legs may give a selective advantage to a hunter of web-based spiders. Another spider hunter, *Rhomphaea fictilium*, also has proportionately long front legs. Heavy spines on the legs of *Mimetus* probably protect it from retaliatory bites from prey spiders.

The eye arrangement in Mimetidae is somewhat similar to the eye arrangement of the orb weavers. Recent phylogenetic analysis indicates that these two groups are closely related to each other.

Male entelegyne spiders, like the pirate spiders, have highly complex pedipalps compared with the pedipalps of mygalomorphs and most haplogyne spiders.

which is highly effective against spiders. The pirate spider then settles down to suck out the contents of the orb weaver, leaving the cuticle of its prey almost completely intact. The web built by the orb weaver to capture food is now the platform for its own consumption.

Slow and stealthy, pirate spiders hardly seem like lethal predators at first glance. But pirate spiders have made a specialty of hunting other spiders, especially the orb weavers and the combfooted spiders (the theridiids), including even black widows. The spiders in the family Mimetidae possess extremely long front legs armed with heavy, slightly curved setae. Another spider hunter, *Rhomphaea*, also has exceptionally long front legs. Perhaps this is a useful adaptation when attacking a spider in its web, giving the attacker a superior reach.

The name Mimetidae is based on the Greek word for "imitator, actor, and impersonator." In fact, the genus *Eros* is notorious for incorporating deception into its hunting repertoire, utilizing what is called aggressive mimicry. Aggressive mimicry is defined by a predator (the mimic) imitating a harmless organism (the model), thereby attracting the prey. Aggressive mimics that target more than one species of prey may have evolved highly complex repertoires of behaviors, and may demonstrate plasticity in the use of these behaviors. Staying near the periphery of the victim's web, *Eros* seductively plucks the silk, imitating a courting male. When the resident spider hurries over to investigate, *Eros* bites her on the leg and kills her. *Eros* then eats not only the spider, but any eggs as well. In 1850, Nicholas Marcellus Hentz wrote, "The *Mimetus* … prefers prowling in the dark, and taking possession of the industrious *Epeira*'s threads and home, or the patient *Theridion*'s web, after murdering the unsuspecting proprietor." Occasionally, though, if *Mimetus* becomes a little careless, the hunter may become the hunted, and the pirate spider may be killed by the resident spider and then eaten.

Although the female *Mimetus* seems to prefer spiders as the predominant prey, male *Mimetus* readily hunt other small arthropods such as gnats. The males may be seen as regular visitors on sliding glass doors at night, feeding on small gnats attracted to the light. It is sometimes difficult to imagine that this innocuous male belongs to the same species as the lethal female.

Corinnidae and Trachelidae: Ant and Wasp Mimics

The family Corinnidae consists of both diurnal and nocturnal wandering hunters. Among the daytime hunters is a group of incredibly fast spiders belonging to the genus *Castianeira*.

Several species in this group are noteworthy for their wasp mimicry. *Castianeira occidens* may be seen running about on the ground even in the heat of a summer day. With its red abdomen and extremely rapid movements, it resembles a wasp flitting from one spot to another. When it pauses, it may wave its front legs in the air in a credible imitation of antennae. Only upon closer examination might the curious observer see that this is actually a spider rather than a wasp. The bright red aposematic colors of this spider along with its wasp-mimicking behavior are an example of Batesian mimicry, in which a harmless animal is avoided because of its resemblance to a potentially dangerous animal.

Other species in the family Corinnidae are ant mimics. Often they have the dark coloration of an ant, frequently including a lighter-colored band circling the abdomen that adds to the illusion of a hard, shiny abdomen. Both their walking gait as well as a tendency to tilt their abdomens upward creates a reasonable facsimile of an ambulatory ant. Ants have few predators, and so the spider benefits from mimicking a hard, unappetizing insect.

While the ant and wasp mimics of the family Corinnidae are frequently diurnal, other members of their family as well as members of the family Trachelidae may be primarily nocturnal. The brown and tan *Meriola decepta* is often found in grassy meadows or grassy stream banks. During the day it may take shelter under rocks or other debris. The genus *Septentrinna* has evolved a different type of mimicry. It is able to infiltrate the nests of the formidable harvester ants (*Pogonomyrmex*) with the chemical protection of its cuticle, which mimics the chemical identity of the ants. Spiders in the genus *Trachelas* are nocturnal hunters, climbing up into bushes in search of prey. During the day, they retreat to a protected shelter under a rock or under bark. Although *Trachelas* are sometimes referred to as ant mimics, these nocturnal spiders are less convincing ants than the day-active *Castianeira* species. Presumably there is greater selection for mimicry in an active, conspicuous diurnal spider that must contend with visual hunters such as

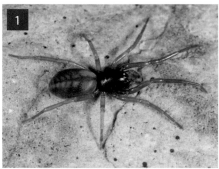

The family Trachelidae contains some commonly encountered ground hunters.

1. *Meriola decepta*. *Meriola decepta* occurs throughout North America, living in grassy meadows, agricultural fields, and other habitats with moisture. In the arid southwestern United States, this species can be found in low-elevation riparian areas, especially along the grassy banks of streams.

2. *Trachelas pacificus* female.

3. *Trachelas pacificus* male. The cephalothorax of the bull-headed sac spider is distinctively textured. The abdomen of the female appears olive green, while the male has a reddish-colored abdomen. These spiders can be found in sheltered areas where they can build a small silken refuge, especially in the nooks and crannies of dead wood lying on the ground.

birds and lizards than in nocturnal species that have little exposure to these predators.

Gnaphosidae: Ground Spiders

Both the scientific name Gnaphosidae, which means "dark" or "obscure" in Greek, and the common name ground spider are apt descriptions of these little hunters. A coat of short, velvety setae in the subdued colors of brown, tan, or gray, relatively short legs, and nocturnal habits paint an image of an inconspicuous little spider. Hiding in a silk nest under rocks or dead wood by day, gnaphosids emerge at twilight or after dark to wander the ground surface and capture their prey in short dashes. Like the nocturnal lycosids, many gnaphosids possess a tapetum in their eyes that reflects light at night like the eyes of a cat or a wolf. But they also utilize polarized light in order to navigate. In one study, it was found that the oval shape and the 90-degree orientation of the posterior median eyes made them good receptors of polarized light, especially at dawn or dusk when polarized light is oriented in one direction. Dawn and dusk are when these crepuscular little spiders are commuting, as it were, to and from their day nest and their nocturnal hunting ground. When these specialized, flat eye lenses were covered,

the spiders had great difficulty navigating their way home to their day nest.

Male gnaphosids may share a retreat with a penultimate female before she undergoes her final molt. The male may build a separate silk nest adjacent to the female's nest, and he strategically mates with her shortly after her final molt, at the point when she is sexually mature but before her exoskeleton has had time to harden completely. In this way, he is safe from attack since the fangs of the female are too soft to be used yet.

After mating, the female constructs an egg sac hidden under debris. The silk for egg sacs comes from specific silk glands that may produce colored silk. Different species of ground spiders may produce egg sacs of red, pink, white, or even apricot-colored silk. Some gnaphosids stay with the egg sac in order to guard it, while others may simply leave it.

Many gnaphosids may be seen hunting at night near doors and windowsills. Others, like the western parson spider, *Herpyllus proquinquus*, are frequent inhabitants of human dwellings, helping to control crickets and other insects that may venture inside.

The genus *Callilepis* specializes in hunting ants. This small, dark gnaphosid has a fixed set of behaviors for capturing its prey. First, it bites the antenna of an ant. Without an antenna to bite, the spider will not initiate an attack. This preliminary bite injures the ant, causing

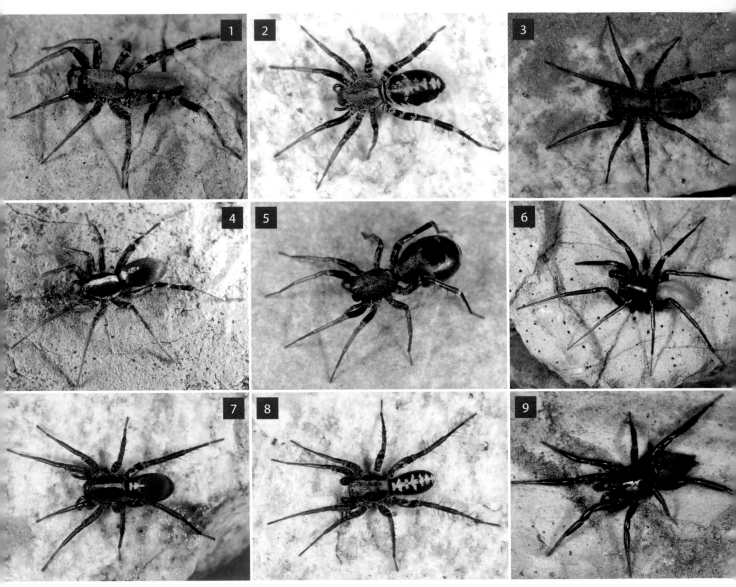

1. *Castianeira dorsata* adult male. Red and black is aposematic (warning) coloration seen in a number of diurnal predators. The model for this Batesian mimic is a wasp.

2 and 3. *Castianeira dorsata* adult females. There can be considerable variation in color patterns within any one species of *Castianeira*.

4. *Castianeira occidens*. This species has both the coloration and the movement of a wasp. It runs so rapidly that it looks almost as if it were flying.

5. *Castianeira* species. Some *Castianeira* mimic ants instead of wasps. This individual moved with a convincingly antlike gait, and the scales on its abdomen gave it a shiny appearance.

6. *Septentrinna bicalcarata*. This genus has been found living in harvester ant colonies (*Pogonomyrmex*). Its cuticle contains chemicals that mimic the ant's own chemical identity.

7. *Castianeira crocata* female. Despite looking similar to *C. occidens*, this female was identified by microscopic structures as *C. crocata*.

8. This male had "staked out" the female shown as number 7 just before she molted into an adult. They readily mated once she had molted, and she produced viable offspring. He resembles *Castianeira longipalpa*.

9. *Zelotes* species, in the family Gnaphosidae.

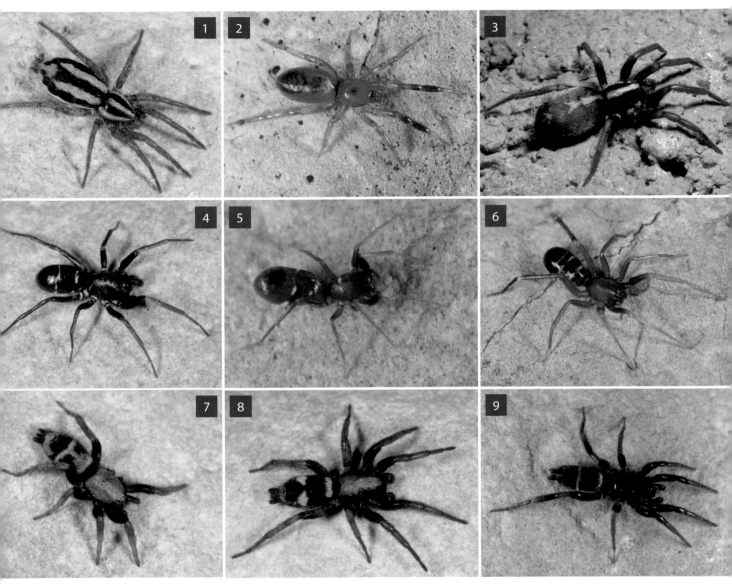

1. *Cesonia classica*. This ground hunter is not only incredibly fast, but it can also run across water and up the sides of a nonstick pan.

2. *Drassyllus lepidus*.

3. *Herpyllus propinquus*. The western parson spider is commonly seen in homes in the arid southwestern United States.

4. *Micaria pulicaria*. This species of ant mimic occurs in a wide area of North America, but in the hot deserts of the arid southwest it is limited to cool, high-elevation forests of the sky island mountains.

5. *Micaria longipes*. A narrow white band enhances the illusion of a shiny ant abdomen in many species of ant mimics, including all three of the species shown here.

6. *Micaria gertschi* (presumptive). The genus *Micaria* typically has shiny scales and an antlike gait and tips its abdomen in an antlike manner, as seen in this image. These spiders are frequently seen in low-elevation desert.

7. *Sergiolus angustus*.

8. *Sergiolus columbianus*. This species and *S. angustus* are found at high, forested elevations in the mountains of the arid southwest.

9. *Gertschosa* species, possibly *G. concinna*. This genus occurs in Mexico and has been only rarely documented in the United States (Texas and Arizona).

it to walk in circles. At this point the spider administers a second bite, which paralyzes the ant. *Callilepis* then grabs the paralyzed ant and runs with it, like a football player with the ball. Other ants attempt to intercept the spider, but she evades them and escapes with her prey.

Like the family Corinnidae, the family Gnaphosidae includes ant mimics—perhaps not surprising given that both families are largely terrestrial hunters. The genus *Micaria* contains many species of extremely convincing ant mimics. These lack the velvety hair of their relatives, instead being covered with fine, flat, iridescent scales that give them a shiny, antlike appearance. Some have a slight constriction in their abdomen accentuated with lighter-colored scales, suggestive of the constriction between an ant's body segments. An antlike walking gait and the characteristic tilt of the abdomen complete the illusion of an ant—especially an ant in a hurry. In fact, *Micaria* prey on ants and are found in close association with them although they may also capture other prey such as crickets if the opportunity occurs. *Micaria* are active during the day, paralleling the ant mimics in the family Corinnidae and in the family Salticidae.

Caponiidae: Lungless Spiders

The family Caponiidae is distinctive with a beautiful deep orange prosoma and legs, a build much like that of a gnaphosid, and eyes clustered closely together. Only one genus in this family has 8 eyes; the other three genera each have only 2 close-set eyes, a very rare arrangement among spiders. Among these 2-eyed spiders is the genus *Orthonops*, found primarily in the southwestern United States in arid environments. During the day this spider seeks refuge under rocks or dead wood, or in pack rat nests. At night, it principally hunts other ground spiders but may take other prey. Its respiratory system is also different from most other spiders of comparable size. Caponiids have no book lungs whatsoever; instead, they

1. *Orthonops icenoglei.* This genus is easy to recognize, having only 2 closely set eyes. The only other two-eyed spiders in North America also belong to the family Caponiidae. *Orthonops* specialize in hunting other spiders.

2. *Escaphiella hespera* female. The female has a sclerotized plate shielding each lateral side of her abdomen. The male has dorsal and ventral plates shielding his abdomen.

This genus has 6 eyes arranged in a cluster.

3. *Yumates* species. This goblin spider was collected in southern Arizona and is in the process of being formally described and named.

breathe by means of respiratory slits or spiracles leading to tubular tracheae that branch into the body. This system, similar to an insect's respiratory system, is more efficient in the transfer of oxygen and gives this spider greater endurance and stamina when running.

Oonopidae: The Goblin Spiders

These tiny spiders may be easily overlooked, especially since they hide under rocks or in leaf litter and are only 0.04 to 0.12 inches (1–3 mm) in body length. But perhaps the charming name "goblin spider" justifies a closer look at these minuscule spiders. One of their most distinctive features is that they have only 6 eyes, frequently clustered tightly together. The vast majority of the 455 described species are tropical, but a few species are found in the southern United States, including *Escaphiella hespera*. The female *Escaphiella hespera* has a plate of tougher, darker, more sclerotized cuticle covering each lateral side of her abdomen, making her easy to recognize. The male *Escaphiella* has sclerotized plates covering the dorsal and ventral surfaces of the abdomen. Like gnaphosids and many other ground hunters, these spiders rapidly chase down their prey instead of using webs for prey capture.

Dysderidae: Sowbug Killers

Spiders in this family are native to Eurasia, with the greatest concentration of species found in the Mediterranean region; however, one species, *Dysdera crocata*, has been introduced to North America and become naturalized in much of the United States, including parts of the southwest. Distinguished by heavy projecting chelicerae and having 6 eyes arranged in an arc, these spiders resemble no other spider in North America.

Armed with exceptionally long fangs, *Dysdera* is specialized to hunt sowbugs and pillbugs. This is no mean feat for a spider, since these little land crustaceans (isopods) are protected by a series of hard calcareous plates covering their bodies. In addition, pillbugs can roll themselves into an armor-plated sphere, looking a bit like a crustacean version of an armadillo. Their defense is almost perfect—the operative word being "almost." *Dysdera* will attempt to find a chink in the isopod's armor, and if it can work a fang into even the narrowest of spaces, the isopod will be killed and eaten.

Only its empty armor exoskeleton will be left behind, eloquent testimony to the determination and skill of the sowbug killer.

Dysdera crocata also energetically hunts other small arthropods such as crickets. Living under rocks or other debris, *Dysdera* excavates a series of small tunnels in the soil and constructs a silk bedroom in which to retreat during the day. At night, *Dysdera* emerges onto the surface of the ground to hunt.

Dysdera also uses the silk retreat for molting and for egg laying. The mother spider guards her eggs until the spiderlings hatch and disperse. The spiderlings may take more than a year to reach maturity and may live for as long as 5 years.

One hypothesis regarding dysderids proposes that human habitations increase the amount of calcium leaching out of concrete or mortar and into the surrounding soil, thus supporting populations of isopod crustaceans that use calcium in their exoskeletons and therefore attracting the dysderids that prey upon them. Although this idea may have merit, the areas surrounding human habitations may provide other aspects of a microenvironment that would serve equally well to support larger populations of isopods and their predators. Most human habitations have moisture somewhere in their vicinity, along with some cultivated plants and possibly areas of compost or other detritus. These microenvironments can provide both the moisture and the food for isopods to thrive. *Dysdera* species have been introduced over much of the world, probably because of humans unwittingly transporting them, possibly in potted plants, which provide a moist, sheltered microenvironment, or tucked into a silken refuge constructed under building materials or other objects transported by humans.

Europe boasts an impressive diversity of dysderids, including some that have very unusual reproductive strategies. *Dysdera hungarica* can reproduce parthenogenetically, producing all-female clones. In an even more surprising and appropriately named species, *Harpactea sadistica*, the male employs a method referred to as traumatic insemination in order to mate with the female. The male first pierces the female's abdomen with his fangs and then uses each needlelike embolus to penetrate the abdominal wall of the female and deposit sperm internally near her ovaries. The female has only ineffective, atrophied seminal receptacles and thus cannot store sperm in the customary spider fashion.

1. Probing for chinks in a pillbug's armor, the sowbug killer *Dysdera crocata* has evolved extremely long fangs for challenging prey. Pillbugs are land crustaceans that have incorporated calcium into their exoskeletons, giving them a tough armor. Adding to this, the pillbug can roll itself into a sphere, leaving almost no opening for a predator.

2. Victorious at last, the determined sowbug killer has succeeded in killing the pillbug after finding a tiny gap in which to insert a fang. The pillbug can then be partially unrolled before the *Dysdera* feeds on it. Only the empty exoskeleton will remain. *Dysdera crocata* is native to Europe but has become established in much of the world.

Although traumatic insemination is well known in insects such as bed bugs, it has never been documented in any spider other than *Harpactea sadistica*.

Miturgidae: Prowling Hunters

Spiders in the family Miturgidae include terrestrial hunters in the genus *Syspira*.

Syspira superficially resemble wolf spiders with their long legs and brown and tan coloration; however, their eye arrangement is very different from that of the lycosids. They can be seen hunting on the ground at night, and they may climb up on doorsills or windowsills to capture their preferred prey of moths. Like a leopard dragging away a gazelle, *Syspira* may straddle a captured moth and carry it away to feed on it in peace and quiet. During the day, these spiders take refuge under rocks or debris on the ground. They are found in the southern United States from west Texas to southern California.

Eutichuridae

Cheiracanthium, on the other hand, specialize on living in vegetation. At night, these spiders prowl about on plants, feeding on caterpillars, flies, and the eggs of insects and spiders. The recognition of eggs as food is unusual in spiders, but this spider appears to be quite partial to eggs, especially the eggs of the green lynx spider. As it cruises around on plants, it may also derive significant energy from drinking nectar from both flowers and extrafloral nectaries.

One common name for *Cheiracanthium inclusum* is "sac spider," for good reason. These spiders construct a retreat by bending over a blade of grass or a leaf which is then used as the framework for a silken chamber. These silken nests are very characteristic of this genus. Care must be taken when handling vegetation with these spiders, since they may bite if they feel threatened. It may be a beneficial spider in agriculture, eating caterpillars and the eggs of insect pests. This species is found throughout the United States.

Liocranidae

Many of the genera in this family used to be in the family Clubionidae, and it is possible that the family will

Syspira analytica (presumptive). The eye arrangement distinguishes these from wolf spiders.

undergo more modifications in future revisions. Most members of this family are tan or brown-colored ground hunters, with little known about their natural history.

Anyphaenidae: Ghost Spiders

Ghost spiders are relatively pale spiders that hunt on foliage or among leaf litter on the ground. Most are nocturnal except for *Hibana*, which may be encountered during the day. This tan and greenish spider depends on cryptic camouflage as its primary defense against sharp-eyed birds. These spiders are extremely active, and they apparently supplement their energy requirements with nectar from plants as they forage for prey. These nectar sources may occur either in flowers or in extrafloral nectaries. Extrafloral nectaries are nectar-producing areas of a plant, such as leaves or petioles, that attract insects such as ants or wasps to the plant. In exchange for the nectar, these protective insects defend the plant against herbivorous insects such as caterpillars. Species of *Hibana* fit neatly into this scenario by ingesting the nectar for energy and then consuming insects such as aphids as well as the eggs or larvae of moths such as the fall webworm, *Hyphantria cunea*.

Ghost spiders have a couple of unusual adaptations for their active lifestyle. Their tarsi (feet) have 2 rows of specialized lamelliform setae that are long and flattened with a widened spatulate tip. These specialized claw tufts probably assist the ghost spider in gripping the surface of leaves as it deftly runs about on foliage. Another adaptation for their energetic lifestyle is a modified respiratory system, compared with that of most other spiders. In ghost spiders (as well as in salticids, thomisids, and dysderids), the tracheae start in the abdomen and extend through the pedicel and into the cephalothorax and even into the legs, thus delivering oxygen directly to the actively moving parts of the body. In many spiders, the tracheae are limited to the abdomen. The spiders with extensive tracheae tend to have smaller hearts and a lower heartbeat rate than those of other groups of spiders. This difference

is probably due to the more efficient oxygen exchange in these spiders. Also, the ghost spiders' tracheae are 3 to 4 times as wide as the tracheae of other spiders. The males have somewhat wider tracheae than the females, probably a selective advantage for their highly energetic courtship.

As many as 500 species of ghost spiders occur around the world, with most found in the American tropics. About 15 species of ghost spiders occur in the southwestern United States.

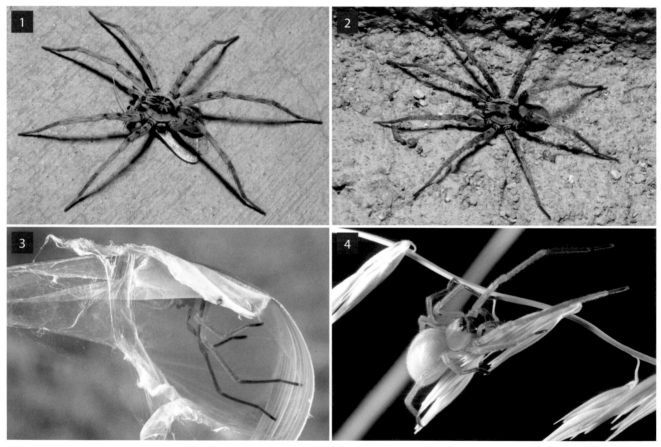

1. *Syspira analytica* (presumptive).

2. *Syspira tigrina*. Spiders in the genus *Syspira* superficially resemble wolf spiders, but they differ in several ways, including their eye arrangement (two straight rows of 4 eyes each, one row above the other). Primarily a desert spider, *Syspira* occurs from Texas to southern California and south into Mexico. More than one species of *Syspira* may occur in any one location; it is not known how they avoid direct competition for resources. *Syspira* are primarily terrestrial hunters.

3 and 4. *Cheiracanthium inclusum*, the long-legged sac spider. Also known as the agrarian sac spider, this species is found in both cultivated areas as well as relatively undisturbed areas throughout the United States and into Central America. This genus constructs a silken refuge using a curved leaf such as a blade of grass as part of the structure. *Cheiracanthium* are primarily foliage hunters, eating caterpillars, eggs of all sorts (including the eggs of green lynx spiders), and a variety of other prey. They also obtain energy and moisture from nectar.

1. *Neoanagraphis chamberlini*, Liocranidae. Little is known about the natural history of this nocturnal spider. Extremely long legs and rapid running speed suggest that this terrestrial hunter must run down fast-moving prey. Two long tarsal claws (INSET) on each third and fourth leg might give this spider better traction. This species is sometimes found in the burrows of other animals. *Neoanagraphis* occurs across the southwest from Texas to California, and south into Mexico. It is a medium-sized spider, from 0.15 to 0.38 inches (3.8–9.7 mm) in body length.

2. *Hibana incursa*, Anyphaenidae. Specialized lamelliform tarsal setae allow this species to grip the surface of leaves as it runs about in foliage. This spider has an unusual dietary preference for aphids, moth eggs, and caterpillars. It also obtains moisture and energy from nectar. This species is found throughout the southwest, from west Texas to California.

3. *Wulfila immaculellus*, Anyphaenidae. Truly a ghost spider, this delicate foliage hunter is almost translucent. Almost nothing is known about the natural history of this genus.

Fishing Spiders:
Pisauridae and Trechaleidae

An impressive predator, this mature female *Tinus peregrinus* fishing spider has captured a fish as large as herself. She has carried it up into vegetation, where she will masticate and predigest the fish. She must feed out of water or the enzymes needed for predigestion will be diluted out.

Resting its front feet on the water's surface, a *Dolomedes* fishing spider waits along the edge of a small, slow-moving stream. It reads every disturbance, however subtle, on the water's surface much the way that an orb weaver spider reads the vibrations within its web. In addition to detecting motion with its feet (specifically with the metatarsal lyriform organ), it can also see quite well; its large eyes are not very different from those of its cousin the wolf spider. The fishing spider's patience is rewarded when an immature grasshopper attempts to leap across the stream and falls onto the surface of the water. Faster than the eye can follow, the fishing spider gallops across the water's surface and grasps the hapless grasshopper between its two impressive fangs. The spider then returns to the edge of the stream to eat the grasshopper on land, where it efficiently masticates its food and sucks down the liquefied portion until all that is left of the grasshopper is an unrecognizable crumb and a few fragments.

Dolomedes belongs to the family Pisauridae, also known as nursery web spiders and fishing spiders. In some ways, the fishing spider is the aquatic analogue to the terrestrial wolf spider. This family includes characteristically large, handsome spiders with good eyesight that depend on their speed and strength in order to capture prey. Many of the family frequent moist habitats, but it is the genus *Dolomedes* that has mastered a lifestyle connected to the water. Despite the fact that some of the species in this genus reach an impressive size (*Dolomedes okefenokensis* has a leg span of 4 to 5 inches, or 10 to 12.7 cm), they can "row" or even rest their bodies on the water's surface without breaking the surface tension. The water simply indents or dimples where their legs and body contact the surface. While the spider is on the surface of the water, it is vulnerable to attack from below by underwater predators such as frogs. In this situation, the spider literally levitates by rapidly pushing all its legs downward against the water's surface to generate the force needed to jump straight up. It then gallops to safety. If the fishing spider becomes startled or frightened by a bird or a wasp, it scrambles underwater, clinging to vegetation so it doesn't pop back up to the surface. A thin layer of air clings to the hydrophobic cuticle and hairs on the spider's body, giving it a lovely silvery appearance. It can remain underwater for a good 40 minutes while waiting for the danger to pass.

Dolomedes spiders must remain vigilant while hunting, because they themselves are hunted. A spider wasp in the pompilid family, *Anoplius depressipes*, preys exclusively on female *Dolomedes* spiders. If a fishing spider sees one of these wasps nearby, it takes evasive action, fleeing from the wasp and diving under water in an attempt to escape. But the wasp does something really extraordinary. It actually dives and then swims underwater in pursuit of the unfortunate spider. Once it finds its prey, the wasp stings and paralyzes the spider. The wasp then surfaces with the paralyzed spider and drags it across the water as it skims across in a low flight trajectory. The spider is installed in the nest burrow of the wasp, and a single egg is laid on it. The wasp larva feeds on the still-living, paralyzed *Dolomedes* until it finally kills the spider. Then the wasp larva pupates.

Unlike the wasp, *Dolomedes* hunts on the surface of the water. Some authors have written that it hunts underwater, but this has yet to be clearly documented. Instead, it captures its prey primarily either at the surface of the water or on land. Despite this limitation, it can readily catch fish as they swim very close to the water's surface. The fangs and venom appear to be highly effective in killing the captured fish almost instantly, making it easier for the spider to carry its prey across the water and onto land or up into vegetation growing at the edge of the water. Because spiders ingest only liquefied, predigested food, the fishing spider must eat its prey above the water or else its digestive fluids will be diluted or lost.

Courtship for males in the family Pisauridae is an extremely hazardous business. Many males are killed by females even before they can mate. Consequently, some species in this family have developed defensive strategies for courting these formidable females. In the species *Tinus peregrinus*, the male captures an insect, wraps it in silk, and offers it to his prospective mate. If she accepts this "nuptial gift," he can usually safely mate with her while she is occupied with eating the offering. This nuptial gift may prove advantageous to the male in two ways. It buys time for the male to mate, and the extra nutrition contributes to the female's fecundity. The male of the genus *Pisaura* also offers a nuptial gift during courtship. In one species of *Pisaurina*, the female is first bound by a "bridal veil" of silk by the male, and then mating takes place with the couple hanging from a dragline, where presumably the female cannot gain traction to rapidly extricate herself from the silk. In the genus *Dolomedes*, the male detects the presence of a female through pheromones in the female's silk or on

1. *Dolomedes triton* can rest on the surface of the water without breaking the surface tension. It rarely stays on the surface for more than a few seconds at a time, presumably because of the danger of being eaten by fish or frogs. If a predator attacks from below, *Dolomedes* can push down against the water's surface with all its feet and leap literally straight up in the air. The typical hunting pose consists of the spider sitting on a rock or vegetation and resting its front legs on the water's surface, reading any surface disturbance much the way an orb weaver reads the vibrations in a silk web.

2. If a small fish disturbs the surface of the water, *Dolomedes* can detect and attack it. The large chelicerae and fangs serve it well in the capture of fish that may be as large as the spider.

If a fishing spider feels threatened, it dives under the water and clings to vegetation or a rock. The small setae covering its body hold tiny air bubbles, giving the spider a lovely silvery appearance. It can remain underwater for as long as 40 minutes before resurfacing. If it releases its hold on the vegetation, the spider pops up to the surface like a cork. Presumably, the layer of trapped air bubbles gives it tremendous buoyancy.

the surface of the water. The male initiates courtship by using the water's surface to send a rhythmic signal to the female with his legs. The receptive female answers the male by drumming on the surface of the water with her pedipalps and waving her front legs in the air slowly. The male exercises caution while approaching the female, and after mating, he departs promptly.

Female *Dolomedes* produce a large brown egg sac that may hold from 1,000 to 2,000 eggs. She carries the egg sac with her chelicerae until just before the spiderlings are ready to emerge, at which point she fastens the egg sac to some vegetation. She may tie down some leaves with silk to fashion a shelter for the egg sac. She then strategically positions herself in a spot nearby to guard the egg sac. Once the spiderlings have emerged, they remain in this protected nursery under the vigilant eye of their mother until they molt and are ready to disperse. In the rich habitat of the water's edge, the young spiders have an abundance of small flies and other arthropods to hunt, and in a year's time, they may have grown into impressively large raft spiders.

Tinus peregrinus produces an egg sac that may contain a couple of hundred eggs. Like the other spiders of this family, she carries the egg sac in her chelicerae until just before the young are ready to emerge. She then builds a nursery consisting of two parallel sheets of silk attached to vegetation and places the egg sac between the sheets. The mother spider stands watch over her young until they are ready to disperse. She becomes quite weakened and emaciated during her vigil, refusing to hunt or eat until her offspring have become independent. After the young disperse, if the female obtains enough food for herself, she may produce successive egg sacs, repeating the entire careful incubation for each in turn. The spiderlings grow fast, reaching maturity in just a few months.

Long-legged raft spiders in the family Trechaleidae carry the egg sac attached to their spinnerets in much the same way that their sister group the wolf spiders do. After the young emerge, they spend a day clustered on the outside of the empty egg sac. The following day they add the surface of their mother's abdomen to their roosting area, in addition to the egg sac. By the following day, they are dispersing away from their mother. It is only after the babies have left that the mother spider releases the empty egg sac.

All three species of southwestern fishing spiders reach an impressive size, all are found in aquatic habitats, and all are generalist predators. Therefore, it is puzzling that *Tinus peregrinus* can be found in the same location as *Dolomedes* or *Trechalea gertschi*. This unexpected sympatry may be possible due to the abundance of prey in these habitats; however, further investigation would be required to see if there is subtle partitioning of habitat between species.

An expert generalist, the fishing spider is able to exploit prey both on land and on the surface of the water. A surprising variety of prey is taken, including invertebrates, small fish, and amphibians. These spiders may even take nonliving food; one captive juvenile was routinely observed feeding on flake fish food. The ability to opportunistically utilize such different media (land and water) and such a variety of prey demonstrates a considerable degree of flexibility and complexity in the behavior of this group of spiders.

Death in the duckweed:

After a struggle, the six-spotted fishing spider (*Dolomedes triton*) has killed a leopard frog many times its own size. It will have to drag its heavy prize to the edge of the pool before it starts to feed. These large fishing spiders routinely capture vertebrate prey, including fish and tadpoles, as well as a variety of invertebrate prey. However, a large frog such as this was a challenge even for such a formidable predator.

This species has a wide distribution, including much of the United States. Populations in Arizona are frequently darker than eastern populations, and the broad white band seen elsewhere may be reduced in size and may be yellow. Genetic analysis may be needed to determine if these populations warrant separate species designation. Photo by Timothy A. Cota.

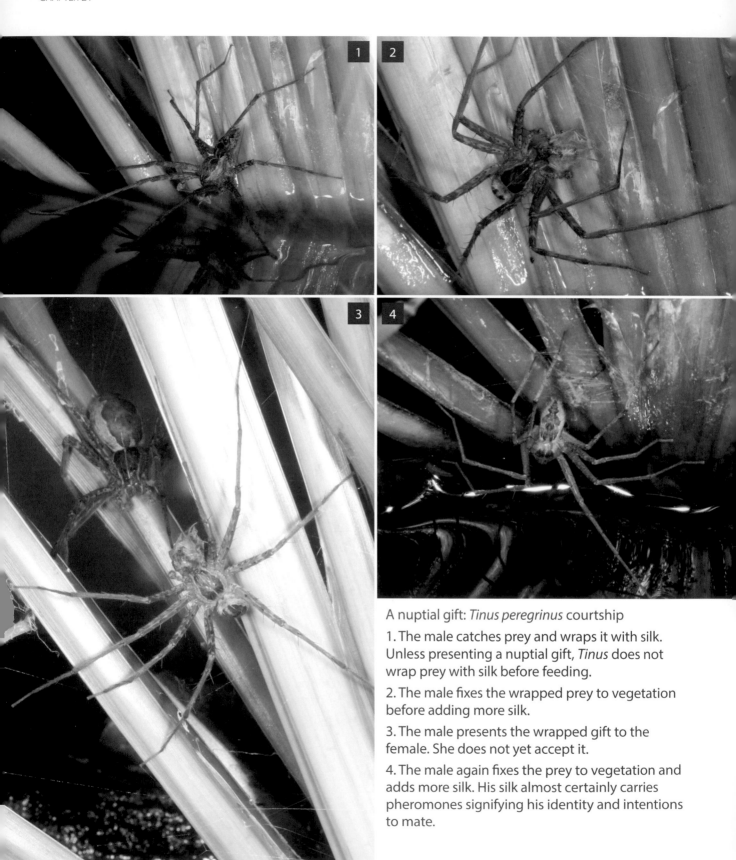

A nuptial gift: *Tinus peregrinus* courtship

1. The male catches prey and wraps it with silk. Unless presenting a nuptial gift, *Tinus* does not wrap prey with silk before feeding.

2. The male fixes the wrapped prey to vegetation before adding more silk.

3. The male presents the wrapped gift to the female. She does not yet accept it.

4. The male again fixes the prey to vegetation and adds more silk. His silk almost certainly carries pheromones signifying his identity and intentions to mate.

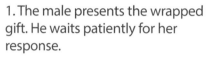

1. The male presents the wrapped gift. He waits patiently for her response.

2. The female leans forward and takes the gift. This signifies that she has accepted the male.

3. The male mates with the female while she feeds on the nuptial gift. The male will repeatedly mate with her while she feeds, and he may leave unscathed; however, the female may occasionally kill and eat the male after he has mated. The nuptial gift buys time for the male to mate and provides nutrition to the female, which may increase the number of eggs she produces.

4. A few days after mating, a well-nourished female will produce an egg sac with about 200 eggs. She carries the egg sac in her chelicerae until just before the spiderlings are ready to emerge. At this point, she attaches the egg sac to vegetation, builds a silk "tent" (or nursery web) over it, and stands guard nearby until the babies have emerged and then dispersed.

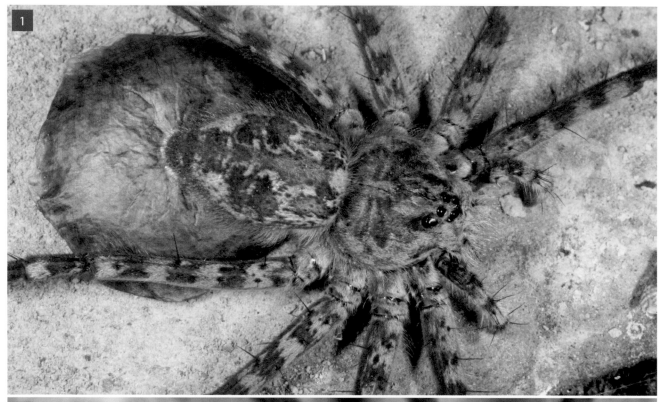

1. Like others in the family Trechaleidae, *Trechalea gertschi* carries the egg sac attached to its spinnerets. The silk is a peculiar silvery teal color.

2. The mother spider continues to carry the empty egg sac after the spiderlings have emerged, and this egg sac provides a surface for the babies to cling to for a day

3. On the second day after their emergence, the babies are distributed on both the empty egg sac as well as their mother's abdomen. Unlike its relative the wolf spider, the mother longlegged raft spider rarely moves around while the babies are still on her. On the third day after emerging, the spiderlings disperse, leaving their mother. She then abandons the empty egg sac.

4. *Trechalea gertschi*, the long-legged raft spider, has a 4-inch (10 cm) leg span in a mature female. This species occurs in Arizona and New Mexico.

Spitting Spiders:
Scytodidae

In a fraction of a second, a spitting spider squirts two narrow streams of a glue/silk mixture from her fangs. This mixture instantly binds her prey to the substrate.

In the absolute darkness of a cave, a predator lies in wait. Her long, spindly front legs are extended in front of her, resting motionless on the rock surface. A cave cricket chances to brush up against the tip of one leg. Sensory setae on the leg send a signal to the predator, alerting her to the presence of the cricket. Almost instantly, two narrow streams of glue zigzag over the hapless cricket, fastening it securely to the wall of the cave. Within a few seconds, the predator approaches her immobilized prey and delivers a killing bite on the leg joint of the cricket, where the cuticle is thin. The predator has once again successfully utilized a hunting technique that may be unique among the animal kingdom—and for this, she has earned the name "spitting spider."

The spitting spider has a distinctive morphology consistent with its hunting method. The high-domed prosoma (cephalothorax) accommodates both a massive glue gland as well as a venom gland. The tiny, 70-micrometer- (0.07 mm) long fangs are used for directing the streams of glue. An open tapering channel in each fang focuses the liquid into a narrow stream as it is squirted out under high pressure. The glue is focused into such a narrow stream that it cannot be seen directly with the human eye. It is seen only in photographic enlargements or indirectly as a pattern of moist contact lines on the substrate surface, which appear as a series of zigzag lines. The chelicerae are fused together where they connect to the prosoma, so they cannot rotate. But the tiny fangs oscillate back and forth at frequencies of up to 1,700 Hz (with an average of 826 Hz). The duration of one entire spitting episode is on average only 25 milliseconds (or one-fortieth of a second). Because of the oscillation of the fangs and the extreme rapidity of the spitting, the average angular velocity of the fangs is calculated to be 123,900 degrees per second, a truly mind-boggling number. The most likely mechanism for this rapid oscillation is hydrodynamic forcing, where fluid flowing past a structure causes it to oscillate. Muscles in the prosoma control the hydraulic pressure that powers the delivery of the glue, which can reach a velocity of 92 feet (28 m) per second.

The spit consists of both a silklike fibrous component as well as a gluelike liquid distributed along the fiber. The fibers contract almost immediately after being squirted out. In combination with the liquid glue, this helps to immobilize the prey. The contraction of the fibers would effectively adjust the bonds so they hold the prey snugly. The venom may be used later, when delivering the killing bite. Because the fangs are so tiny, the killing bite must be delivered where the cuticle is thin, such as a leg joint.

Many spiders such as wolf spiders, jumping spiders, and tarantulas use their large fangs to hold prey as well as to deliver venom. The fangs of the spitting spider are far too small and weak to hold prey directly, but by using them to squirt glue, the fangs can be used indirectly to immobilize prey. In fact, the fangs' ability to rapidly oscillate back and forth is contingent upon small size; very large fangs could not achieve this rapid oscillation rate. Likewise, the long, spindly legs of the spitting spider appear barely adequate to support the weight of the spider as it walks, in contrast to the sturdy, robust legs of many other spiders; however, they do assist the spider in detecting and capturing prey. In fact, although spitting spiders have 6 eyes (arranged in 3 pairs), a cricket can be mere fractions of an inch (millimeters) in front of the spider and still elicit no response, until it touches one of the legs. Then the spider reacts unhesitatingly.

After administering the killing bite, the spitting spider fastidiously cleans her front legs. The front legs may get some glue on them, since they are in the vicinity of the prey as the glue is squirted. These legs very likely give the spider information regarding both direction and distance to her target.

Once her legs are clean, the spitting spider again approaches her prey and wraps some very fine silk from her spinnerets around it. She then grasps the victim with her chelicerae and carries it to a secluded spot where she can feed without being disturbed.

She introduces digestive fluids into the body of the captured animal and then sucks out the liquefied contents with her powerful pumping stomach, leaving the exoskeleton of the prey almost completely intact. This should come as no surprise, since her tiny fangs would be inadequate for masticating an arthropod into mush.

Spitting spiders use their glue for defense as well as for hunting. If a spitting spider feels threatened by an approaching insect, it may squirt a small amount of the glue and escape while the insect extricates itself. This seems especially effective as long as the antennae of the approaching insect are glued to the substrate. After a few minutes, the insect can usually free itself and seems to suffer no ill effects aside from having to groom itself. Usually the amount of glue used for defense is relatively small compared with the amount used for capturing prey. It would certainly make sense that the sole ammunition used for securing food would be conserved for that primary purpose, rather than "wasting" it on a threat. This may be analogous to the occasional "dry" defensive bite of a rattlesnake.

Some species of spitting spiders do build small irregular snare webs that assist in the detection of prey. One species from the Philippines, *Scytodes pallidus*, builds webs on the leaves of trees and shrubs where it is most likely to encounter its jumping spider prey. Of course, sometimes the jumping spider kills the spitting spider first, especially if the jumping spider happens to be *Portia*, a spider whose specialty is hunting other spiders.

The majority of spitting spider species are solitary for most of their lives; however, one species of spitting spider from Madagascar is social, living in a colony consisting of both adults and juveniles. *Scytodes socialis* constructs a fist-sized communal web which is collectively maintained by the colony's residents. Adult spiders cooperatively kill prey and then share the kill with all members of the colony. It is probable that in this species sociality derives from an extension of maternal care of the young and tolerance between individuals of the same species.

In solitary species of spitting spiders, when the male matures, he stops hunting for food and instead goes wandering in search of a mate. When he finds a mature female, he may engage in a brief round of sparring. The couple may strike each other rapidly with their front legs. If the female is receptive to the male's advances, she suddenly falls over backward and lies on her back, allowing the male to mate with her. He uses both pedipalps simultaneously to transfer sperm to the female. After mating, he may recharge his pedipalps using a modified sperm web consisting of a single strand of silk to convey the sperm drop to each pedipalp.

After mating, the female produces an egg sac that has a loose and open network of silk holding the relatively large eggs; this sac somewhat resembles a mesh bag of onions. She holds this egg sac in her chelicerae for the entire duration of the incubation period, during which time she does not hunt or feed. Because the eggs are visible through the open meshwork of silk, the development of the embryos is easy to observe. The legs appear and grow in definition and length, the prosoma and opisthosoma become distinguishable, the eyes appear, and finally even

The high-domed cephalothorax accommodates a massive glue gland as well as a smaller venom gland. The 6 eyes are arranged in three pairs, similar to the eye arrangement of a related family of spiders, the highly venomous Sicariidae. However, spitting spiders pose no risk to humans; their fangs are far too small to penetrate our skin. This species is *Scytodes univittata*. A number of other species of *Scytodes* occur in the southwestern United States, including several undescribed species.

the speckled markings on the baby spiders' abdomens become visible. The spiderlings remain in the egg sac for a day or two after they hatch out of the eggs. Even after they emerge from the egg sac, they continue to develop for a few more days while living off the stored yolk in their bodies. Finally, after their first molt, they are ready to start their lives as hunters, capturing prey in exactly the same manner as their parents—squirting two tiny streams of glue in the blink of an eye.

1. Spitting spiders rely on nonvisual sensory structures in order to detect and identify prey. In this case, although the cricket was only fractions of an inch (millimeters) from the spider, there was no reaction from the spider until the cricket's antenna touched the front leg of the spider. Sensory setae on the spider's legs detect the cricket, enabling the spider to identify the cricket as prey. Once contact is made, the reaction is immediate.

2 and 3. Each fang oscillates back and forth rapidly, resulting in two narrow streams of glue squirting out over the prey. The two streams are almost parallel in photo number 2, whereas in photo number 3, each stream has changed its direction. The cricket is securely fastened to the substrate in a 40th of a second.

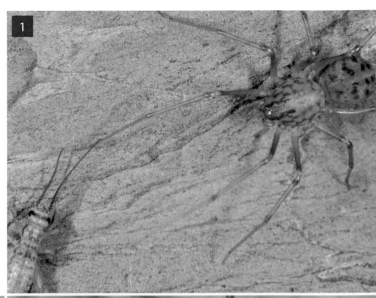

3400 15.0kV 16.3mm x230 SE 2/8/2010   200um

S3400 15.0kV 16.0mm x550 SE 2/8/2010   1C

1 and 2. Electron micrographs reveal the structure of the tiny fangs. Each fang consists of an open, tapering funnel that channels the liquid mixture of glue and silk into a narrow stream. The fangs oscillate at an extremely rapid frequency that averages 826 Hz, but may reach a rate of 1,700 Hz. Such high rates of oscillation can be achieved only with tiny fangs. The fangs on this adult spitting spider are only 70 microns (0.07 mm) in length, comparable to the width of a human hair.

3. The chelicerae of the spitting spider are fused where they connect to the cephalothorax. Therefore, the majority of the cheliceral structure is fixed and cannot rotate; however, the fangs oscillate rapidly back and forth from hydrodynamic forcing as the muscles within the cephalothorax exert pressure on the massive glue gland and the fluid is forced out via the fangs. The liquid shoots out at a velocity of 92 feet (28 m) per second. Two streams of glue/silk mixture can be seen as they are channeled by the tiny fangs.

The pedipalps are held out of the way as the two streams of liquid are squirted out. A silklike fibrous component of the mixture starts to contract almost instantly, helping to securely bind the targeted prey.

1. The rapid oscillation of the fangs results in many lines of liquid glue and silk deposited on the prey and on the substrate. The entire spitting sequence lasts for only a 40th of a second.

2. After securely fastening the prey to the substrate, the spider delivers the killing bite at the joint of a leg where the cuticle is thin enough for the tiny fangs to penetrate.

3. The spitting spider then fastidiously cleans any glue from its own front legs.

4. After wrapping the prey in a bit of silk from its spinnerets, the spider carries it to a quiet spot to feed. The spitting spider almost never feeds where it kills the prey.

Facing page:

1. After a brief sparring match using their front legs, the female spitting spider signals that she has accepted a courting male by falling over onto her back. He then uses both pedipalps simultaneously to transfer sperm to the female.

2, 3, and 4. Shortly after mating, a well-fed female produces an egg sac. The female carries the egg sac in her chelicerae during the 2–3 weeks of incubation time. She does not hunt or feed during this period of time. Details of the developing embryos can easily be observed through the loose mesh of the enveloping silk. Legs appear and grow in length, and even the speckled markings of the babies are visible.

5. A tiny baby spitting spider has just emerged from the egg sac. Even the babies must hunt prey the same way the adults do, squirting out two streams of glue in the blink of an eye.

Scytodes lugubris is found across the southern United States, including the arid southwestern states.

In the Company of Arachnids

Indra's Net was an ancient Indian metaphor symbolizing the interconnectedness of all the universe. An orb web connects all things, and each and every dewdrop reflects the entire universe.

Seattle, July 16, 2014. "A Seattle house was set on fire Tuesday by a man trying to kill a spider with a makeshift blowtorch," CBS affiliate KIRO-TV reported.

Arachnids evoke strong emotions in people. Throughout centuries of human civilization, arachnids have inspired fear and disgust, reverence and wonder. In contemporary Western culture, arachnids are usually depicted in a negative connotation; however, in many ancient civilizations arachnids were venerated.

The concept of a spider web symbolizing the interconnectedness of the universe is both ancient and modern. Dating from third century India, the Avatamsaka Sutra introduces the metaphor of Indra's Net. Indra's Net is illustrated by an orb web hung with dewdrops. This web connects all things, and in each and every dewdrop is reflected the entire universe. In our modern world, technology has provided a tool to connect humans from all over the world. The terms "Internet" and "World Wide Web" convey the theme of the web as a symbol of universal interconnection.

Humans derive material advantages from arachnids as well as symbolic benefits. From ancient times to the present, people from around the world have adapted silk and venom for their own purposes. The antimicrobial properties of spider silk made it useful as a wound dressing even from the time of Alexander the Great. Spider silk has been used for catching fish by Pacific Islanders, and in Madagascar the silk from the golden orb weaver (Nephila sp.) was harvested and fashioned into shimmering golden cloth. Arachnids even played a crucial role for the Allies during World War II. Until the 1960s, spider silk was used for cross hairs and sighting marks in a variety of scientific instruments and weapon sights. Spider silk was especially useful for optical instruments such as periscopes, telescopes, and microscopes.

Spider silk may also be used in medical applications. It shows promise in someday helping victims of traumatic spinal cord injury. The spider silk functions as a guide for the regenerating nerves. Bundles of nerve fibers grow together along the silk path, rebuilding neural connections.

Scorpion venom also has some surprising medical applications. James Olson of the Seattle Children's Hospital, in partnership with the University of Washington, has been developing a "tumor paint" made from a component of venom derived from the deathstalker scorpion (Leiurus quinquestriatus).

It binds to matrix metalloproteinase-2 expressed in gliomas and some other cancers but not found in normal brain tissue. The venom peptide is labeled with the fluorescent molecular marker indo-cyanine green. Once the peptide has had a chance to circulate through the brain, surgeons can see cancerous tissue glow under a certain wavelength of light, permitting them to remove the tumor tissue while sparing the healthy tissue. The fluorescent paint permits the detection of even small tumors consisting of as few as 2,000 cells. This is 500 times as sensitive as an MRI. This technology is still in a clinical-trial phase both in human and in veterinary medicine. However, it is now a possibility that many lives might be saved thanks to the unlikely partnership of a lethal scorpion and an impassioned doctor.

Certainly, a few arachnids may be problematic for humans. Crops may be damaged by spider mites, rust and gall mites, or penthaleid mites. Ticks transmit a variety of diseases, including Lyme disease, Rocky Mountain spotted fever, relapsing fever, Colorado tick fever, babesiosis, and erhlichiosis. Chiggers and scabies mites can cause severe itching. Finally, a small but significant number of arachnids from around the world have potentially dangerous stings or bites. Included in their ranks are the Arizona brown spiders (Loxosceles species), black widows (Latrodectus species), and bark scorpions (Centruroides sculpturatus).

However, out of the tens of thousands of species of arachnids found on Earth, fewer than 1 percent could be viewed as significantly hazardous to humans. The vast majority of arachnids are quietly and inconspicuously going about their lives as predators, herbivores, and detritivores, doing their part as they contribute to the dynamic balance of life on our planet.

For more than 400 million years arachnids have been surviving and evolving within a complex and changing world. During the late Silurian Period about 420 million years ago, arachnids began to colonize land. During the following Devonian and Carboniferous periods ending about 300 million years ago, more arachnids appeared. They were contemporary with forests of huge horsetail plants and giant clubmoss.

They have survived several mass extinctions, including the "Great Dying" at the end of the Permian Period 252 million years ago, in which about 85 percent of species became extinct. They also survived the mass extinction triggered by the Chicxulub impact off the Yucatan Peninsula, resulting in the end of the Age of the Dinosaurs. They have survived the "Ice Age" of the Pleistocene Epoch, with its cycles of warm and freezing conditions. They have adapted to millions of climatic, geologic, and biological changes within their world.

Throughout all, arachnids have persevered. Some are almost unchanged from their ancestors, living relics from the deep past. But over hundreds of thousands of millennia, these small, unobtrusive creatures have evolved a stunning array of survival tactics. Their anatomy, physiology, and behavior have become adapted to their environment, allowing them to survive in every conceivable niche despite the most severe challenges.

The study of arachnids provides insights into universal principles governing life on Earth, many of which directly concern human existence. However, after having survived endless challenges over eons of time, these little creatures face a relatively new threat: *Homo sapiens*. Our species has so completely changed our planet that the current geologic time has been named the Anthropocene Epoch. *Homo sapiens* may be the catalyst for yet another mass extinction. It remains to be seen whether arachnids, or for that matter humans, survive this extinction event. Masters in the art of patience, perseverance, endurance, and adaptation, arachnids have much to teach us regarding the art of survival in a changing world.

Perhaps Walt Whitman said it best in the poem "A Noiseless Patient Spider":

A noiseless patient spider,
I mark'd where on a little promontory
  it stood isolated,
Mark'd how to explore the vacant vast
  surrounding,
It launch'd forth filament, filament,
  filament, out of itself,
Ever unreeling them, ever tirelessly
  speeding them.

And you O my soul where you stand,
Surrounded, detached, in measureless
  oceans of space,
Ceaselessly musing, venturing,
  throwing, seeking the spheres to
  connect them,
Till the bridge you will need be
  form'd, till the ductile anchor hold,
Till the gossamer thread you fling
  catch somewhere, O my soul.

Thank you, arachnids.

Photo by Bill Savary

Acknowledgments and Photo Credits

I would like to thank the following people for their generous permission for the use of their amazing photographs:

Alice Abela: 65, 66, 67

Matt Coors: 27, 187

Timothy A. Cota: 111, 170, 176, 177, 301

Scott Justis: 147

Philip Kline: 106

Bill Savary: 316

Michael Seiter: 97

Bruce D. Taubert: 68, 69, 239

I would like to thank Steven Hernandez and Paul Wallace and the University of Arizona for their assistance in obtaining scanning electron micrographs.

I would like to thank Alex Yelich for his help in photographing a palpigrade.

I would like to thank the following arachnologists for all their patience in answering my questions and providing identifications over the past 10 years: Luis de Armas, Leticia Aviles, Greta Binford, Gita Bodner, Sarah Crews, Paula Cushing, Bruce Cutler, Rebecca Duncan, G. B. Edwards, Ray Fisher, Oscar Franke, Mark Harvey, Marshal Hedin, Brent Hendrixson, Bernhard Huber, Garrett Brady Hughes, Hans Klompen, Joseph Lapp, Joel Ledford, Wayne Maddison, Jaime Mayoral, Kari McWest, Norman Platnick, Heather Proctor, David Richman, Mark Rowland, Michael Seiter, Jeffrey Shultz, W. David Sissom, Darrell Ubick, David Walter, Cal Welbourn, and Peter Weygoldt.

I would like to thank all the folks who contributed so much of their time, specimens, expertise, and encouragement: Carel Brest van Kempen, Marcus Bullock, Don Cadle, Joe Cicero, Andrea and Timothy A. Cota, Vince Dolan, Joel Hallan, Mark Heitlinger, Dennis Hoburg, J. D. Mizer, Wendy Moore, June Olberding, Andrew Olson, Carl Olson, Eskild Petersen, Wanda and Ken Petty, John Rhodes, Fran Rome, Matt and Ashlee Rowe, Justin Schmidt, and Bruce Taubert.

And most of all, I would like to thank my long-suffering spouse, Bill Savary. Without him, this book would never have been possible.

Paramarpissa female.

References

Adams, R. J. 2014. *Field Guide to the Spiders of California and the Pacific Coast States.* University of California Press, Berkeley, U.S.A.

Agnarsson, I. 2002. Sharing a web—on the relation of sociality and kleptoparasitism in theridiid spiders (Theridiidae: Araneae). *Journal of Arachnology* 30: 181–188.

Agnarsson, I., L. Avilés, and W. P. Maddison. 2013. Loss of genetic variability in social spiders: genetic and phylogenetic consequences of population subdivision and inbreeding. *Journal of Evolutionary Biology* 26: 27–37.

Alvarado-Castro, J. A., and M. L. Jiménez. 2011. Reproductive behavior of *Homalonychus selenopoides* (Araneae: Homalonychidae). *Journal of Arachnology* 39: 118–127.

Armas, L. F. de. 2012. Nueva especie de *Paraphrynus* Moreno, 1940 (Amblypygi: Phrynidae) de México y el suroeste de los EE.UU. de América. *Revista Ibérica de Aracnología* 21: 27–32.

Avilés, L., and G. Gelsey. 1998. Natal dispersal and demography of a subsocial *Anelosimus* species and its implications for the evolution of sociality in spiders. *Canadian Journal of Zoology* 76: 2137–2147.

Barrantes, G., and M. J. Ramírez. 2013. Courtship, egg sac construction, and maternal care in *Kukulcania hibernalis*, with information on the courtship of *Misionella medensis* (Araneae, Filistatidae). *Arachnology* 16: 72–80.

Beavis, A. S., D. M. Rowell, and T. Evans. 2007. Cannibalism and kin recognition in *Delena cancerides* (Araneae: Sparassidae), a social huntsman spider. *Journal of Zoology* 271: 233–237.

Beccaloni, J. 2009. *Arachnids.* University of California Press, Berkeley, U.S.A. 320 pages.

Becker, N., E. Oroudjev, S. Mutz, J. S. Cleveland, P. K. Hansma, C. Hayashi, D. E. Makarov, and H. G. Hansma. 2003. Molecular nanosprings in spider capture-silk threads. *Nature Materials* 2: 278–283.

Binford, G. J., and M. A. Wells. 2003. The phylogenetic distribution of sphingomyelinase D activity in venoms of Haplogyne spiders. *Comparative Biochemistry and Physiology Part B: Biochemistry and Molecular Biology* 135: 25–33.

Binford, G. J., M.H.J. Cordes, and M. A. Wells. 2005. Sphingomyelinase D from venoms of *Loxosceles* spiders: evolutionary insights from gene sequence and structure. *Toxicon* 45: 547–560.

Binford, G. J., and A. Rypstra. 1992. Foraging behavior of the communal spider, *Philoponella republicana* (Araneae: Uloboridae). *Journal of Insect Behavior* 5(3): 321–335.

Boulton, A. M., and G. A. Polis. 1999. Phenology and life history of the desert spider, *Diguetia mojavea* (Araneae, Diguetidae). *Journal of Arachnology* 27: 513–521.

Boulton, A. M., and G. A. Polis. 2002. Brood parasitism among spiders: interactions between salticids and *Diguetia mojavea*. *Ecology* 83(1): 282–287.

Bradley, R. A. 2013. *Common Spiders of North America.* University of California Press, Berkeley, U.S.A..

Brady, A. R. 2007. *Sosippus* revisited: review of a web-building wolf spider genus from the Americas (Araneae, Lycosidae). *Journal of Arachnology* 35: 54–83.

Brownell, P., and G. Polis (eds.). 2001. *Scorpion Biology and Research.* Oxford University Press, Cambridge, U.K. 448 pages.

Carico, J. E. 1973. The Nearctic species of the genus *Dolomedes* (Araneae: Pisauridae). *Bulletin of the Museum of Comparative Zoology* 144(7): 435–488.

Carico, J. E. 1976. The spider genus *Tinus* (Pisauridae). *Psyche* 83: 63–78.

Carico, J. E. 1993. Revision of the genus *Trechalea* Thorell (Araneae, Trechaleidae) with a review of the taxonomy of the Trechaleidae and Pisauridae of the Western Hemisphere. *Journal of Arachnology* 21: 226–257.

Clos, L. M. 2008. *North America through Time: A Paleontological History of Our Continent.* Fossil News, Colorado, U.S.A. 284 pages.

Coyle, F. A. 1985. Ballooning behavior of *Ummidia* spiderlings (Araneae, Ctenizidae). *Journal of Arachnology* 13: 137–138.

Coyle, F. A., and W. R. Icenogle. 1994. Natural history of the Californian trapdoor spider genus *Aliatypus* (Araneae, Antrodiaetidae). *Journal of Arachnology* 22: 225–255.

Cushing, P. E. 1997. Mymecomorphy and myrmecophily in spiders: A review. *Florida Entomologist* 80(2): 165–193.

Cushing, P. E., J. O. Brookhard, H-J. Kleebe, G. Zito, and P. Payne. 2005. The suctorial organ of the Solifugae (Arachnida, Solifugae). *Arthropod Struct Dev.* 34: 397–406.

Cushing, P. E., and P. Castro. 2012. Preliminary survey of the setal and sensory structures on the pedipalps of camel spiders (Arachnida: Solifugae). *Journal of Arachnology* 40: 123–127.

Derkarabetian, S., J. Ledford, and M. Hedin. 2011. Genetic diversification without obvious genitalic morphological divergence in harvestmen (Opiliones, Laniatores, *Sclerobunus robustus*) from montane sky islands of western North America. *Molecular Phylogenetics and Evolution* 61: 844–853.

DiDomenico, A., and M. Hedin. 2016. New species in the *Sitalcina sura* species group (Opiliones, Laniatores, Phalangodidae), with evidence for a biogeographic link between California desert canyons and Arizona sky islands. *ZooKeys* (586): 1–36.

Domínguez, K., and M. Jiménez. 2005. Mating and self-burying behavior of *Homalonychus theologus* Chamberlin (Araneae, Homalonychidae) in Baja California Sur. *Journal of Arachnology* 33: 167–174.

Duncan, R. P., K. Autumn, and G. J. Binford. 2007. Convergent setal morphology in sand-covering spiders suggests a design principle for particle capture. *Proceedings of the Royal Society of London B*, 274(1629): 3049–3056.

Dunlop, J. A. 2010. Geological history and phylogeny of Chelicerata. *Arthropod Structure and Development* 39: 124–142.

Eberhard, W. G., and R.D.L. Briceño. 1983. Chivalry in pholcid spiders. *Behavioral Ecology and Sociobiology* 13: 189–195.

Eisner, T. E. 2003. *For Love of Insects*. Belknap Press of Harvard University Press, Cambridge, MA, U.S.A. 448 pages.

Eisner, T., M. Eisner, and M. Siegler. 2005. *Secret Weapons: Defenses of Insects, Spiders, Scorpions, and Other Many-Legged Creatures*. Belknap Press of Harvard University Press, Cambridge, MA, U.S.A. 372 pages.

Elias, D., A. C. Mason, W. P. Maddison, and R. R. Hoy. 2003. Seismic signals in a courting male jumping spider (Araneae: Salticidae). *Journal of Experimental Biology* 206: 4029–4039.

Evans, T. A., E. Wallis, and M. Elgar. 1995. Making a meal of mother. *Nature* 376: 299.

Fabre, J. H. 1912. *The Life of the Spider*. Blue Ribbon Books, New York, U.S.A. 404 pages.

Fink, L. 1984. Venom spitting by the green lynx spider, *Peucetia viridans* (Araneae, Oxyopidae). *Journal of Arachnology* 12: 372–373.

Fink, L. S. 1987. Green lynx spider egg sacs: sources of mortality and the function of female guarding (Araneae, Oxyopidae). *Journal of Arachnology* 15: 231–239.

Foelix, R. F. 2011. *Biology of Spiders*, Third Edition. Oxford University Press, New York, U.S.A. 419 pages.

Garwood, R. J., and J. Dunlop. 2014. Three-dimensional reconstruction and the phylogeny of extinct chelicerate orders. *PeerJ* 2: e641; DOI 10.7717/peerj.641.

Gertsch, W. J. 1949. *American Spiders: A Guide to the Life and Habits of the Spider World*. D. Van Nostrand, New York, U.S.A. 285 pages.

Gosline, J. M., P. A. Guerette, C. S. Ortlepp, and K. N. Savage. 1999. The mechanical design of spider silks: from fibroin sequence to mechanical function. *Journal of Experimental Biology* 202: 3295–3303.

Hadley, N. F. 1974. Adaptational biology of desert scorpions. *Journal of Arachnology* 2: 11–23.

Hamilton, C., B. Hendrixson, and J. Bond. 2016. Taxonomic revision of the tarantula genus *Aphonopelma* Pocock, 1901 (Araneae, Mygalomorphae, Theraphosidae) within the United States. *ZooKeys* 560: 1–340.

Harmer, A.M.T., T. A. Blackledge, J. S. Madin, and M. E. Herberstein. 2011. High-performance spider webs: integrating biomechanics, ecology and behaviour. *Journal of the Royal Society Interface* 8(57): 457–471.

Harvey, M. S. 2003. *Catalogue of the Smaller Arachnid Orders of the World: Amblypygi, Uropygi, Schizomida, Palpigradi, Ricinulei and Solifugae*. CSIRO, Colllingwood, Australia. 385 pages.

Hayashi, C. Y., and R. V. Lewis. 1998. Evidence from flagelliform silk cDNA for the structural basis of elasticity and molecular nature of spider silks. *Journal of Molecular Biology* 275(5): 773–784.

Haynes, K. F., C. Gemeno, K. V. Yeargan, J. G. Millar, and K. M. Johnson. 2002. Aggressive chemical mimicry of moth pheromones by a bolas spider: how does this specialist predator attract more than one species of prey? *Chemoecology* 12(2): 99–105.

Hedin, M. C. 1997. Molecular phylogenetics at the population/species interface in cave spiders in the southern Appalachians (Araneae: Nesticidae: *Nesticus*). *Molecular Biology and Evolution* 14(3): 309–324.

Herberstein, M. E. (ed.). 2011. *Spider Behaviour*. Cambridge University Press, New York, U.S.A. 391 pages.

Hillyard, P. 2008. *The Private Life of Spiders*. Princeton University Press, Princeton, NJ, U.S.A. 160 pages.

Hinman, M. B., and R. V. Lewis. 1992. Isolation of a clone encoding a second dragline silk fibroin. *Nephila clavipes* dragline silk is a two-protein fiber. *Journal of Biological Chemistry* 267(27): 19320–19324.

Jackson, R. R., and F. R. Cross. 1987. A cognitive perspective on aggressive mimicry. *Journal of Zoology* 290(3): 161–171.

Jackson, R. R., X. J. Nelson, and K. Salm. 2008. The natural history of *Myrmarachne melanotarsa*, a social ant-mimicking jumping spider. *New Zealand Journal of Zoology* 35: 225–235.

Kim, K. W., and C. Roland. 2000. Trophic egg laying in the spider, *Amaurobius ferox*: mother-offspring interactions and functional value. *Behavioural Processes* 50: 31–42.

Kim, K. W., C. Roland, and A. Horel. 2001. Functional value of matriphagy in the spider *Amaurobius ferox*. *Ethology* 106: 729–742.

Kim, K. W., B. Krafft, and J. C. Choe. 2005. Cooperative prey capture by young subsocial spiders: I. Functional value. *Behavioral Ecology and Sociobiology* 59: 92–100.

Kim, K. W., B. Krafft, and J. C. Choe. 2005. Cooperative prey capture by young subsocial spiders: II. Behavioral mechanism. *Behavioral Ecology and Sociobiology* 59: 101–107.

Kim, K. W., and A. Horel. 2010. Matriphagy in the spider *Amaurobius ferox* (Araneae Amaurobiidae): an example of mother-offspring interactions. *Ethology* 104: 1021–1037.

Klingel, H. 1963. Mating and maternal behaviour in *Telyphonus caudatus* L. (Pedipalpi, Holopeltidia, Uropygi). *Treubia* 26: 65–69.

Krantz, G. W., and D. E. Walter (eds.). 2009. *A Manual of Acarology,* third edition. Texas Tech University Press, Lubbock, U.S.A. 816 pages.

Ledford, J. M., and C. E. Griswold. 2010. A study of the subfamily Archoleptonetinae (Araneae, Leptonetidae) with a review of the morphology and relationships for the Leptonetidae. *Zootaxa* 2391: 1–32.

Ledford, J., P. Paquin, J. Cokendolpher, J. Campbell, and C. Griswold. 2012. Systematics, conservation and morphology of the spider genus *Tayshaneta* (Araneae, Leptonetidae) in central Texas caves. *ZooKeys* 167: 1–102.

Lee, V. F. 1979. The maritime pseudoscorpions of Baja California, Mexico (Arachnida: Pseudoscorpionida). *Occasional Papers of the California Academy of Sciences* 131: 1–38.

Levi, H. W., and L. R. Levi. 1987. *Spiders and Their Kin*. Golden Press, New York, U.S.A. 160 pages.

Lubin, Y., and T. Bilde. 2007. The evolution of sociality in spiders. *Advances in the Study of Behavior* 37: 83–145.

Marija, M., I. Agnarsson, J. Svenning, and T. Bilde. 2013. Social spiders of the genus *Anelosimus* occur in wetter, more productive environments than non-social species. *Naturwissenschaften* 100: 1031–1040.

Meehan, C. J., E. J. Olson, and R. L. Curry. 2008. Exploitation of the *Pseudomyrmex-Acacia* mutualism by a predominantly vegetarian jumping spider (*Bagheera kiplingi*). *93rd ESA Annual Meeting*, PS 62–107.

Miller, J. A. 2006. Web-sharing sociality and cooperative prey capture in a Malagasy spitting spider (Araneae: Scytodidae). *Proceedings of the California Academy of Sciences* 57: 739–750.

Montaño-Moreno, H., and O. F. Franke. 2013. Observations on the life history of *Eukoenenia chilanga* Montaño (Arachnida: Palpigradi). *Journal of Arachnology* 41: 205–212.

Moya-Laraño, J., D. Vinković, E. De Mas, G. Corcobado, and E. Moreno. 2008. Morphological evolution of spiders predicted by pendulum mechanics. *PLoS ONE* 3(3): e1841. doi: 10.1371/journal.pone.0001841.

Nelson, X. J., and R. R. Jackson. 2006. A predator from East Africa that chooses malaria vectors as preferred prey. *PLoS ONE* 1(1): e132.

Newell, I. M., and R. E. Ryckmann. 1966. Species of *Pimeliaphilus* (Acari: Pterygosomidae) attacking insects, with particular reference to the species parasitizing triatominae (Hemiptera: Reduviidae). *Hilgardia* 37(12): 403–436.

Penney, D. 2009. A new spider family record for Hispaniola—a new species of *Plectreurys* (Araneae: Plectreuridae) in Miocene Dominican amber. *Zootaxa* 2144: 65–68.

Pepato, A. R., C.E.F. da Rocha, and J. A. Dunlop. 2010. Phylogenetic position of the acariform mites: sensitivity to homology assessment under total evidence. *BMC Evolutionary Biology* 10: 235.

Pierce, W. D. 1951. Fossil arthropods from onyx-marble. *Bulletin of the Southern California Academy of Sciences* 50: 34–49.

Pinto-da-Rocha, R., G. Machado, and G. Giribet (eds.). 2007. *Harvestmen: The Biology of Opiliones*. Harvard University Press, Cambridge MA., U.S.A. 597 pages.

Platnick, N. I. 1986. On the tibial and patellar glands, relationships, and American genera of the spider family Leptonetidae (Arachnida, Araneae). *American Museum Novitates* 2855: 1–16.

Platnick, N. I., and R. R. Forster. 1982. On the Micromygalinae, a new subfamily of Mygalomorph spiders (Araneae, Microstigmatidae). *American Museum Novitates* 2734: 1–13.

Polis, G. A. (ed.). 1990. *The Biology of Scorpions*. Stanford University Press, California, U.S.A. 587 pages.

Porter, S. D., and D. A. Eastmond. 1982. *Euryopis coki* (Theridiidae), a spider that preys on *Pogonomyrmex* ants. *Journal of Arachnology* 10: 275–277.

Powers, K. S., and L. Avilés. 2003. Natal dispersal patterns of a subsocial spider *Anelosimus* cf. *jucundus* (Theridiidae). *Ethology* 109: 725–737.

Preston-Mafham, R. 1996. *The Book of Spiders and Scorpions*. Barnes and Noble Books, New York, U.S.A. 144 pages.

Punzo, F. 1994. Changes in brain amine concentrations associated with postembryonic development in the solpugid *Eremobates palpisetulosus* Fichter (Solpugida: Eremobatidae). *Journal of Arachnology* 22: 1–4.

Punzo, F. 1998. *The Biology of Camel-Spiders (Arachnida, Solifugae)*. Kluwer Academic, Dordrecht, Netherlands. 301 pages.

Ramirez, M. J. 2014. The morphology and phylogeny of dionychan spiders (Araneae: Araneomorphae). *Bulletin of the AMNH*, 390. 374 pages.

Raynor, L. S., and L. A. Taylor. 2006. Social behavior in amblypygids, and a reassessment of arachnid social patterns. *Journal of Arachnology* 34: 399–421.

Reddell, J. R., and J. Cokendolpher. 1995. *Catalogue, Bibliography, and Generic Revision of the Order Schizomida (Arachnida)*. Texas Memorial Museum Speleological Monographs No. 4. 170 pages.

Riechert, S. E., R. Roeloffs, and A. C. Echternacht. 1986. The ecology of the cooperative spider *Agelena consociata* in equatorial Africa (Araneae, Agelenidae). *Journal of Arachnology* 14: 175–191.

Řezáč, M., J. Král, and S. Pekár. 2008. The spider genus *Dysdera* (Araneae, Dysderidae) in central Europe: revision and natural history. *Journal of Arachnology* 35: 432–462.

Roble, S. M. 1985. Submergent capture of *Dolomedes triton* (Araneae, Pisauridae) by *Anoplius depressipes* (Hymenoptera, Pompilidae). *Journal of Arachnology* 13: 391–392.

Ross, K., and R. L. Smith. 1979. Aspects of the courtship and behavior of the black widow spider, *Latrodectus hesperus* (Araneae: Theridiidae), with evidence for the existence of a contact sex pheromone. *Journal of Arachnology* 7: 69–77.

Rovner, J. S. 1980. Morphological and ethological adaptations for prey capture in wolf spiders (Araneae, Lycosidae). *Journal of Arachnology* 8: 201–215.

Rowe, A. H., Y. Xiao, M. P. Rowe, T. R. Cummins, and H. Zakon. 2013. Voltage-gated sodium channel in grasshopper mice defends against bark scorpion toxin. *Science* 342: 441–446.

Rowland, J. M. 1971. A new *Trithyreus* from a desert oasis in southern California (Arachnida: Schizomida: Schizomidae). *Pan-Pacific Entomologist* 47: 304–309.

Rowland, J. M. 1972. Origins and distribution of two species groups of schizomida (Arachnida). *Southwestern Naturalist* 17: 153–160.

Rowland, J. M. 1972. The brooding habits and early development of *Trithyreus pentapeltis* (Cook) (Arachnida: Schizomida). *Entomological News* 83: 69–74.

Rowland, M., and D. Sissom. 1980. Report on a fossil palpigrade from the Tertiary of Arizona, and a review of the morphology and systematics of the order (Arachnida: Palpigradida). *Journal of Arachnology* 8: 69–86.

Rowell, D. M., and L. Avilés. 1995. Sociality in a bark-dwelling huntsman spider from Australia, *Delena cancerides* Walckenaer (Araneae: Sparassidae). *Insectes Sociaux* 42: 287–302.

Sandoval, C. P. 1994. Plasticity in web design in the spider *Parawixia bistriata*: A response to variable prey type. *Functional Ecology* 8: 701–707.

Schmalhofer, V. R. 2000. Diet induced and morphological color changes in juvenile crab spiders (Araneae, Thomisidae). *Journal of Arachnology* 28: 56–60.

Schmidt, J. O. 2003. Vinegaroon (*Mastigoproctus giganteus*) life history and rearing. *Proceedings of the 2003 Invertebrates in Captivity Conference*: 73–80.

Schmidt, J. O. 2009. Vinegaroons, in V. Resh and R. Carde (eds.), *Encyclopedia of Insects,* second edition. Academic Press, Boston, U.S.A., pp. 1038–1041.

Schultz, S. A., and M. J. Schultz. 2009. *The Tarantula Keeper's Guide*. Barron's, New York, U.S.A. 376 pages.

Sharpton, V. L., G. B. Dalrymple, L. E. Marin, G. Ryder, B. C. Schuraytz, and J. Urrutia-Fucugauchi. 1992. New links between the Chicxulub impact structure and the Cretaceous/Tertiary boundary. *Nature* 359: 819–821.

Shear, W. 2009. Harvestmen. *American Scientist* 97(6): 468–475.

Shook, R. S . 1978. Ecology of the wolf spider *Lycosa carolinensis* Walckenaer (Araneae: Lycosidae) in a desert community. *Journal of Arachnology* 6: 53–64.

Selden, P. A., and D. Huang. 2010. The oldest haplogyne spider (Araneae: Plectreuridae), from the Middle Jurassic of China. *Naturwissenschaften* 97(5): 449–459.

Simmons, A. H., C. A. Michal, and L. W. Jelinski. 1996. Molecular orientation and two-component nature of the crystalline fraction of spider dragline silk. *Science* 271(5245): 84–87.

Smith, D. R. 1982. Reproductive success of solitary and communal *Philoponella oweni* (Araneae: Uloboridae). *Behavioral Ecology and Sociobiology* 11: 149–154.

Smrž, J., Ĺ. Kováč, J. Mikeš, and A. Lukešová. 2013. Microwhip scorpions (Palpigradi) feed on heterotrophic cyanobacteria in Slovak caves—a curiosity among Arachnida. *PLoS ONE* 8(10): e75989.

Sponner, A., W. Vater, S. Monajembashi, E. Unger, F. Grosse, and K. Weisshart. 2007. Composition and hierarchical organisation of spider silk. *PLoS ONE* 2(10): e998.

Starrett, J., and E. R. Waters. 2007. Positive natural selection has driven the evolution of the Hsp70s in *Diguetia* spiders. *Biology Letters* 3(4): 439–444.

Stockmann, R., and E. Ythier. 2010. *Scorpions of the World*. N.A.P. France. 568 pages.

Stowe, M. K., T.C.J. Turlings, J. H. Loughrin, W. J. Lewis, and J. H. Tumlinson. 1995. The chemistry of eavesdropping, alarm and deceit. *Proceedings of the National Academy of Sciences USA* 92: 23–28.

Suter, R. B., and G. E. Stratton. 2009. Spitting performance parameters and their biomechanical implications in the spitting spider, *Scytodes thoracica*. *Journal of Insect Science* 9(1): 62.

Taylor, R. M., and R. A. Bradley. 2009. Plant nectar increases survival, molting, and foraging in two foliage wandering spiders. *Journal of Arachnology* 37: 232–237.

Tizo-Pedroso, E., and K. Del-Claro. 2007. Parental care and social behavior in neotropical pseudoscorpions: the example of *Paratemnoides nidificator* (Pseudoscorpiones: Atemnidae). *17th International Congress of Arachnology Abstracts*, p. 45.

Ubick, D., P. Paquin, P. E. Cushing, and V. Roth (eds.). *Spiders of North America: An Identification Manual.* American Arachnological Society. 377 pages.

van Beek, J. D., S. Hess, F. Vollrath, and B. H. Meier. 2002. The molecular structure of spider dragline silk: folding and orientation of the protein backbone. *PNAS* 99(16): 10266–10271.

Vetter, R. S., L. S. Vincent, A. A. Itnyre, D. E. Clarke, K. I. Reinker, D.W.R. Danielson, L. J. Robinson, and J. N. Kabashima. 2012. Predators and parasitoids of egg sacs of the widow spiders, *Lactodectus geometricus* and *Lactrodectus hesperus* (Araneae: Theridiidae) in southern California. *Journal of Arachnology* 40: 209–214.

Waloszek, D., and J. Dunlop. 2002. A larval sea spider (Arthropoda: Pycnogonida) from the upper Cambrian 'Orsten' of Sweden, and the phylogenetic position of pycnogonids. *Palaeontology* 45: 421–446.

Walter, D. E., and H. C. Proctor. 2013. *Mites: Ecology, Evolution and Behaviour: Life at a Microscale,* second edition. Springer, Dordrecht, Netherlands. 494 pages.

Welbourn, W. C. 1999. Invertebrate cave fauna of Kartchner Caverns, Kartchner Caverns, Arizona. *Journal of Cave and Karst Studies* 61(2): 93–101.

Weygoldt, P. 1969. *The Biology of Pseudoscorpions.* Harvard University Press, Cambridge, MA, U.S.A. 145 pages.

Weygoldt, P. 2000. *Whip Spiders (Chelicerata: Amblypygi).* Apollo Books, Stenstrup, Denmark. 163 pages.

Witt, P. N., and J. S. Rovner (eds.). 1982. *Spider Communication: Mechanisms and Ecological Significance.* Princeton University Press, Princeton, NJ, U.S.A. 440 pages.

Wolfe, J. A. 1991. Palaeobotanical evidence for a June 'impact winter' at the Cretaceous/Tertiary boundary. *Nature* 352: 420–422.

Wolff, J. O., A. Schönhofer, C. F. Schaber, and S. N. Gorb. 2014. Gluing the 'unwettable': soil-dwelling harvestmen use viscoelastic fluids for capturing springtails. *Journal of Experimental Biology* 217: 3535–3544.

Yeargan, K. V., and L. W. Quate. 1996. Juvenile bolas spiders attract psychodid flies. *Oecologia* 106(2): 266–271.

Yip, E. C., D. M. Rowell, and L. S. Raynor. 2012. Behavioural and molecular evidence for selective immigration and group regulation in the social huntsman spider, *Delena cancerides. Biological Journal of the Linnean Society* 106: 749–762.

Zeh, J. A., and D. Zeh. 1990. Cooperative foraging for large prey by *Paratemnus elongatus* (Pseudoscorpionida, Atemnidae). *Journal of Arachnology* 18: 307–311.

Index of Common and Scientific Names

Numbers in bold refer to pages where there are pictures of the species